Engineering Rock Mass Classification

Tunneling, Foundations, and Landslides

Bhawani Singh
Former Professor of Civil Engineering
Indian Institute of Technology
Roorkee - 247 667 (India)

R. K. Goel
Scientist G
Central Institute of Mining and Fuel Research
Regional Centre, CBRI Campus
Roorkee - 247 667 (India)

Edinburgh London New York Oxford Philadelphia St Louis Sydney Toronto 2011

Butterworth-Heinemann is an imprint of Elsevier
225 Wyman Street, Waltham, MA 02451, USA
The Boulevard, Langford Lane, Kidlington, Oxford, OX5 1GB, UK

Library of Congress Cataloging-in-Publication Data

Singh, Bhawani.

Engineering rock mass classification : tunneling, foundations, and landslides / Bhawani Singh & R. K. Goel.

p. cm.

Includes bibliographical references and index.

ISBN 978-0-12-810364-7 (alk. paper)

1. Engineering geology. 2. Tunneling. 3. Foundations. 4. Landslides–Prevention. 5. Rocks–Classification. 6. Rock mechanics. I. Goel, R. K. II Title.

TA705.S53638 2011

625.1′22–dc22 2011006029

British Library Cataloguing-in-Publication Data

A catalogue record for this book is available from the British Library.

For information on all Butterworth-Heinemann publications visit our Web site at *www.elsevierdirect.com*

11 12 13 14 15 10 9 8 7 6 5 4 3 2 1

Engineering Rock Mass Classification

Dedicated to
Researchers and Readers

Contents

Preface

The urgent need for this book, *Engineering Rock Mass Classification: Tunneling, Foundations, and Landslides*, was our motivation to write it. Many questions went through our minds: Is Classification reasonably reliable? Can it be successful in crisis management of geohazards? Can a single classification system be general enough for rock structures? Is classification a scientific approach? Laborious field research was needed to find answers to these vital questions.

By God's grace, scientists of the Central Institute of Mining and Fuel Research (CIMFR), IIT Roorkee, Central Soil and Material Research Station (CSMRS), Irrigation Research Institute (IRI), and the Norwegian Geotechnical Institute (NGI) came together. These God-gifted ideas and the reliable field data made our task of interpretation less tortuous. Consequently, several improvements in correlations have been possible and practical doubts were erased. At this point, consultancy works were started in the previously mentioned institutions. The success in consultancy further boosted our morale. Finally, the research work for this book was systematically compiled to help a new confident generation. The aim of this book is to generate more creative confidence and interest among civil, mining, and petroleum engineers and geologists. This book is a comprehensive revision of our book, *Rock Mass Classification—A Practical Approach in Civil Engineering*, and includes rock mass characterization, examples, and modern classifications.

Based on research, many classification approaches are scientific. Nevertheless, the scientific spirit of prediction, check, and cross-check should be kept alive; thus, many alternative classification systems have been presented here for particular rock structures. In feasibility designs of major projects, the suggested correlations in this book may be used. For final designs of complex openings, rational approaches are recommended. In the design of minor projects, field correlations may be used. The notation for uniaxial compressive strength of rock material in this book is q_c instead of σ_c. The engineering rock mass classification is an amazingly successful approach because it is simple, reliable, and time-tested for more than three decades.

Today the rational approach is becoming popular in consultancy on major projects. Our goal should be a reliable engineering strategy/solution of geological problems and not rigorous analysis. This should remove the prevailing dissatisfaction from the minds of designers. Thus, computer modeling may be the future trend of research at this time.

It appears that field testing and monitoring may always be the key approach to use in rock engineering projects, because all practical knowledge has been gained from interpretations of field observations.

The Himalayas provide the best field laboratory to learn rock mechanics and engineering geology because of complex geological problems. Further, the hypnotic charm of the upper Himalayas is very healing especially to concerned engineers and geologists. Natural oxygenation on hill tracking charges our whole nervous system and gives a marvelous feeling of energy and inner healing. So, working in the majestic Himalayas is a twin boon.

Acknowledgments

Our foremost wish is to express deep gratitude to Professor Charles Fairhurst, University of Minnesota; Dr. N. Barton; Professor J. A. Hudson, Imperial College of Science and Technology, London; Professor E. Hoek, International Consulting Engineer; Professor J.J.K. Daemen, University of Nevada; Dr. E. Grimstad, NGI; Professor G. N. Pandey, University of Swansea; Professor J. Nedoma, Academy of Sciences of the Czech Republic; Professor S. Sakurai; Professor Z. T. Bieniawski; Professor Jian Zhao, LMR, EPFL, Switzerland; Professor T. Ramamurthy, IIT Delhi; Professor V. D. Choubey; Dr. B. Singh, Banaras Hindu University; Professor B. B. Dhar; Professor Jagdish Narain, Former Vice Chancellor, University of Roorkee; Dr. N. M. Raju; Dr. A. K. Dube; Dr. J. L. Jethwa; Dr. Amalendu Sinha, Director, CIMFR; Dr. V. M. Sharma, ATES; Professor Gopal Ranjan; Professor P. K. Jain; Professor M. N. Viladkar; Professor A. K. Srivastava; Professor N. K. Samadhiya; Professor Mahendra Singh; Professor R. Anbalagan; Dr. J. P. Narayan and Dr. Daya Shankar, IIT Roorkee; Dr. T. N. Singh, IIT Mumbai; Dr. V. K. Mehrotra; Dr. Subhash Mitra, IRI, Roorkee; Dr. Bhoop Singh, DST; Dr. Surya Prakash, NIDM; Dr. Rajbal Singh, CSMRS; Dr. S. K. Jain, J. P. University, Solan; and Mr. H. S. Niranjan, HBTI, for constant moral support and vital suggestions and free sharing of precious field data.

We are also grateful to the scientists at CIMFR, CSMRS, IRI, and IIT Roorkee and all project authorities for supporting field research. For whole-hearted moral support, we are grateful to Mr. N. P. Aterkar; Mr. Sandesh Aterkar, Soilex Consultant, Roorkee; Mr. Phillip C. Helwig, Canada; and Mr. A. K. Bajaj, ceramic engineer, Roorkee.

We are very thankful to A. A. Balkema, Rotterdam, Netherlands; the American Society of Civil Engineers (ASCE), Reston; Ellis Horwood, Chichester, UK; the Institution of Mining & Metallurgy, London; John Wiley & Sons, New York; Springer-Verlag, Berlin, Germany; TransTech Europe, Oldenburg, Germany; Taylor & Francis; Maney Publishing; ICIMOD; Van Nostrand Reinhold, New York; ISO; ISRM; and all other publishers whose work has been referred to in this book. We appreciate their kind permission to use excerpts from their publications. In addition, we thank all the eminent researchers whose work is mentioned here. The authors are deeply grateful to Elsevier/Butterworth-Heinemann for the editing, production, and publication of this book.

We are also deeply grateful to our beloved families for their sacrifice, love, deep moral support, and suggestions; and to all of our friends and students. We also thank Holy Teacher Dr. B. K. Saxena, former scientist, CBRI, Roorkee, for his kind blessings.

We wish to encourage all enlightened engineers and geologists to kindly send their important suggestions for improving this book to us.

Chapter 1

Philosophy of Engineering Classifications

When you can measure what you are speaking about, and express it in numbers, you know something about it, but when you cannot measure it, when you cannot express it in numbers, your knowledge is of a meagre and unsatisfactory kind; it may be the beginning of knowledge, but you have scarcely in your thoughts, advanced to the stage of science.

Lord Kelvin

THE CLASSIFICATION

The science of classification is called "taxonomy"; it deals with the theoretical aspects of classification, including its basis, principles, procedures, and rules. Knowledge tested in projects is called the "practical knowledge." Surprisingly the rating and ranking systems have become popular in every part of life in the twenty first century.

Rock mass classifications form the backbone of the empirical design approach and are widely employed in rock engineering. Engineering rock mass classifications have recently been quite popular and are used in feasibility designs. When used correctly, a rock mass classification can be a powerful tool in these designs. On many projects the classification approach is the only practical basis for the design of complex underground structures. The Gjovik Underground Ice Hockey Stadium in Norway was designed by the classification approach.

Engineering rock mass classification systems have been widely used with great success in Austria, South Africa, the United States, Europe, and India for the following reasons:

1. They provide better communication between planners, geologists, designers, contractors, and engineers.
2. An engineer's observations, experience, and judgment are correlated and consolidated more effectively by an engineering (quantitative) classification system.
3. Engineers prefer numbers in place of descriptions; hence, an engineering classification system has considerable application in an overall assessment of the rock quality.
4. The classification approach helps in the organization of knowledge and is amazingly successful.
5. An ideal application of engineering rock mass classification occurs in the planning of hydroelectric projects, tunnels, caverns, bridges, silos, building complexes, hill roads, rail tunnels, and so forth.

The classification system, in the last 60 years of its development, has been cognizant of the new advances in rock support technology starting from steel rib supports to the latest supporting techniques such as rock bolts and steel fiber reinforced shotcrete (SFRS).

PHILOSOPHY OF CLASSIFICATION SYSTEM

In any engineering classification system, the minimum rating is called "poor rock mass" and the maximum rating is called "excellent rock mass." Thus, every parameter of a classification plays a more dominant role as overall rating decreases, and many classifications are accurate in both excellent and poor rock conditions. Reliability may decrease for medium rock conditions. No single classification is valid for assessment of all rock parameters. Selection of a classification for estimating a rock parameter is, therefore, based on experience. The objective should be to classify the undisturbed rock mass beyond excavated faces. *Precaution should be taken to avoid the double-accounting of joint parameters in the classification and in the analysis. Thus, joint orientation and water seepage pressure should not be considered in the classification if these are accounted for in the analysis.*

It is necessary to account for fuzzy variation of rock parameters after allowing for uncertainty; thus, it is better to assign a range of ratings for each parameter. There can be a wide variation in the engineering classifications at a location. When designing a project, the average of rock mass ratings (RMR) and geological strength index (GSI) should be considered in the design of support systems. For rock mass quality (Q), a geometric mean of the minimum and the maximum values should also be considered in the design.

A rigorous classification system may become more reliable if uncertain parameters are dropped and considered indirectly. An easy system's approach (Hudson, 1992) is very interesting and tries to sequence dominant parameters at a site (see Chapter 27). This classification is a holistic (whole) approach, considering all parameters.

Hoek and Brown (1997) realized that a classification system must be non-linear to classify poor rock masses realistically. In other words, the reduction in strength parameters with classification should be non-linear, unlike RMR in which strength parameters decrease linearly with decreasing RMR. (Mehrotra, 1993, found that strength parameters decrease non-linearly with RMR for dry rock masses.) More research is needed on the non-linear correlations for rock parameters and rock mass characterization.

Sound engineering judgment evolves out of long-term, hard work in the field.

NEED FOR ENGINEERING GEOLOGICAL MAP

Nature tends to be heterogeneous, which makes it easy to predict its weakest link. More attention should be focused on the weak zones (joints, shear zones, fault zones, etc.) in the rock mass that may cause wedge failures and/or toppling. Rock failure is localized and three dimensional in heterogeneous rock mass and not planar, as in homogeneous rock mass.

First, a geological map on macro-scale (1:50,000) should be prepared before tunneling or laying foundations. Then an engineering geological map on micro-scale (1:1000) should be prepared soon after excavation. This map should highlight geological details for an excavation and support system. These include Q, RMR, all the shear zones, faults, dip and dip directions of all joint sets (discontinuities), highest ground water table

(GWT), and so forth along tunnel alignment. The engineering geological map helps civil engineers immensely. Such detailed maps prepared based on thorough investigation are important for tunnel excavations. If an engineering geological map is not prepared then the use of a tunnel boring machine (TBM) is not advisable, because the TBM may get stuck in the weak zones, as experienced in Himalayan tunneling. An Iraqi proverb eloquently illustrates this idea:

Ask 100 questions, but do not make a single mistake.

MANAGEMENT OF UNCERTAINTIES

Empirical, numerical, or analytical and observational approaches are various tools for engineering designs. The empirical approach, based on rock mass classifications, is the most popular because of its simplicity and ability to manage uncertainties. Geological and geotechnical uncertainties can be tackled effectively using proper classifications. Moreover, this approach allows designers to make on-the-spot decisions regarding supporting measures if there is a sudden change in the geology. The analytical approach, on the other hand, is based on assumptions and obtaining correct values of input parameters. This approach is both time-consuming and expensive. The observational approach, as the name indicates, is based on monitoring the efficiency of the support system.

Classifications are likely to be invalid in areas where there is damage due to blasting and weathering such as in cold regions, during cloudbursts, and under oceans. If the rock has extraordinary geological occurrence (EGO) problems, then these should be solved under the guidance of national and international experts.

According to Fairhurst (1993), designers should develop design solutions and design strategies so that support systems are ductile and robust, that is, able to perform adequately even in unknown geological conditions. For example, shotcreted and reinforced rock arch is a robust support system. The Norwegian Method of Tunneling (NMT) after 30 years, has evolved into a successful strategy that can be adopted for tunnel supporting in widely different rock conditions.

PRESENT-DAY PRACTICE

Present-day practice is a combination of all of the previously described approaches. This is basically a "design as you go" approach. Experience led to the following strategy of refinement in the design of support systems.

1. In feasibility studies, empirical correlations may be used for estimating rock parameters.
2. At the design stage, in situ tests should be conducted for major projects to determine the actual rock parameters. It is suggested that in situ triaxial tests (with σ_1, σ_2, and σ_3 applied on sides of the cube of rock mass) should be conducted extensively, because σ_2 is found to affect both the strength and deformation modulus of rock masses in tunnels. This is the motivation for research, and its presentation in this book is likely to prove an urgent need for in situ polyaxial tests.
3. At the initial construction stage, instrumentation should be carried out in drifts, caverns, intersections, and other important locations with the objective of acquiring field data on displacements both on the supported excavated surfaces and within the rock mass. Instrumentation is also essential for monitoring construction quality. Experience confirms that instrumentation in a complex geological environment is the key to success

for a safe and steady tunneling rate. These data should be utilized in computer modeling for back analysis of both the model and its parameters (Sakurai, 1993).

4. At the construction stage, forward analysis of rock structures should be carried out using the back analyzed model and the parameters of rock masses. Repeated cycles of back analysis and forward analysis (BAFA) may eliminate many inherent uncertainties in geological mapping and knowledge of engineering behavior of rock masses. Where broken/plastic zones are predicted, the borehole extensometers should reveal a higher rate of displacement in the broken zone than in the elastic zone. The predicted displacements are very sensitive to the assumed model, parameters of rock masses and discontinuities, in situ stresses, and so forth.
5. The principle of dynamic programming should be adopted. Construction strategy will evolve with time in every step to reach the goal quickly; for example, grouting may improve ground conditions significantly. Dynamic programming is essentially a "re-design while you go" evolutionary approach.
6. The aim of computer modeling should be to design site-specific support systems and not just analysis of the strains and stresses in the idealized geological environment. In a non-homogeneous and complex geological environment, which is difficult to predict, slightly conservative rock parameter values may be assumed for the purpose of designing site-specific remedial measures (lines of defenses) and for accounting inherent uncertainties in geological and geotechnical investigations.
7. Be prepared for the worst and hope for the best.

SCOPE OF THE BOOK

This book presents an integrated system of classifications and their applications for tunnels, foundations, and landslides in light of the field research conducted in India and Europe during the last three decades. This revised edition offers an integrated practical knowledge on the rock mass characterization for use in software packages along with extensive tables.

This text is a specialized book on rock mass classifications and is written for civil engineers and geologists who have basic knowledge of these classifications. The analysis and design of rock slopes is beyond the scope of this book (see Singh & Goel, 2002). There are several types of popular software for non-linear analysis, but they need an approximate solution to be useful, which is provided by the engineering rock mass classification.

This book is written to help civil engineers and geologists working on civil engineering jobs such as hydroelectric projects, foundations, tunnels, caverns, and rapid landslide hazard zonation.

Some engineers work under the assumption that a rock mass is homogeneous and isotropic, but this may not always be correct as shear zones are encountered frequently. Because of this, shear zone treatment is discussed in Chapter 2.

REFERENCES

Fairhurst, C. (1993). Analysis and design in rock mechanics—The general context. In *Comprehensive rock engineering* (Vol. 3, Chap. 1, pp. 1–29). New York: Pergamon.

Hoek, E., & Brown, E. T. (1997). Practical estimation of rock mass strength. *International Journal of Rock Mechanics and Mining Sciences*, *34*(8), 1165–1186.

Hudson, J. A. (1992). *Rock engineering systems — Theory and practice* (p. 185). Chichester, UK: Horwood Ltd.

Mehrotra, V. K. (1993). *Estimation of engineering properties of rock mass* (p. 267). Ph.D. thesis. Uttarakhand, India: IIT Roorkee.

Sakurai, S. (1993). Back analysis in rock engineering. *ISRM News Journal*, *2*(2), 4–16.

Singh, B., & Goel, R. K. (2002). *Software for engineering control of landslide and tunnelling hazards* (p. 344). Rotterdam, A. A. Balkema (Swets & Zeitlinger).

Chapter 2

Shear Zone Treatment in Tunnels and Foundations

Nature is different everywhere, and she does not follow the text books.

Stini

SHEAR ZONE

A shear zone is a zone in which shearing has occurred so that the rock mass is crushed and brecciated. A shear zone is the outcome of a fault where the displacement is not confined to a single fracture, but is distributed through a fault zone. Shear zones vary in thickness from a fraction of meters to hundreds of meters. Depending upon the thickness, the shear zone has a variable effect on the stability of underground openings and foundations. The thicker the shear zone, the higher chance it will be unstable. Clay-like gouge in shear zones is generally highly over-consolidated and shows high cohesion. Similarly, weak zones, fault zones, and thrust zones can also cause instability.

TREATMENT FOR TUNNELS

Rock mass classifications consider only the homogeneous units, so downgrading the rock quality adjacent to shear zones may be difficult. It is envisaged that the rock mass affected by a shear zone is much larger than the shear zone. Hence, this rock mass must be downgraded to the quality of the shear zone so that a heavier support system can be installed. A method has been developed at the Norwegian Geotechnical Institute (NGI) for assessing support requirements using the Q-system for rock masses affected by shear zones (Grimstad & Barton, 1993). In this method, weak zones and the surrounding rock mass are allocated their respective Q-values from which a mean Q-value can be determined, taking into consideration the width of the weak zone. Equation (2.1) may be used in calculating the weighted mean Q-value (Bhasin et al., 1995).

$$\log Q_m = \frac{b.\ \log Q_{wz} + \log Q_{sr}}{b+1} \tag{2.1}$$

where Q_m = mean value of rock mass quality Q for deciding the support; Q_{wz} = Q-value of the weak zone; Q_{sr} = Q-value of the surrounding rock; and b = width of the weak zone in meters.

The strike direction (θ) and thickness of the weak zone (b) in relation to the tunnel axis is important for the stability of the tunnel; therefore, the following correction factors have been suggested for the value of b in Eq. (2.1).

if $\theta = 90°–45°$ to the tunnel axis, then use 1b
if $\theta = 45°–20°$, x3 then use 2b in place of b
if $\theta = 10°–20°$, then use 3b in place of b
if $\theta < 10°$, then use 4b in place of b

Equation (2.1) may also be used for estimating the weighted average value of the joint roughness number (J_{rm}) after appropriately replacing the log Q by J_r. Similarly, the weighted mean of joint alteration number (J_{am}) may also be estimated.

Further, when multiplying Eq. (2.1) by 25 in the numerator and replacing 25 log Q by E (E = 25 log Q; Barton et al., 1980), the average value of modulus of deformation E_m can be estimated as follows:

$$E_m = \frac{b.\ E_{wz} + E_{sr}}{b + 1} \tag{2.2}$$

where E_{wz} = modulus of deformation of the weak zone or the shear zone and E_{sr} = modulus of deformation of the surrounding rock mass.

Thus, E_m, Q_m, and J_{rm} may also be used to design support systems for shear zones or weak zones by using the semi-empirical method discussed in Chapter 12 or TM software (Singh & Goel, 2002).

A 3D finite element analysis of the underground powerhouse of the Sardar Sarovar Hydroelectric Project in India shows that the maximum deformations of walls are increased near the shear zone (b = 2 m) by a factor of E_{sr}/E_m. The predicted support pressure on shotcrete near the shear zone is increased to about $0.2\, Q_m^{-1/3}/J_{rm}$ (MPa) and the support pressures in the surrounding rock away from the shear zone are approximately $0.2\, Q_{sr}^{-1/3}/J_{rsr}$ (MPa), in which J_{rsr} is the joint roughness number of the surrounding rock mass (Samadhiya, 1998). These computations are quite encouraging.

If the surrounding rock mass near a shear zone is downgraded by using Eqs. (2.1) and (2.2), a heavier support should be chosen for the whole area instead of just the weak zone.

Figure 2.1 shows a typical treatment method for shear zones (Lang, 1971). First the shear zone is excavated with caution up to some depth. Immediately after excavation one thin layer of steel fiber reinforced shotcrete (SFRS) is sprayed. The weak zone is then

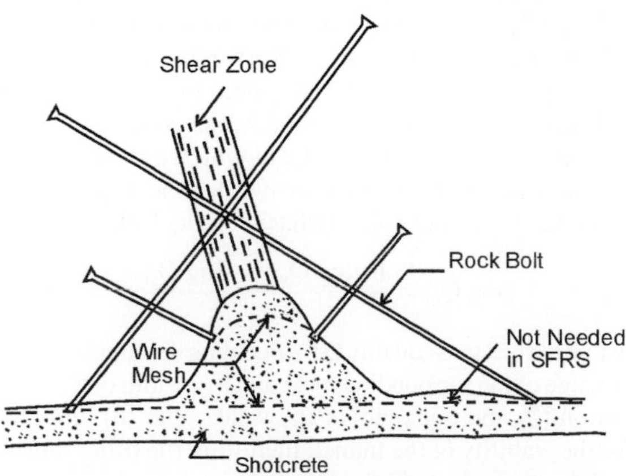

FIGURE 2.1 Shear zone treatment in an underground opening. *(From Lang, 1971)*

reinforced with inclined rock bolts, and shotcrete (preferably SRFS) is sprayed ensuring its proper thickness in the weak zones. This methodology is urgently needed if the New Austrian Tunneling Method (NATM) or Norwegian Method of Tunneling (NMT) is to be used in the tunnels of the Himalayan region, as seams, shear zones, faults, thrusts, and thin intra-thrust zones are frequently found along tunnels and caverns there. "Stitching" is perhaps the terminology that best suits this requirement.

In a thick shear zone (b >> 2m) with sandy gouge, umbrella grouting or rock bolting is used to enhance the strength of the roof and walls before tunneling. The excavation is made manually. Steel ribs are placed closely and shotcreted until the shear zone is crossed. Each round of advancement should be limited to 0.5 m or even smaller depending upon the stand-up time of the material and be fully supported before starting another round of excavation.

In the Himalayan tunnels the rock mass above the shear zone is often water charged. This may be because of the presence of impermeable gouge material in the shear zone. Hence, engineers should be prepared to tackle this problem from the start of the project.

TREATMENT FOR DAM FOUNDATIONS

Treatment of a shear zone in a concrete dam foundation consists of dental treatment, as shown in Figure 2.2. The vertical depth d of excavation of the weak zone and backfilling by (dental) concreting is recommended by the USBR (1976) as follows:

$$\begin{aligned} d &= 0.00656\ b\ H + 1.53,\ (m)\ \text{for}\ H < 46m \\ &= 0.3\ b + 1.52,\ (m)\ \text{for}\ H \geq 46m \\ &> 0.1\ H\ \text{in seams with clayey gouge} \end{aligned} \tag{2.3}$$

where H = height of dam above general foundation level in meters; b = width of weak zone in meters; and d = depth of excavation of weak zone below surface adjoining the sound rock in meters.

The infilling and crushed weathered rock is oozed out by water jet at very high pressure and then backfilled by rich concrete. No blasting is used to avoid damage to the rock mass. Sharma (personal communication with Bhawani Singh) designed

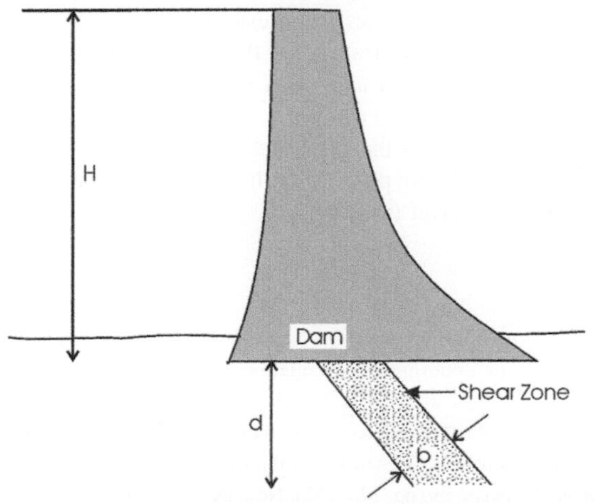

FIGURE 2.2 Shear zone treatment below dam foundations.

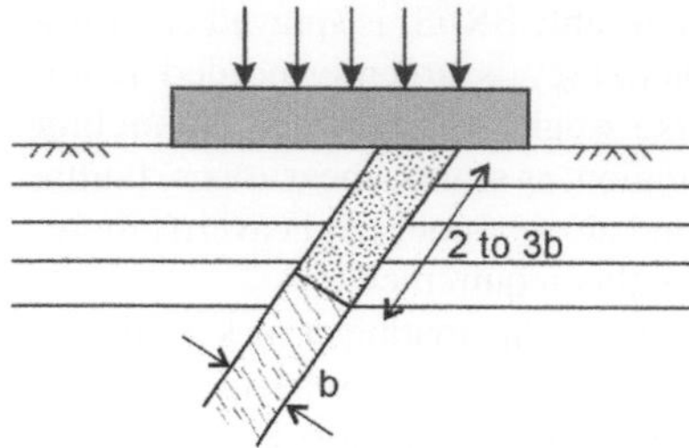

FIGURE 2.3 Weak seams under foundation less than 20% of the area.

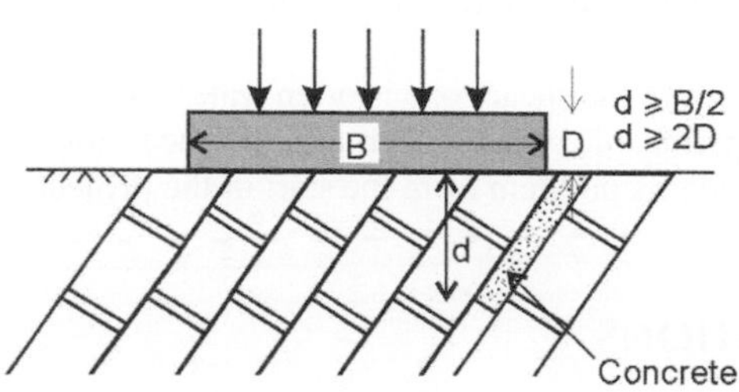

FIGURE 2.4 Foundation on steeply dipping clay seam (D = foundation depth).

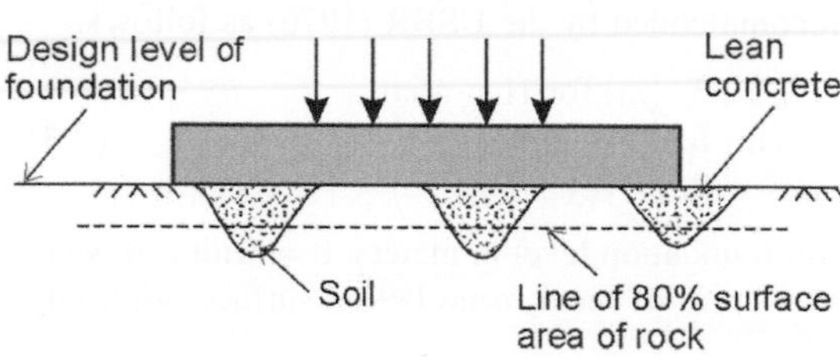

FIGURE 2.5 Foundation on undulating rock surface.

reinforcement at the bottom of the gravity dam by cable anchors to rest over a wide shear zone with five branches in the foundation area in the Himalayas by using the computer program FLAC^{3D}. Unfortunately, it was too late to change the site of this dam.

The treatment of shear zones, joints, solution cavities in limestone, and so forth is essential for the long life of building foundations. The strategy for their treatment should be the same as that adopted for dam foundations and as shown in Figures 2.3 to 2.5 as per Indian Standard code (IS13063, 1991).

Undulating rock profiles are a major problem in the construction of footings, well foundations, and piles. However, massive rocks do not pose instability problems, because their behavior is similar to that of the rock material (intact rock).

REFERENCES

Barton, N., Loset, F., Lien, R., & Lunde, J. (1980). Application of Q-system in design decisions concerning dimensions and appropriate support for underground installations. In *Subsurface Space* (pp. 553–561). Oxford: Pergamon.

Bhasin, R., Singh, R. B., Dhawan, A. K., & Sharma, V. M. (1995). Geotechnical evaluation and a review of remedial measures in limiting deformations in distressed zones in a powerhouse cavern. In *Conference on Design and Construction of Underground Structures* (pp. 145–152). New Delhi, India.

Grimstad, E., & Barton, N. (1993). Updating of the Q-system for NMT. In *Proceedings of the International Symposium on Sprayed Concrete—Modern Use of Wet Mix Sprayed Concrete for Underground Support*. Oslo: Fagernes, Norwegian Concrete Association.

IS13063. (1991). *Structural safety of buildings on shallow foundations on rock—Code of practice* (p. 15). New Delhi: Bureau of Indian Standards.

Lang, T. A. (1971). Theory and practice of rock bolting. *AIME Trans*. 220.

Samadhiya, N. K. (1998). *Influence of shear zone on stability of cavern* (p. 334). Ph.D. Thesis. Uttarakhand, India: Dept. of Civil Engineering, IIT Roorkee.

Singh, B., & Goel, R. K. (2002). *Software for engineering control of landslides and tunnelling hazards* (p. 344). Rotterdam: A. A. Balkema (Swets & Zeitlinger).

USBR. (1976). *Design of gravity dams* (pp. 97–105). U.S. Bureau of Reclamation.

Chapter 3

Rock Material

In all things of nature there is something of the marvelous.

Aristotle

ROCK MATERIAL

The term "rock material" refers to the intact rock within the framework of discontinuities. In other words, this is the smallest element of rock block not cut by any fracture. There are always some micro-fractures in the rock material, but these should not be treated as fractures. Rock material differs from "rock mass," which refers to in situ rock together with its discontinuities and weathering profile. Rock material has the characteristics shown in Figure 3.1.

HOMOGENEITY AND INHOMOGENEITY

Bray (1967) demonstrated that if a rock contains ten or more sets of discontinuities (joints), then its behavior can be approximated to the behavior of a homogeneous and isotropic mass with only 5% error due to assumed homogeneity and isotropic condition. Also, if a rock is massive and contains very little discontinuity, it could ideally behave as a homogeneous medium. Hoek and Brown (1980) showed that homogeneity is a characteristic dependent on the sample size. If the sample size is considerably reduced, the most heterogeneous rock will become a homogeneous rock (Figure 3.2). In the figure s is a constant that depends on rock mass characteristics as discussed in Chapter 26. Deere et al. (1969) suggested that if the ratio between fracture spacing and opening size is equal to or less than 1/100, the rock should be considered discontinuous and beyond this range it should be considered a continuum and possibly anisotropic.

An inhomogeneous rock is more predictable than a homogeneous rock because the weakest rock gives distress signals before final collapse of the rock structure.

CLASSIFICATION OF ROCK MATERIAL

Ancient Shilpshastra in India classified rocks on the basis of color, sound, and heaviness. ISO14689-1 (2003) proposed classification of rock material based on uniaxial compressive strength (UCS) as shown in Table 3.1. It is evident that rock material may show a large scatter in strength, say of the order of 10 times; hence, the need for a classification system based on strength and not mineral content.

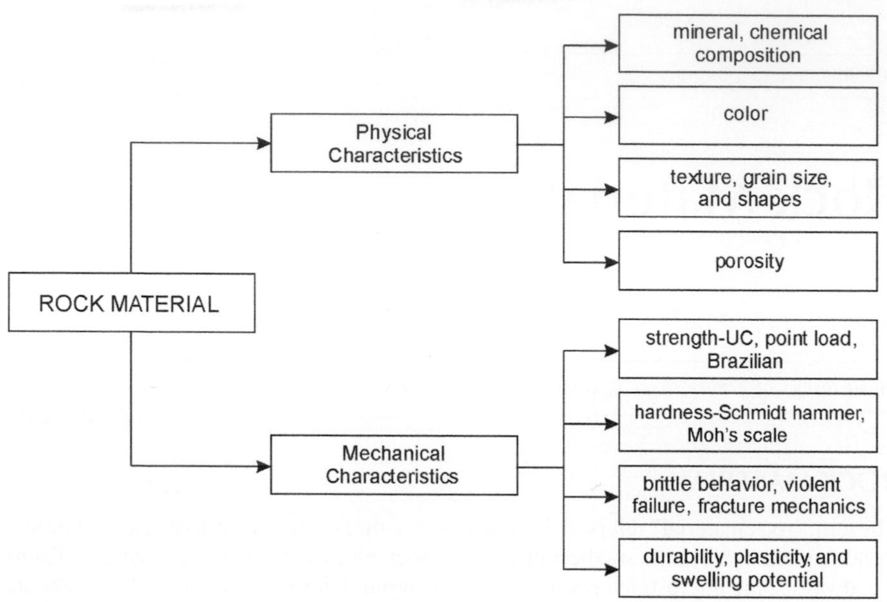

FIGURE 3.1 Material characteristics of rocks.

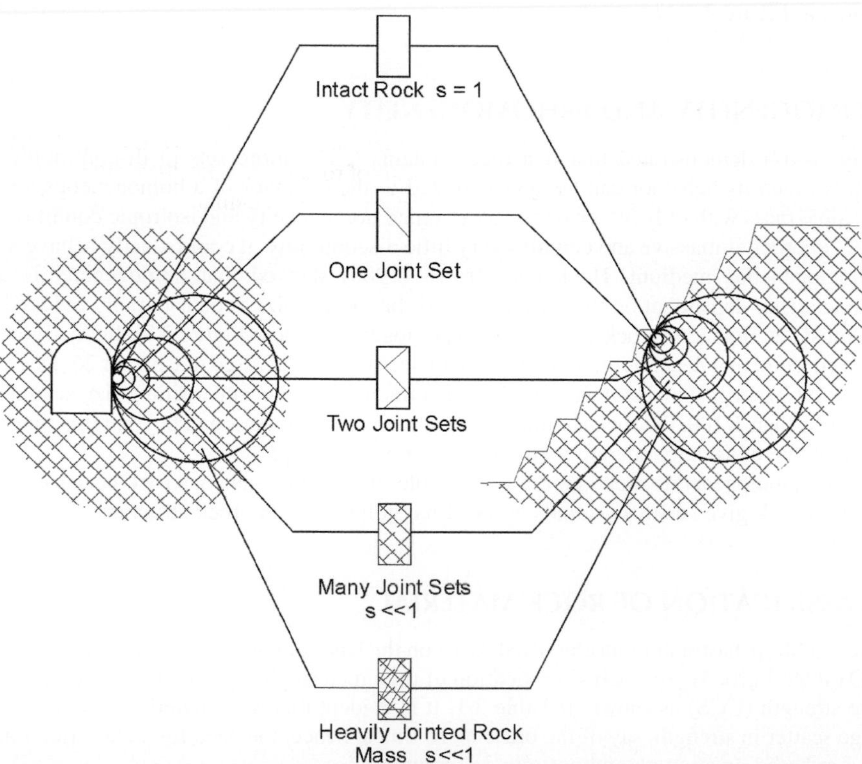

FIGURE 3.2 Rock mass conditions under the Hoek-Brown failure criterion. *(From Hoek, 1994)*

TABLE 3.1 Classification of Rock Material Based on Unconfined Compressive Strength

Term for uniaxial compressive strength	Symbol	Strength (MPa)	Ranges for common rock materials: *Granite, basalt, gneiss, quartzite, marble*	*Schist, sandstone*	*Limestone, siltstone*	*Slate*	*Concrete*
Extremely weak*	EW	<1		**	**		
Very weak	VW	1–5		**	**	**	**
Weak	W	5–25		**	**	**	**
Medium strong	MS	25–50	**		**	**	
Strong	S	50–100	**				
Very strong	VS	100–250	**				
Extremely Strong	ES	>250	**				

**Some extremely weak rocks behave as soils and should be described as soils.*
***Indicates the range of strength of rock material.*
Source: ISO 14689-1, 2003.

The UCS can be easily predicted from point load strength index tests on rock cores and rock lumps right at the drilling site because ends of rock specimens do not need to be cut and lapped. UCS is also found from Schmidt's rebound hammer (see Chapter 15). Table 8.13 lists typical approximate values of UCS.

There are frequent legal disputes on soil-rock classification. The International Standard Organization (ISO) classifies geological material having a UCS less than 1.0 MPa as soil.

Deere and Miller (John, 1971) suggested another useful classification system based on the modulus ratio, which is defined as the ratio between elastic modulus and UCS. Physically, a modulus ratio indicates the inverse of the axial strain at failure. Thus, brittle materials have a high modulus ratio and plastic materials exhibit a low modulus ratio.

CLASS I AND II BRITTLE ROCKS

Rock material has been divided into two classes according to their post-peak stress-strain curve (Wawersik, 1968).

Class I: Fracture propagation is stable because each increment of deformation beyond the point of maximum load-carrying capacity requires an increment of work to be done on the rock.

Class II: Rocks are unstable or self-sustaining; elastic energy must be extracted from the material to control fracture.

The introduction of partial confinement, as in short samples when end constraint becomes prominent, is likely to have a satisfactory effect. If end restraint becomes severe, it is possible that a Class II rock might behave like a Class I material.

Wawersik (1968) conducted experiments on six rock types to demonstrate the features of Class I and II rocks (Figure 3.3). Typical S-shape stress-strain curves may be obtained for rocks with micro-fractures. Further, the post-peak curve for Class II rocks shows reduction of strain after failure. The lateral strain increases rapidly after peak stress in Class II rocks. Brittle rocks, therefore, may be kept in the Class II category.

A deep tunnel within dry, massive, hard Class II and laminated rocks may fail because of rock bursts due to uncontrolled fracturing where tangential stress exceeds the strength of the rock material (see Chapter 13). Hence, it is necessary to test rock material in a Servo-controlled closed loop testing machine to get the post-peak curve.

UNIAXIAL COMPRESSION

Rock failure in uniaxial compression occurs in two modes: (1) local (axial) splitting or cleavage failure parallel to the applied stress, and (2) shear failure.

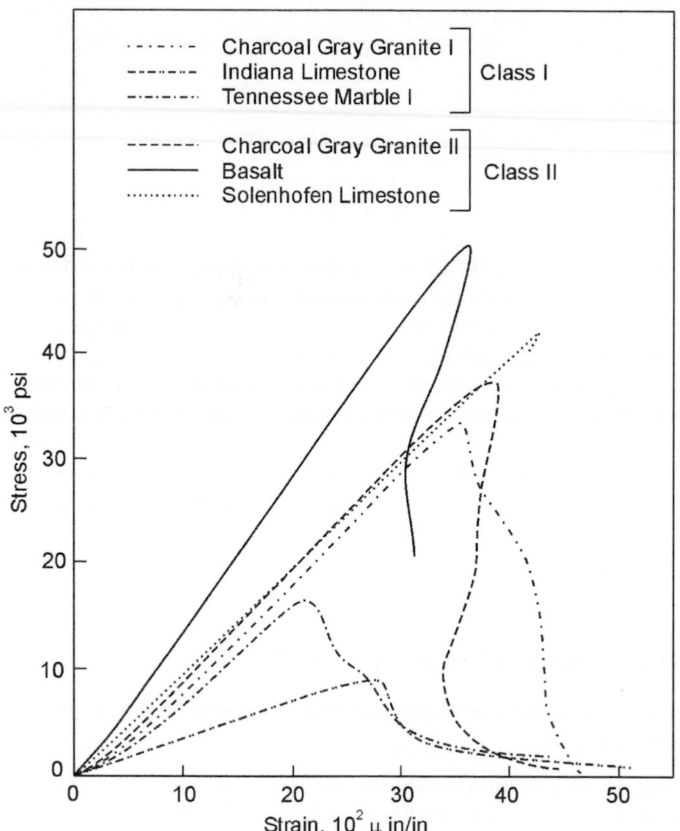

FIGURE 3.3 Stress-strain curves for six representative rocks in uniaxial compression. *(From Wawersik, 1968)*

Local cleavage fracture characterizes fracture initiation at 50 to 95% of the compressive strength and is continuous throughout the entire loading history. Axial cleavage fracture is a local stress-relieving phenomenon that depends on the strength anisotropy and brittleness of the crystalline aggregates as well as on the grain size of the rock. Local axial splitting is virtually absent in fine-grained materials at stress levels below their compressive strength.

Shear failure manifests in the development of boundary faults (followed by interior fractures), which are oriented at approximately 30 degrees to the sample axis. In fine-grained materials where the inhomogeneity of the stress distribution depends only on the initial matching of the material properties at the loading platen interfaces, boundary and interior faults are likely to develop simultaneously and appear to have the same orientation for any rock type within the accuracy of the measurements on the remnant pieces of collapsed specimens (basalts, etc.).

Local axial fracturing governs the maximum load-carrying ability of coarse-grained, locally inhomogeneous Class I and II rock types. Thus, in coarse-grained rocks the ultimate macroscopic failure mode of fully collapsed samples in uniform uniaxial compression cannot be related to peak stress. In fine-grained, locally homogeneous rock types, which most likely are Class II, the peak stress is probably characterized by the development of shear fractures seen in continuous failure planes. In controlled fracture experiments on very fine-grained rocks, the final appearance of a collapsed rock specimen probably correlates with its compressive strength. However, if rock fracture is uncontrolled, then the effects of stress waves produced by the dynamic release of energy may override the quasi-elastic failure phenomenon to such an extent that the latter may no longer be recognizable.

The extent of the development of the two basic failure modes, local axial splitting and slip or shear failure, determines the shape of the stress-strain curve for all rocks subjected to unidirectional or triaxial loading. Partially failed rocks still exhibit elastic properties. However, the sample stiffness decreases steadily with increasing deformation and loss of strength.

Macroscopic cleavage failure (e.g., laboratory samples splitting axially into two or more segments) was never observed in the experiments on Class I and II rocks. An approximate theoretical analysis of the "sliding surface" model, which was proposed by Fairhurst and Cook (1966), revealed qualitatively that unstable axial cleavage fracture is an unlikely failure mode of rocks in uniaxial compression.

The dynamic tensile strength of rocks (granite, diorite, limestone, and grigen) is found to be about four to five times the static tensile strength (Mohanty, 2009). Brazilian tensile strength of laminated rocks and other argillaceous weak rocks like marl do not appear to be related to the UCS of rock material (Constantin, personal communication).

STABILITY IN WATER

In hydroelectric projects, rocks are charged with water. The potential for disintegration of rock material in water can be determined by immersing rock pieces in water for up to one week. Their stability can be described using the terms listed in Table 3.2 (ISO 14689-1, 2003).

Ultrasonic pulse velocity in a saturated rock is higher than in a dry rock because it is easier for pulse to travel through water than in air voids. However, the UCS and modulus of elasticity are reduced significantly after saturation, particularly in rocks with water sensitive minerals. On the other hand, the post-peak stress-strain curve becomes flatter in the case of undrained UCS tests on saturated samples because increasing fracture porosity after failure creates negative pore water pressure.

TABLE 3.2 Rock Material Stability in Water

Term	Description (after 24 h in water)	Grade
Stable	No changes	1
Fairly stable	A few fissures are formed or specimen surface crumbles slightly	2
	Many fissures are formed and broken into small lumps or specimen surface crumbles	3
Unstable	Specimen disintegrates or nearly the whole specimen surface crumbles	4
	The whole specimen becomes muddy or disintegrates into sand	5

Source: ISO 14689-1, 2003.

CLASSIFICATION ON THE BASIS OF SLAKE DURABILITY INDEX

Based upon his tests on representative shales and clay stones for two 10-minute cycles after drying, Gamble (1971) found the slake durability index varied from 0 to 100%. There are no visible connections between durability and geological age, but durability increased linearly with density and inversely with natural water content. Based on his results, Gamble proposed a classification of slake durability as seen in Table 3.3. The slake durability classification is useful when selecting rock aggregates for road, rail line, concrete, and shotcrete.

Rock in field is generally jointed. It was classified by core recovery in the past and later in the 1960s by modified core recovery (RQD), which will be discussed in Chapter 4.

TABLE 3.3 Slake Durability Classification

Group name	% retained after one 10-minute cycle (dry weight basis)	% retained after two 10-minute cycles (dry weight basis)
Very high durability	>99	>98
High durability	98–99	95–98
Medium high durability	95–98	85–95
Medium durability	85–95	60–85
Low durability	60–85	30–60
Very low durability	<60	<30

Source: Gamble, 1971, 2003.

REFERENCES

Bray, J. W. (1967). A study of jointed and fractured rock. Part I. *Rock Mechanics and Engineering Geology, 5–6*(2–3), 117–136.

Deere, D. U., Peck, R. B., Monsees, J. E., & Schmidt, B. (1969). *Design of tunnel liners and support system* (Final Report, University of Illinois, Urbana, for Office of High Speed Transportation, Contract No. 3-0152, p. 404). Washington, D.C.: U.S. Department of Transportation.

Fairhurst, C., & Cook, N. G. W. (1966). The phenomenon of rock splitting parallel to the direction of maximum compression in the neighborhood of a surface. In: *Proceedings 1st Congress, International Society of Rock Mechanics,* Lisbon, pp. 687–692.

Gamble, J. C. (1971). *Durability—Plasticity classification of shales and other argillaceous rocks* (p. 159). Ph.D. Thesis. University of Illinois.

Hoek, E. (1994). Strength of rock and rock masses. *ISRM News Journal, 2*(2), 4–16.

Hoek, E., & Brown, E. T. (1980). *Underground excavations in rocks. Institution of Mining and Metallurgy* (p. 527). London: Maney Publishing.

ISO 14689-1 (2003). (E). *Geotechnical investigation and testing—Identification and classification of rock—Part 1: Identification and description* (pp. 1–16). Geneva: International Organization for Standardization.

Mohanty, B. (2009). Measurement of dynamic tensile strength in rock by means of explosive-driven Hopkinson bar method. In *Workshop on Rock Dynamics, ISRM Commission on Rock Dynamics.* Lausanne, Switzerland: EPFL, June.

Wawersik, W. R. (1968). *Detailed analysis of rock failure in laboratory compression tests* (p. 165). Ph.D. Thesis. University of Minnesota.

Chapter 4

Rock Quality Designation

Strength and weaknesses go together both in matter and life. If nature has given weakness, nature will compensate. No one is perfect.

IIT Roorkee

ROCK QUALITY DESIGNATION

Rock quality designation (RQD) was introduced by Deere in 1964 as an index of assessing rock quality quantitatively. It is more sensitive as an index of the core quality than the core recovery.

The RQD is a modified percent core recovery that incorporates only sound pieces of core that are 100 mm (4 in.) or greater in length along the core axis

$$\text{RQD} = \frac{\text{sum of core pieces} \geq 10\ \text{cm}}{\text{total drill run}}.\ 100,\ \%$$

RQD is found to be a practical parameter for core logging, but it is not sufficient on its own to provide an adequate description of rock mass (Bieniawski, 1984). The following methods are used for obtaining RQD.

DIRECT METHOD

For RQD determination, the International Society for Rock Mechanics (ISRM) recommends a core size of at least NX (54.7 mm) drilled with double-tube core barrel using a diamond bit. Artificial fractures can be identified by close fitting cores and unstained surfaces. All of the artificial fractures should be ignored while counting the core length for RQD. A slow rate of drilling will also give better RQD.

The relationship between RQD and the engineering quality of the rock mass as proposed by Deere (1968) is seen in Table 4.1.

The correct procedure for measuring RQD is shown in Figure 4.1. RQD is perhaps the most common method for characterizing the degree of jointing in borehole cores, although this parameter may also implicitly include other rock mass features like weathering and "core loss" (Bieniawski, 1989).

TABLE 4.1 Correlation between RQD and Rock Quality

S. No.	RQD (%)	Rock quality
1	<25	Very poor
2	25–50	Poor
3	50–75	Fair
4	75–90	Good
5	90–100	Excellent

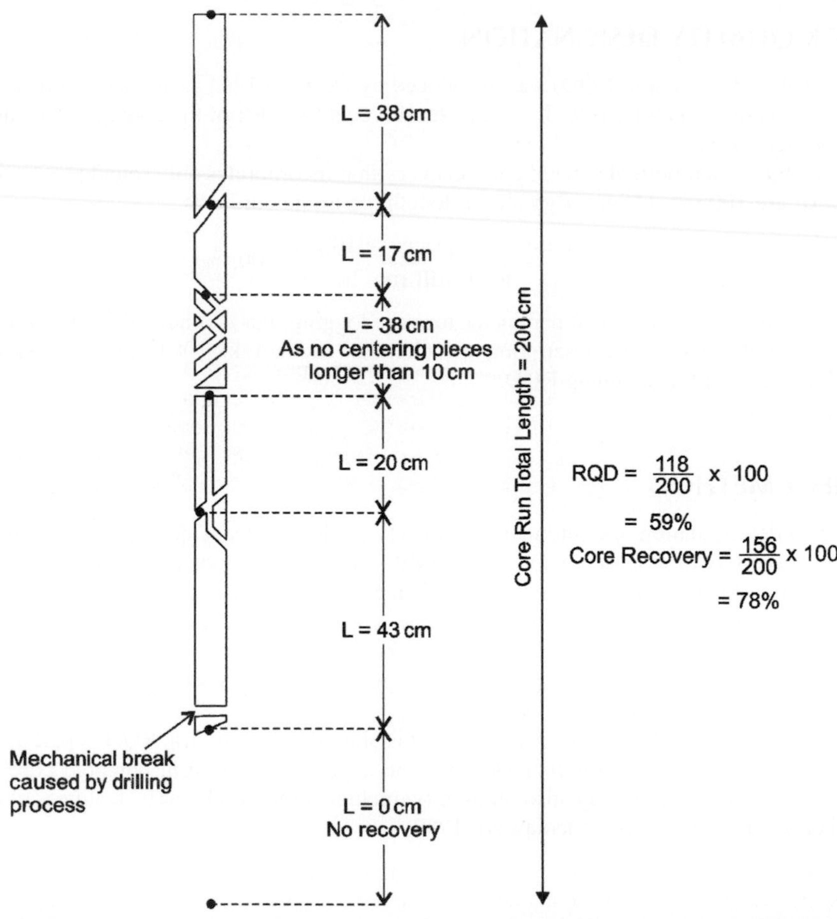

FIGURE 4.1 Procedure for measurement and calculation of rock quality designation (RQD). *(From Deere, 1989)*

INDIRECT METHODS

Seismic Method

The seismic survey method uses the variation of elastic properties of the strata that affect the velocity of the seismic waves traveling through them, thus providing useful information about the subsurface strata. This method is relatively cheap and rapid to apply and is helpful when studying a large volume of rock masses. The following information regarding rock masses is obtained from these tests: a (1) location and configuration of bedrock and geological structures in the subsurface, and (2) the effect of discontinuities in rock mass may be estimated by comparing the in situ compressional wave velocity with laboratory sonic velocity of intact drill core obtained from the same rock mass:

$$\mathrm{RQD} = (V_F/V_L)^2 \,.\, 100 \tag{4.1}$$

where V_F is in situ compressional wave velocity and V_L is compressional wave velocity in intact rock core. For details of a seismic method, any textbook dealing with this topic may be useful.

Volumetric Joint Count

When cores are not available, RQD may be estimated from the number of joints (discontinuities) per unit volume (J_v). A relationship used to convert J_v into RQD for clay-free rock masses is (Palmstrom, 1982)

$$\mathrm{RQD} = 115 - 3.3\, J_v \tag{4.2a}$$

where J_v represents the total number of joints per cubic meter or the volumetric joint count. Palmstrom (2005) proposed a new equation (Eq. 4.2b):

$$\mathrm{RQD} = 110 - 2.5\, J_v \tag{4.2b}$$

The new correlation (Eq. 4.2b) probably gives a more appropriate average correlation than the existing Eq. (4.2a), which may be representative for the long or flat blocks, while Eq. (4.2b) is better used for blocks of a cubical (bar) shape (Palmstrom, 2005).

The volumetric joint count (J_v) has been described by Palmstrom (1982, 1985, 1986) and Sen and Eissa (1992). It is a measure for the number of joints within a unit volume of rock mass defined by

$$J_v = \sum_{i=1}^{J} \left(\frac{1}{S_i}\right) \tag{4.3}$$

where S_i is the average joint spacing in meters for the ith joint set and J is the total number of joint sets except the random joint set.

Random joints may also be considered by assuming a "random spacing." Palmstrom (1982) presented an approximate rule of thumb correction for this with a spacing of 5 m for each random joint (Palmstrom, 2005):

$$J_v = \sum_{i=1}^{J} \left(\frac{1}{S_i}\right) + \frac{N_r}{5\sqrt{A}} \tag{4.4}$$

where N_r is the number of random joints in the actual location and A is the area in m^2. N_r can be estimated from joint observations, because it is based on measurements of random frequencies. In cases where random or irregular jointing occurs, J_v can be found by counting all of the joints observed in an area of known size. Table 4.2 shows the classification of J_v.

TABLE 4.2 Classification of Volumetric Joint Count (J_v)

S. No.	Degree of jointing	J_v
2	Very low	< 1.0
3	Low	1–3
4	Moderately	3–10
5	High	10–30
6	Very high	30–60
7	Crushed	>60

Source: Palmstrom, 2005.

Palmstrom (2002) reported that Eq. (4.2a) may be inaccurate for several situations. Eq. (4.2a) generally gives values of RQD that are too low. However, when cores are not available, Eq. (4.2a,b) has been found to be an alternative for estimating RQD.

Although RQD is a simple and inexpensive index, when considered alone it is not sufficient to provide an adequate description of a rock mass because it disregards joint orientation, joint condition, type of joint filling, and stress condition.

Correlation between J_v and V_b

As has been shown by Palmstrom (2005), the correlation between the block volume (V_b) and the volumetric joint count (J_v) is

$$V_b = \beta (J_v)^{-3} \tag{4.5a}$$

where β is the block shape factor, having the following characterization:

- For equidimensional (cubical or compact) blocks β = 27
- For slightly long (prismatic) and for slightly flat (tabular) blocks β = 28–32
- For moderately long and for moderately flat blocks β = 33–59
- For long and for flat blocks β = 60–200
- For very long and for very flat blocks β > 200.

A common value for β = 36.

Palmstrom (2005) has shown that the block shape factor (β) may crudely be estimated from

$$\beta = 20 + 7a3/a1 \tag{4.5b}$$

where a1 and a3 are the shortest and longest dimensions of the block.

WEIGHTED JOINT DENSITY

The weighted joint measurement method, proposed by Palmstrom (1996), was developed to achieve better information from borehole and surface observations. In principle, it is based on the measurement of the angle between each joint and the surface or the drill hole (Figure 4.2). The weighted joint density (wJd) is defined as

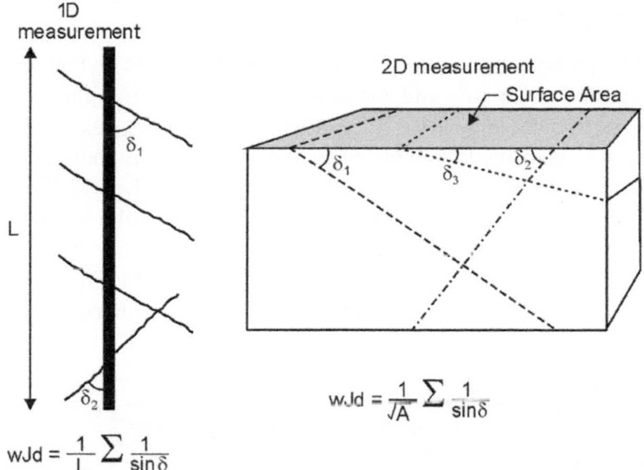

FIGURE 4.2 The intersection between joints and a drill core hole (left) and between joints and a surface (right). *(From Palmstrom, 1996)*

$$\text{for measurements in rock surface: } \text{wJd} = \frac{1}{\sqrt{A}}\sum f_i \tag{4.6}$$

$$\text{for measurements along a drill core or scan line: } \text{wJd} = \frac{1}{L}\sum f_i \tag{4.7}$$

where δ is the intersection angle, that is, the angle between the observation plane or drill hole and the individual joint (Fig. 4.2); A is the size of the observed area in m^2; L is the length of the measured section along the core or scan line (Figure 4.2); and f_i is a rating factor (Table 4.3).

To solve the problem of small intersection angles and to simplify the observations, the angles have been divided into intervals for which a rating of f_i has been selected, as shown in Table 4.3. The selection of intervals and the rating of f_i have been determined from a simulation.

To make the approach clear, examples are given in the next section for both surface and drill hole measurements.

TABLE 4.3 Angle Intervals and Rating of the Factor f_i

Angle interval (between joint and borehole or surface)	1/sinδ	Chosen rating of the factor f_i
$\delta > 60°$	<1.16	1
$\delta = 31–60°$	1.16–1.99	1.5
$\delta = 16–30°$	2–3.86	3.5
$\delta < 16°$	>3.86	6

Source: Palmstrom, 2005.

Surface Measurement

Two examples of jointing seen on a surface are shown in Figure 4.3. The observation area in both the examples is 25 m^2, and the results from the observations are given in Table 4.4. In the second example all of the joints belong to joint sets and there is no random joint. Thus, it is possible to calculate the volumetric joint count ($J_v = 3.05$) from the joint spacings of 0.85 m, 1.0 m, and 1.1 m. As observed, the weighted joint density measurement produces values that are somewhat higher than the known value for the volumetric joint count (Palmstrom, 1996).

The rock block shape should be described according to the terms in Table 4.5, such as tabular blocks, columnar blocks, and so forth. The shape of rock blocks should be correlated to the joint spacing.

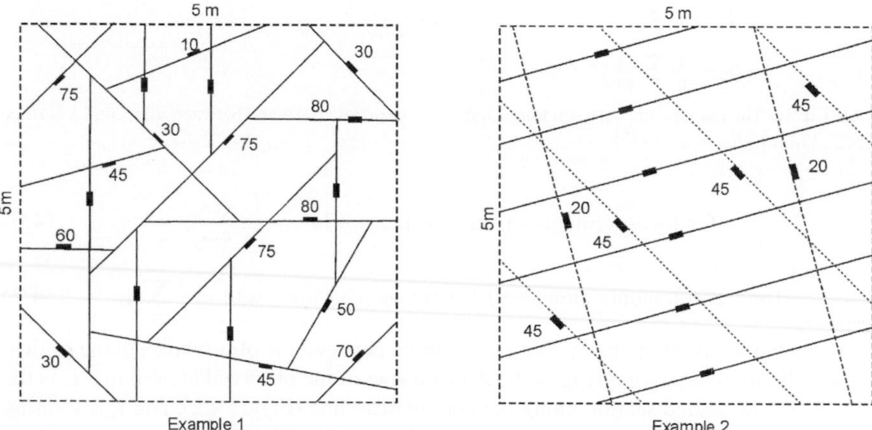

FIGURE 4.3 Two examples of jointing on a surface. *(From Palmstrom, 1996)*

TABLE 4.4 Calculation of Weighted Joint Density from Analysis of Jointing Shown for the Surfaces in Figure 4.3

	Area (A)	Number of joints (n) within each interval				Total number	Number of weighted joints		
Location	m^2	>60°	31–60°	16–30°	<16°	of joints	$N_w = \sum n \times f_i$	wJd = $(1/\sqrt{A})\, N_w$	Jv
Example 1	25	12	4	3	1	20	34.5	6.9	
Example 2	25	6	4	2	0	12	19	3.8	3.05
Rating of f_i =		1	1.5	3.5	6				

Source: Palmstrom, 1996.

TABLE 4.5 Terms to Describe the Main Rock Mass Structures and Block Shapes

S. No.	Terms	Figure	Description
1	Polyhedral blocks	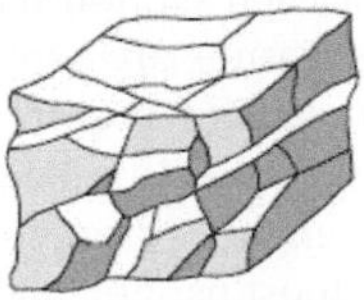	Irregular discontinuities without arrangement into distinct sets, and of small persistence
2	Tabular blocks	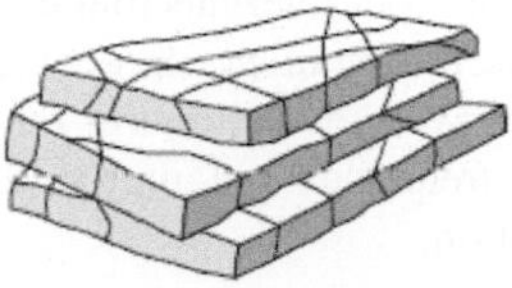	One dominant set of parallel discontinuities, for example, bedding planes, with other noncontinuous joints; thickness of blocks much less than length or width
3	Prismatic blocks	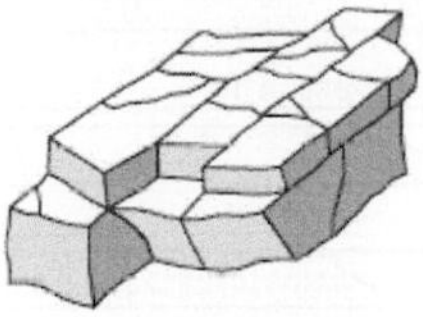	Two dominant sets of discontinuities, approximately orthogonal and parallel, with a third irregular set; thickness of blocks much less than length or width
4	Equidimensional blocks	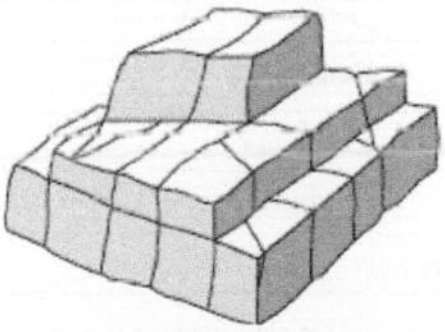	Three dominant sets of discontinuities, approximately orthogonal, with occasional irregular joints, giving equi-dimensional blocks
5	Rhombohedral blocks	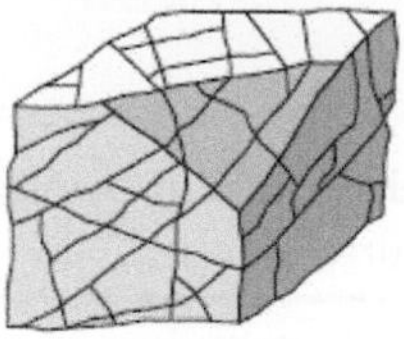	Three (or more) dominant, mutually oblique, sets of joints, giving oblique-shaped, equidimensional blocks
6	Columnar blocks	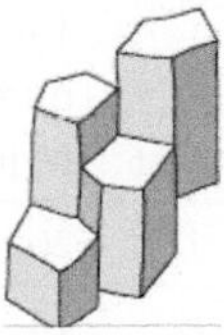	Several, usually more than three, sets of discontinuities; parallel joints usually crossed by irregular joints; length much greater than other dimensions

Sources: ISO 14689-1, 2003; Palmstrom, 2005.

Drill Hole Measurements

An example from core logging is shown in Figure 4.4. The 5 m long part of the core has been divided into the following three sections with similar density of joints: 50.0–52.17 m, 52.17–53.15 m, and 53.15–55.0 m. For each section the number of joints within each angle interval has been counted and the results are shown in Table 4.6.

The evaluation of weighted joint density requires small additional effort over currently adopted logging practices. The only additional work is to determine the number of joints within each angle interval. The angles chosen for the intervals between the joint and the drill hole should be familiar to most people, and this should make the observations for wJd quick. The use of only four intervals makes the registration simple and easy. Eventually, wJd may prove a useful parameter to accurately measure the joint density.

Priest and Hudson (1976) derived the following relationship (Eq. 4.8) between the RQD and linear discontinuity frequency per meter (λ) where discontinuity spacing follows an exponential distribution.

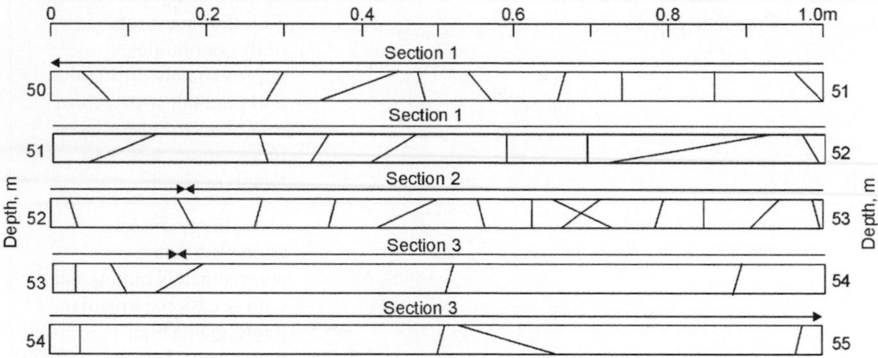

FIGURE 4.4 Example of jointing along part of a borehole. *(From Palmstrom, 1996)*

TABLE 4.6 Calculation of the Weighted Joint Density from Registration of Jointing in the Borehole in Figure 4.4

Depth	Length (L)	Number of joints (n) within each interval				Total number of joints	Number of weighted joints	wJd =
m	*m*	*>60°*	*31–60°*	*16–30°*	*<16°*		$N_w = \sum n \times f_i$	(1/L) N_w
50–52.17	2.17	11	6	2	1	20	33	15
52.17–53.15	0.98	9	3	2	0	14	20.5	20.9
53.15–55.0	1.85	5	0	1	0	6	8.5	4.6
Rating of f_i =		1	1.5	3.5	6			

Source: Palmstrom, 1996.

$$\text{RQD} = 100(0.1\lambda + 1)e^{-0.1\lambda} \tag{4.8}$$

Romana (1993) validated Eq. (4.8) for RQD > 50%. For $6 < \lambda < 16$ a simplification is

$$\text{RQD} = -3.68\lambda + 110.4 \tag{4.9}$$

RED-FLAG EFFECT OF LOW RQD

As suggested by Deere and Deere (1988), a low RQD value should be considered a "red flag" for further action. The reason for low RQD values must be determined: poor drilling techniques, core breakage upon handling, stress-relief or air staking, thinly bedded or closely jointed zone, or zone of poor rock conditions with shearing, weathering, and so forth. It is the last condition that would be of most concern. If these conditions were found to exist, additional borings or other types of explorations might be required to assess the orientation and characteristics of the weak zone and its potential effect on the engineering structure to be built.

Deere and Deere (1988) highlighted the "red-flag" zones by plotting both the total percentage of core recovery and the RQD as a function of depth on the same graphical column of the boring log; this plot is easy to draft as the RQD. RQD value is always equal to or less than the core recovery. To highlight RQD values less than 50%, the areas that are included between the line representing the low RQD value and the 50% line are colored red.

A zone of RQD of 45% would have only a narrow colored band (5%), while a zone of very poor rock represented by, say, 12% would have a wide colored band (38%). Thus the zone would be adequately red-flagged; the worse the rock, the larger the red flag. By use of this simple technique a quick comparison can be made among boring logs in various parts of the site and, upon occasion, a weak structural feature can be followed from boring to boring.

The depth of weathering and its general decrease in severity with depth as indicated by the RQD is successfully depicted with the red-flag concept. The depth of required foundation excavation often can be determined early with a quick study of the red-flag display.

The RQD is sensitive to the orientation of joint sets with respect to the orientation of the core; that is, a joint set parallel to the core axis will not intersect the core unless the drill hole happens to run along the joint. A joint set perpendicular to the core axis will intersect the core axis at intervals equal to the joint spacing. For intermediate orientations, the spacing of joint intersections with the core will be a cosine function of angle between joints and the core axis. *Thus, RQD is a directionally dependent parameter and its value may change significantly, depending upon the borehole orientation. The use of the volumetric joint count can be useful in reducing this directional dependence.*

An RQD of less than 70% indicates that the rock mass will be more susceptible to blast damage (Singh, 1992). RQD values less than 50% would require close spacing, light loading, and relief holes to produce acceptable results. Laubscher and Taylor (1976) proposed modifications in RQD values because of poor blasting practices. Accordingly, the maximum reduction in the RQD value is 20% for "poor conventional blasting."

Apart from the reduction in the weathering effects, the joints, fractures, and other discontinuities become tighter as they go deeper and deeper. Therefore, in a same rock mass, the RQD may tend to increase with depth.

Several researchers have investigated the influence of RQD in the rock mass classification schemes and discussed problems associated with its use and the RQD's sensitivity to measurement conditions and the experience of the person who classifies RQD. According to Hack (2002), typical problems with RQD are

- The limiting length of 10 cm is arbitrary
- The limiting length of 10 cm is an "abrupt boundary." Hack (2002) gave a simple yet insightful example: A core in a rock mass that includes an ideally uniformly distributed joint spacing of 9 cm shows an RQD of 0% (drilled perpendicular to the joints); if the spacing is just above 10 cm RQD is 100%. The limit of 10 cm is based on extensive experience.
- RQD is biased by orientation of measurement. Some approximate corrections are available to remove these effects.
- RQD is influenced by drilling equipment, size of equipment, handling of core, experience of the personnel, and so forth.

APPLICATION OF RQD

RQDs has been extensively used in engineering classifications of the rock mass as discussed in subsequent chapters of this book.

In addition, RQD has also been used to estimate the deformation modulus of the rock mass. Zhang and Einstein (2004) studied a wider range of rock masses with RQD values ranging from 0 to 100% and proposed the following mean correlation between RQD and modulus ratio:

$$\frac{E_d}{E_r} = 10^{0.0186\ RQD-1.91} \tag{4.10}$$

where E_d and E_r are the deformation moduli of the rock mass and the intact rock, respectively.

Cording and Deere (1972) attempted to relate the RQD index to Terzaghi's rock load factors. They found that Terzaghi's rock load theory should be limited to tunnels supported by steel sets, as it does not apply to openings supported by rock bolts. Chapter 5 deals with Terzaghi's rock load theory.

REFERENCES

Bieniawski, Z. T. (1984). *Rock mechanics design in mining and tunneling* (p. 272). Rotterdam: A. A. Balkema.

Bieniawski, Z. T. (1989). *Engineering rock mass classifications* (p. 251). New York: John Wiley.

Cording, E. J., & Deere, D. U. (1972). Rock tunnel support and field measurements. In *Proceedings of the rapid excavation tunnelling conference* (pp. 601–622). New York: AIME.

Deere, D. U. (1968). *Geological considerations, rock mechanics in engineering practice* (pp. 1–20). In R. G. Stagg & D. C. Zienkiewicz (Eds.). New York: Wiley.

Deere, D. U. (1989). *Rock quality designation (RQD) after twenty years* (U.S. Army Corps of Engineers Contract Report GL-89-1, p. 67). Vicksburg, MS: Waterways Experiment Station.

Deere, D. U., & Deere, D. W. (1988). *The rock quality designation (RQD) index in practice—Rock classification systems for engineering practice*. In L. Kirkaldie (Ed.). (pp. 91–101). ASTM STP 984. Philadelphia: American Society for Testing and Materials.

Hack, R. (2002). An evaluation of slope stability classification. In C. D. Gama et al. (Eds.). *EUROCK 2002, Proceedings of the ISRM International Symposium on Rock Engineering for Mountainous*

Regions (pp. 3–32). Portugal, Madeira, Funchal, 25–28 November. Lisboa: Sociedade Portuguesa de Geotecnia.

ISO 14689-1. (2003). (E). *Geotechnical investigation and testing—Identification and classification of rock—Part 1: Identification and description* (pp. 1–16). Geneva, Switzerland: International Organization for Standardization.

Laubscher, D. H., & Taylor, H. W. (1976). The importance of geomechanics classification of jointed rock masses in mining operations. In *Proceedings of the Symposium of Exploration for Rock Engineering* (pp. 119–128). Johannesburg, South Africa.

Palmstrom, A. (1982). The volumetric joint count—A useful and simple measure of the degree of jointing. In *IVth International Congress IAEG* (pp. V221–V228). New Delhi, India.

Palmstrom, A. (1985). Application of the volumetric joint count as a measure of rock mass jointing. In *Proceedings of the International Symposium on Fundamentals of Rock Joints* (pp. 103–110). Bjorkliden, Sweden.

Palmstrom, A. (1986). A general practical method for identification of rock masses to be applied in evaluation of rock mass stability conditions and TBM boring progress. In *Proceedings of the Conference on Fjellsprengningsteknikk, Bergmekanikk, Geoteknikk* (pp. 31.1–31.31). Oslo, Norway.

Palmstrom, A. (1996). RMi—A system for characterising rock mass strength for use in rock engineering. *Journal of Rock Mechanics and Tunnelling Technology, 1*(2), 69–108.

Palmstrom, A. (2002). *Measurement and characterization of rock mass jointing, in situ characterization of rocks*. In V. M. Sharma & K. R. Saxena (Eds.), Chap. 2, p. 358. New Delhi: Oxford & IBH Publishing Co. Pvt. Ltd., and Rotterdam: A. A. Balkema.

Palmstrom, A. (2005). Measurements of and correlations between block size and rock quality designation (RQD). *Tunnelling and Underground Space Technology, 20*, 362–377.

Priest, S. D., & Hudson, J. A. (1976). Discontinuity spacings in rock. *International Journal of Rock Mechanics and Mining Sciences—Geomechanics Abstracts, 13*, 135–148.

Romana, M. R. (1993). *A geomechanical classification for slopes: Slope mass rating in comprehensive rock engineering, principles—Practice and projects*, J. A. Hudson, Ed., Chap. 3 (pp. 575–600). New York: Pergamon.

Sen, Z., & Eissa, E. A. (1992). Rock quality charts for log-normally distributed block sizes. *International Journal of Rock Mechanics and Mining Sciences—Geomechanics Abstracts, 29*(1), 1–12.

Singh, S. P. (1992). Mining industry and blast damage. *Journal of Mines, Metals and Fuels*, December, 465–471.

Zhang, L., & Einstein, H. H. (2004). Using RQD to estimate the deformation modulus of rock masses. *International Journal of Rock Mechanics and Mining Sciences, 41*, 337–341.

Chapter 5

Terzaghi's Rock Load Theory

The geotechnical engineer should apply theory and experimentation but temper them by putting them into the context of the uncertainty of nature. Judgement enters through engineering geology.

Karl Terzaghi

INTRODUCTION

This was probably the first successful attempt at classifying rock masses for engineering purposes. Terzaghi (1946) proposed that the rock load factor (H_p) is the height of the loosening zone over the tunnel roof, which is likely to load the steel arches. These rock load factors were estimated by Terzaghi from a 5.5-m-wide steel-arch supported railroad tunnel in the Alps during the late 1920s. In these investigations wooden blocks of known strengths were used for blocking the steel arches to the surrounding rock masses. Rock loads were estimated from the known strengths of the failed wooden blocks. Terzaghi used these observations to back analyze rock loads acting on the supports. Subsequently, he conducted "trap-door" experiments on different sands and found that the height of loosened arch above the roof increased directly with the opening width in the sand.

ROCK CLASSES

Terzaghi (1946) considered the structural discontinuities of the rock masses and classified them qualitatively into nine categories as described in Table 5.1. Extensive experience from tunnels in the lower Himalayas showed that "squeezing rock" is really a squeezing ground condition, because a jointed and weak rock mass fails at high overburden stress and squeezes into the tunnels.

ROCK LOAD FACTOR

Terzaghi (1946) combined the results of his trap-door experiments and the estimated rock loads from Alpine tunnels to compute rock load factors (H_p) in terms of tunnel width (B) and tunnel height (H_t) of the loosened rock mass above the tunnel crown (Figure 5.1), which loads the steel arches. Rock load factors for all the nine rock classes are listed in Table 5.2.

TABLE 5.1 Definitions of Rock Classes of Terzaghi's Rock Load Theory

Rock class	Type of rock	Definition
I.	Hard and intact	The rock is unweathered. It contains neither joints nor hair cracks. If fractured, it breaks across intact rock. After excavation the rock may have some popping and spalling failures from the roof. At high stresses spontaneous and violent spalling of rock slabs may occur from the sides or the roof. The unconfined compressive strength is equal to or more than 100 MPa.
II.	Hard stratified and schistose	The rock is hard and layered. The layers are usually widely separated. The rock may or may not have planes of weakness. In this type of rock, spalling is quite common.
III.	Massive moderately jointed	A jointed rock. The joints are widely spaced. The joints may or may not be cemented. It may also contain hair cracks, but the huge blocks between the joints are intimately interlocked so that vertical walls do not require lateral support. Spalling may occur.
IV.	Moderately blocky and seamy	Joints are less spaced. Blocks are about 1 m in size. The rock may or may not be hard. The joints may or may not be healed, but the interlocking is so intimate that no side pressure is exerted or expected.
V.	Very blocky and seamy	Closely spaced joints. Block size is less than 1 m. It consists of almost chemically intact rock fragments that are entirely separated from each other and imperfectly interlocked. Some side pressure of low magnitude is expected. Vertical walls may require supports.
VI.	Completely crushed but chemically intact	Comprises chemically intact rock having the character of a crusher-run aggregate. There is no interlocking. Considerable side pressure is expected on tunnel supports. The block size could be a few centimeters to 30 cm.
VII.	Squeezing rock—moderate depth	Squeezing is a mechanical process in which the rock advances into the tunnel opening without perceptible increase in volume. Moderate depth is a relative term and could be 150 to 1000 m.
VIII.	Squeezing rock—great depth	The depth may be more than 150 m. The maximum recommended tunnel depth is 1000 m.
IX.	Swelling rock	Swelling is associated with volume change and is due to chemical change of the rock usually in the presence of moisture or water. Some shales absorb moisture from air and swell. Rocks containing swelling minerals such as montmorillonite, illite, kaolinite, etc., can swell and exert heavy pressure on rock supports.

Source: Sinha, 1989.

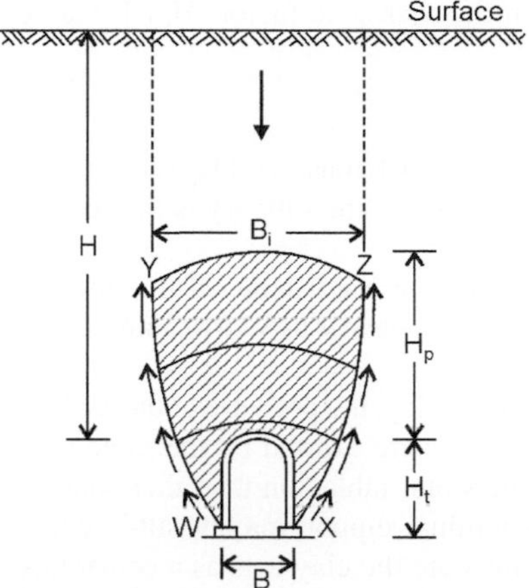

FIGURE 5.1 Terzaghi's (1946) rock load concept in tunnels.

TABLE 5.2 Rock Load in Tunnels within Various Rock Classes

Rock class	Rock condition	Rock load factor H_p	Remarks
I	Hard and intact	Zero	Light lining required only if spalling or popping occurs.
II	Hard stratified or schistose	0 to 0.5 B	Light support mainly for protection against spalling. Load may change erratically from point to point.
III	Massive, moderately jointed	0 to 0.25 B	
IV	Moderately blocky and seamy	0.25 B to 0.35 ($B + H_t$)	No side pressure
V	Very blocky and seamy	(0.35 to 1.10) ($B + H_t$)	Little or no side pressure
VI	Completely crushed but chemically intact	1.10 ($B + H_t$)	Considerable side pressure. Softening effects of seepage toward bottom of tunnel requires either continuous support for lower ends of ribs or circular ribs.
VII	Squeezing rock—moderate depth	(1.10 to 2.10) ($B + H_t$)	Heavy side pressure, invert struts required. Circular ribs are recommended.
VIII	Squeezing rock—great depth	(2.10 to 4.50) ($B + H_t$)	
IX	Swelling rock	Up to 250 ft. (80 m), irrespective of the value of ($B + H_t$)	Circular ribs are required. In extreme cases, use of yielding support recommended.

B = tunnel span in meters; H_t = height of the opening in meters; and H_p = height of the loosened rock mass above tunnel crown developing load (Figure 5.1).

Source: Terzaghi, 1946.

For obtaining the vertical support pressure from the rock load factor (H_p) Terzaghi suggested the following equation:

$$p_v = \gamma . H_p \tag{5.1}$$

where p_v is the support pressure, γ is the unit weight of the rock mass, and H_p is the height of loose overburden above the tunnel roof (Figure 5.1). Terzaghi's theory is limited because it may not be applicable for tunnels wider than 6 m.

The roof of a tunnel is assumed to be located below the water table. If it is located permanently above the water table, the values given for Classes IV to VI in Table 5.2 can be reduced by 50% (Rose, 1982).

If the joints in a blocky and seamy rock do not contain clay, the pressure of the rock on the tunnel support may be as high as one-half of the pressure exerted by the same rock on the same tunnel at a considerable depth below the water table. On the other hand, if the joints are partially or entirely filled with clay, a nominal support may be sufficient to hold up the roof during the dry season; in a dried-out state the clay acts as a cementing material. However, during long wet spells the clay ceases to act as an effective binder and the pressure on the tunnel support becomes as heavy as if the joints were lubricated (Proctor & White, 1946).

Because of this, several large tunnels, which were mined and supported during the dry season, caved in soon after the rains. If it is uncertain whether or not the rock located above the tunnel will remain dry throughout the year, it is advisable to design the tunnel supports on the basis of the values obtained by the equations given in Table 5.2 regardless of the appearance of the rock during mining operations.

Deere et al. (1970) modified Terzaghi's classification system by introducing the rock quality designation (RQD) as the lone measure of rock quality (Table 5.3). They have distinguished between blasted and machine excavated tunnels and proposed guidelines for selection of steel set, rock bolts, and shotcrete supports for 6- to 12-m diameter tunnels in rock. These guidelines are listed in Table 5.4.

Deere et al. (1970) also considered the rock mass as an integral part of the support system; Table 5.4 is only applicable if the rock mass is not allowed to loosen and disintegrate extensively. They assumed that machine excavation reduced rock loads by approximately 20 to 25%.

Limitations

Terzaghi's approach was successfully used when conventional drill and blast methods of excavation and steel-arch supports were employed in tunnels of comparable size. This practice lowered the strength of the rock mass and permitted significant roof convergence that mobilized a zone of loosened rock mass above the tunnel roof. The height of this loosened rock mass, called "coffin cover," acted as dead load on the supports. Cecil (1970) concluded that Terzaghi's classification provided no quantitative information regarding the rock mass properties. Despite these limitations, the immense practical value of Terzaghi's approach cannot be denied, and his method is still applied under conditions similar to those for which it was developed.

With the advent of the New Austrian Tunnelling Method (NATM) and Norwegian Method of Tunnelling (NMT), increasing use is made of controlled blasting and machine excavation techniques and support systems employing steel fiber reinforced shotcrete (SFRS) and rock bolts. Even in steel-arch supported tunnels, wooden struts have been

TABLE 5.3 Terzaghi's Rock Load Concept as Modified by Deere

Rock class and condition	RQD (%)	Rock load (H_p)	Remarks
I. Hard and intact	95–100	Zero	Same as Table 5.2
II. Hard stratified or schistose	90–99	0–0.5 B	Same as Table 5.2
III. Massive moderately jointed	85–95	0–0.25 B	Same as Table 5.2
IV. Moderately blocky and seamy	75–85	0.25 B–0.35 ($B + H_t$)	Types IV, V, and VI reduced by about 50% from Terzaghi values because water table has little effect on rock load (Terzaghi, 1946; Brekke, 1968)
V. Very blocky and seamy	30–75	(0.2–0.6) ($B + H_t$)	Same as above
VI. Completely crushed	3–30	(0.6–1.10) ($B + H_t$)	Same as above
VIa. Sand and gravel	0–3	(1.1–1.4) ($B + H_t$)	Same as above
VII. Squeezing rock at moderate depth	NA	(1.10–2.10) ($B + H_t$)	Same as Table 5.2
VIII. Squeezing rock at great depth	NA	(2.10–4.50) ($B + H_t$)	Same as Table 5.2
IX. Swelling rock	NA	Up to 80 m irrespective of the value of ($B + H_t$)	Same as Table 5.2

B = tunnel span; H_t = height of the opening; and H_p = height of the loosened rock mass above the tunnel crown developing load (Figure. 5.1).

Source: Deere et al., 1970.

replaced by pneumatically filled lean concrete. These improvements in tunneling technology preserve the pre-excavation strength of the rock mass and use it as a load-carrying structure to minimize roof convergence and restrict the height of the loosening zone above the tunnel crown.

Consequently, support pressure does not increase directly with the opening width. Based on this argument, Barton, Lien, and Lunde (1974) advocated that the support pressure is independent of opening width in rock tunnels. Rock mass-tunnel-support-interaction analysis of Verman (1993) also suggested that the support pressure is practically independent of the tunnel width, provided support stiffness is not lowered. Goel, Jethwa, and Dhar (1996) also studied the effect of tunnel size on support pressure and found a negligible effect of tunnel size on support pressure in non-squeezing ground

TABLE 5.4 Guidelines for Selection of Steel Sets for 6- to 12-m-Diameter Tunnels in Rock

		Steel sets		Rock bolt		Conventional shotcrete		
						Total thickness (cm)		
Rock quality	Construction method	*Weight of steel sets*	*Spacing*	*Spacing of pattern bolt*	*Additional requirements*	*Crown*	*Sides*	Additional supports
Excellent RQD >90	Tunnel boring machine	Light	None to occasional	None to occasional	Rare	None to occasional	None	None
	Drilling and blasting	Light	None to occasional	None to occasional	Rare	None to occasional	None	None
Good RQD 75 to 90	Boring machine	Light	Occasional or 1.5 to 1.8 m	Occasional or 1.5 to 1.8 m	Occasional mesh and straps	Local application 5 to 7.5 cm	None	None
	Drilling and blasting	Light	1.5 to 1.8 m	1.5 to 1.8 m	Occasional mesh and straps	Local application 5 to 7.5 cm	None	None
Fair RQD 50 to 75	Boring machine	Light to medium	1.5 to 1.8 m	1.2 to 1.8 m	Mesh and straps as required	5 to 10 cm	None	Rock bolts
	Drilling and blasting	Light to medium	1.2 to 1.5 m	0.9 to 1.5 m	Mesh and straps as required	10 cm or more	10 cm or more	Rock bolts

Poor RQD 25 to 50	Boring machine	Medium circular	0.6 to 1.2 m	0.9 to 1.5 m	Anchorage may be hard to obtain; considerable mesh and straps required	10 to 15 cm	10 to 15 cm	Rock bolt as required (1.2 to 1.8 m center to center)
	Drilling and blasting	Medium to heavy circular	0.2 to 1.2 m	0.6 to 1.2 m	As above	15 cm or more	15 cm or more	As above
Very poor RQD <25	Boring machine	Medium to heavy circular	0.6 m	0.6 to 1.2 m	Anchorage may be impossible; 100% mesh and straps required	15 cm or more on whole section		Medium sets as required
	Drilling and blasting	Heavy circular	0.6 m	0.9 m	As above	15 cm or more on whole section		Medium to heavy sets as required
Very poor squeezing and swelling ground	Both methods	Very heavy circular	0.6 m	0.6 to 0.9 m	Anchorage may be impossible; 100% mesh and straps required	15 cm or more on whole section		Heavy sets as required

Source: Deere et al., 1970.

TABLE 5.5 Recommendations of Singh et al. (1995) on Support Pressure for Rock Tunnels and Caverns

Terzaghi's classification			Classification of Singh et al. (1995)				
					Recommended support pressure (MPa)		
Category	*Rock condition*	*Rock load factor* $(H)_p$	*Category*	*Rock condition*	p_v	P_h	Remarks
(1)	(2)	(3)	(4)	(5)	(6)	(7)	(8)
I	Hard and intact	0	I	Hard and intact	0	0	—
II	Hard stratified or schistose	0 to 0.5 B	II	Hard stratified or schistose	0.04–0.07	0	—
III	Massive, moderately jointed	0 to 0.25 B	III	Massive, moderately jointed	0.0–0.04	0	—
IV	Moderately blocky, seamy, and jointed	0.25 B to 0.35 $(B + H_t)$	IV	Moderately blocky, seamy, very jointed	0.04–0.1	0–0.2 p_v	Inverts may be required
V	Very blocky and seamy, shattered arched	0.35 to 1.1 $(B + H_t)$	V	Very blocky and seamy, shattered highly jointed, thin shear zone or fault	0.1–0.2	0–0.5 p_v	Inverts may be required, arched roof preferred
VI	Completely crushed but chemically intact	1.1 $(B + H_t)$	VI	Completely crushed but chemically unaltered, thick shear and fault zone	0.2–0.3	0.3–1.0 p_v	Inverts essential, arched roof essential

VII	Squeezing rock at moderate depth	1.1 to 2.1 $(B + H_t)$	VII	Squeezing rock condition			
			VIIA	Mild squeezing (u_a/a up to 3%)	0.3–0.4	Depends on primary stress values, p_h may exceed p_v	Inverts essential. In excavation flexible support preferred. Circular section with struts recommended
			VIIB	Moderate squeezing (u_a/a = 3 to 5%)	0.4–0.6	-do-	-do-
VIII	Squeezing rock at great depth	2.1 to 4.5 $(B + H_t)$	VIIC	High squeezing (u_a/a >5%)	6.0–1.4	-do-	-do-
IX	Swelling rock	Up to 80 m	VIII	Swelling rock			
			VIIIA	Mild swelling	0.3–0.8	Depends on type and content of swelling clays, p_h may exceed p_v	Inverts essential in excavation, arched roof essential
			VIIIB	Moderate swelling	0.8–1.4	-do-	-do-
			VIIIC	High swelling	1.4–2.0	-do-	-do-

p_v = vertical support pressure; p_h = horizontal support pressure; B = width or span of opening; H_t = height of opening; u_a = radial tunnel closure; a = B/2; thin shear zone = up to 2 m thick.

Source: Singh et al., 1995.

conditions, but the tunnel size could have considerable influence on the support pressure in squeezing ground conditions. For more in-depth coverage on this subject, see Chapter 9.

The estimated support pressures from Table 5.2 have been compared with the measured values with the following conclusions:

1. Terzaghi's method provides reasonable support pressure for small tunnels (B < 6 m).
2. It provides over-safe estimates for large tunnels and caverns (diameter 6–14 m).
3. There is a very large range of estimated support pressure values for squeezing and swelling ground conditions.

MODIFIED TERZAGHI'S THEORY FOR TUNNELS AND CAVERNS

Singh, Jethwa, and Dube (1995) compared support pressure measured from tunnels and caverns with estimates from Terzaghi's rock load theory and found that the support pressure in rock tunnels and caverns does not increase directly with excavation size as assumed by Terzaghi (1946) and others. This is due mainly to the dilatant behavior of rock masses, joint roughness, and prevention of rock mass loosening by improved tunneling technology. They have subsequently recommended ranges of support pressures as listed in Table 5.5 for both tunnels and caverns for those who still want to use Terzaghi's rock load approach. They observed that the support pressures are nearly independent of size of opening.

It is interesting to note that the recommended roof support pressures turn out to be the same as those obtained from Terzaghi's rock load factors when B and H_t are substituted by a tunnel width of 5.5 m. The estimated roof support pressures from Table 5.5 were found to be comparable with the measured values irrespective of the opening size and the rock conditions (Singh et al., 1995). These authors have further cautioned that the support pressure is likely to increase directly with the excavation width for tunnel sections through slickensided shear zones, thick clay-filled fault gouges, weak clay shales, and running or flowing ground conditions where interlocking blocks are likely to be missing or where joint strength is lost and rock wedges are allowed to fall due to excessive roof convergence because of delayed supports beyond stand-up time. It should be noted that wider tunnels require reduced spacing of bolts or steel arches and thicker linings since rock loads increase directly with the excavation width, even if the support pressure does not increase with the tunnel size.

REFERENCES

Barton, N., Lien, R., & Lunde, J. (1974). *Engineering classification of rock masses for the design of tunnel support* (NGI Publication No. 106, p. 48). Oslo: Norwegian Geotechnical Institute.

Brekke, T. L. (1968). Blocky and seamy rock in tunnelling. *Bulletin of the Association of Engineering. Geologists*, *5*(1), 1–12.

Cecil, O. S. (1970). *Correlation of rock bolt—Shotcrete support and rock quality parameters in Scandinavian tunnels* (p. 414). Ph.D. Thesis. Urbana: University of Illinois.

Deere, D. U., Peck, R. B., Parker, H., Monsees, J. E., & Schmidt, B. (1970). Design of tunnel support systems. *Highway Research Record*, No. 339, 26–33.

Goel, R. K., Jethwa, J. L., & Dhar, B. B. (1996). Effect of tunnel size on support pressure, Technical Note. *International Journal of Rock Mechanics and Mining Sciences—Geomechanics Abstracts, 33*(7), 749–755.

Proctor, R. V., & White, T. L. (1946). *Rock tunnelling with steel supports* (p. 271). Youngstown, OH: The Commercial Shearing and Stamping Company.

Rose, D. (1982). Revising Terzaghi's tunnel rock load coefficients. In *Proceedings of the 23rd U.S. Symposium on Rock Mechanics* (pp. 953–960). New York: AIME.

Singh, B., Jethwa, J. L., & Dube, A. K. (1995). A classification system for support pressure in tunnels and caverns. *Journal of Rock Mechanics and Tunnelling Technology, 1*(1), 13–24.

Sinha, R. S. (1989). *Underground structures—Design and instrumentation* (p. 480). Oxford: Elsevier Science.

Terzaghi, K. (1946). Introduction to tunnel geology. In R. V. Proctor & T. L. White (Eds.), *Rock tunnelling with steel supports* (p. 271). Youngstown, OH: Commercial Shearing & Stamping Co.

Verman, M. K. (1993). *Rock mass—Tunnel support interaction analysis* (p. 258). Ph.D. Thesis. Uttarakhand, India: IIT Roorkee.

Chapter 6

Rock Mass Rating

Effectiveness of knowledge through research (E) is $E = mc^2$; where m is mass of knowledge and c is communication of knowledge by publications.

Z.T. Bieniawski

INTRODUCTION

The geomechanics classification or the rock mass rating (RMR) system was initially developed at the South African Council of Scientific and Industrial Research (CSIR) by Bieniawski (1973) on the basis of his experiences in shallow tunnels in sedimentary rocks (Kaiser, MacKay, & Gale, 1986). Since then the classification has undergone several significant evolutions: in 1974, reduction of classification parameters from 8 to 6; in 1975, adjustment of ratings and reduction of recommended support requirements; in 1976, modification of class boundaries to even multiples of 20; in 1979, adoption of ISRM (1978) rock mass description, and so forth. Therefore, it is important to state which version is used when RMR values are quoted. The geomechanics classification reported by Bieniawski (1984) can be found in the section Rock Mass Excavability Index for TBM.

To apply the geomechanics classification system, a given site should be divided into a number of geological structural units in such a way that each type of rock mass is represented by a separate geological structural unit. The following six parameters (representing causative factors) are determined for each structural unit:

1. Uniaxial compressive strength (UCS) of intact rock material
2. Rock quality designation (RQD)
3. Joint or discontinuity spacing
4. Joint condition
5. Groundwater condition
6. Joint orientation

COLLECTION OF FIELD DATA

The ratings of six parameters of the RMR system are given in Tables 6.1 to 6.6. For reducing doubts due to subjective judgments, the ratings for different parameters should be given a range rather than a single value. These six parameters are discussed in the following paragraphs. Beginners do not always understand the value of RMR, Q, and so forth, at a location, and they get confused transitioning from one category to another (Tables 6.4 and 6.5). Usually approximate average RMR is good enough. ISO 14689 describes internationally accepted definitions for rock materials, joints, and rock masses.

TABLE 6.1 Strength of Intact Rock Material

Qualitative description	Compressive strength (MPa)	Point load strength (MPa)	Rating
Extremely strong*	>250	8	15
Very strong	100–250	4–8	12
Strong	50–100	2–4	7
Medium strong*	25–50	1–2	4
Weak	5–25	Use of UCS is preferred	2
Very weak	1–5	-do-	1
Extremely weak	<1	-do-	0

At compressive strength of rock material less than 1.0 MPa, many rock materials would be regarded as soil.
**Terms redefined according to ISO 14689.*

Sources: Bieniawski, 1979, 1984; ISO14689-1, 2003.

TABLE 6.2 Rock Quality Designation

Qualitative description	RQD (%)	Rating
Excellent	90–100	20
Good	75–90	17
Fair	50–75	13
Poor	25–50	8
Very poor	<25	3

Source: Bieniawski, 1979.

TABLE 6.3 Spacing of Discontinuities

Description	Spacing (m)	Rating
Very wide	>2	20
Wide	0.6–2	15
Moderate	0.2–0.6	10
Close	0.06–0.2	8
Very close	<0.06	5

If more than one discontinuity set is present and the spacing of discontinuities of each set varies, consider the unfavorably oriented set with lowest rating. ISO 14689 uses the term "extremely close" for joint spacing less than 0.02 m.

Sources: Bieniawski, 1979; ISO 14689-1, 2003.

TABLE 6.4 Condition of Discontinuities

Description	Joint separation (mm)	Rating
Very rough and unweathered, wall rock tight and discontinuous, no separation	0	30
Rough and slightly weathered, wall rock surface separation <1 mm	<1	25
Slightly rough and moderately to highly weathered, wall rock surface separation <1 mm	<1	20
Slickensided wall rock surface, or 1–5 mm thick gouge, or 1–5 mm wide continuous discontinuity	1–5	10
5 mm thick soft gouge, 5 mm wide continuous discontinuity	>5	0

Source: Bieniawski, 1979.

TABLE 6.5 The RMR System: Guidelines for Classification of Discontinuity Conditions

Parameter*	Ratings				
Discontinuity length (persistence/continuity)	<1 m	1–3 m	3–10 m	10–20 m	>20 m
	6	4	2	1	0
Separation (aperture)	None	<0.1 mm	0.1–1.0 mm	1–5 mm	>5 mm
	6	5	4	1	0
Roughness of discontinuity surface	Very rough	Rough	Slightly rough	Smooth	Slickensided
	6	5	3	1	0
Infillings (gouge)		Hard filling		Soft filling	
	None	<5 mm	>5 mm	<5 mm	>5 mm
	6	4	2	2	0
Weathering discontinuity surface	Unweathered	Slightly weathered	Moderately weathered	Highly weathered	Decomposed
	6	5	3	1	0

**Some conditions are mutually exclusive. For example, if infilling is present, it is irrelevant what the roughness may be, since its effect will be overshadowed by the influence of the gouge. In such cases use Table 6.4 directly.*

Source: Bieniawski, 1993.

TABLE 6.6 Groundwater Condition

Inflow per 10 m tunnel length (L/min)	None	<10	10–25	25–125	>125
Ratio of joint water pressure to major principal stress	0	0–0.1	0.1–0.2	0.2–0.5	>0.5
General description	Completely dry	Damp	Wet	Dripping	Flowing
Rating	15	10	7	4	0

Source: Bieniawski, 1979.

When mixed quality rock conditions are encountered at the excavated rock face, such as when "good quality" and "poor quality" are present in one exposed area, it is essential to identify the "most critical condition" for the assessment of the rock strata. This means that the geological features that are most significant for stability purposes will have an overriding influence. For example, a fault or a shear zone in a high quality rock face will play a dominant role, irrespective of the high rock material strength in the surrounding strata (Bieniawski, 1993).

Uniaxial Compressive Strength of Intact Rock Material (q_c)

The strength of the intact rock material should be obtained from rock cores in accordance with site conditions. The ratings based on both UCS (which is preferred) and point load strength index are given in Table 6.1. UCS may also be obtained from the point load strength index tests on rock lumps at the natural moisture content. See Table 8.13 for average UCS values of a variety of rocks. The pH value of groundwater may affect the UCS in saturated conditions.

Rock Quality Designation

RQD should be determined from rock cores or volumetric joint count (Chapter 4). It is the percentage of rock cores (equal to or more than 10 cm long) in one meter of drill run. The fresh broken cores are fitted together and counted as one piece. The details of RQD rating are given in Table 6.2.

Spacing of Discontinuities

The term "discontinuity" covers joints, beddings or foliations, shear zones, minor faults, and other surfaces of weakness. The linear distance between two adjacent discontinuities should be measured for all sets of discontinuities. Ratings are shown in Table 6.3 for the most critically oriented discontinuity or the lowest rating (Edelbro, 2003). It is widely accepted that spacing of joints is very important when appraising a rock mass structure. The very presence of joints reduces the strength of a rock mass and their spacing governs the degree of such a reduction (Bieniawski, 1973).

Condition of Discontinuities

This parameter includes roughness of discontinuity surfaces, their separation, length of continuity, weathering of the wall rock or the planes of weakness, and infilling (gouge) material. Tables 6.4 and 6.5 illustrate the ratings for discontinuities. The joint set, which is oriented unfavorably with respect to a structure (tunnel or cavern), should be considered along with spacing of the discontinuities.

Groundwater Condition

For tunnels, the rate of inflow of groundwater in liters per minute per 10 m length of the tunnel should be determined, or a general condition may be described as completely dry, damp, wet, dripping, or flowing. If actual water pressure data are available, these should be stated and expressed in terms of the ratio of the seepage water pressure to the major principal stress. The ratings according to the water condition are shown in Table 6.6.

Ratings of the above five parameters (seen in Tables 6.1 to 6.6) are added to obtain the basic rock mass rating, RMR_{basic}.

Orientation of Discontinuities

Orientation of discontinuities refers to the strike and dip of discontinuities. The strike should be recorded with reference to magnetic north. The dip angle is the angle between the horizontal and discontinuity plane taken in a direction in which the plane dips. The value of the dip and the strike should be recorded as shown in Table 6.7. The orientation of tunnel axis or slope face or foundation alignment should also be recorded.

The influence of the strike and dip of discontinuities is considered with respect to the direction of tunnel drivage, slope face orientation, or foundation alignment. To decide whether or not the strike and dip are favorable, reference should be made to Tables 6.8 and 6.9, which provide a quantitative assessment of critical joint orientation effect regard ing tunnels and dam foundations, respectively. Once the rating for the effect of the critical discontinuity is known, as shown in Table 6.9, the sum of the joint adjustment rating and the RMR_{basic} can be obtained. This number is called the "final RMR."

Keep in mind that the effect of orientation in a rough-dilatant joint is not as important in tunnels, according to Table 6.10. That is why the orientation of joints is ignored in the Q-system of the Norwegian Geotechnical Institute (NGI; Chapter 8). The effect of orientation of joints is more important for rafts. It is most important in rock slopes for which slope mass rating (SMR) is recommended (Chapter 18). The cut slopes of the trench before the tunnel should be classified according to SMR and not RMR or Q.

Research is needed to devise a new table to assess joint orientation for shafts not included in Table 6.8. Research should also be done for the assessment of joint orientation for foundations of buildings and silos and so forth on the basis of Figure 20.1, because Table 6.9 is only valid for dam foundations, which are subjected to a high horizontal hydraulic force.

TABLE 6.7 Orientation of Discontinuities

A. Orientation of tunnel/slope/foundation axis
B. Orientation of discontinuities:
Set-1 Average strike........(from.......to.......) Dip/Dip direction..........
Set-2 Average strike........(from.......to.......) Dip/Dip direction..........
Set-3 Average strike........(from.......to.......) Dip/Dip direction..........

TABLE 6.8 Assessment of Joint Orientation Effect on Tunnels

Strike perpendicular to tunnel axis				Strike parallel to tunnel axis		Irrespective of strike
Drive with dip		*Drive against dip*				
Dip 45°–90°	Dip 20°–45°	Dip 45°–90°	Dip 20°–45°	Dip 20°–45°	Dip 45°–90°	Dip 0°–20°
Very favorable	Favorable	Fair	Unfavorable	Fair	Very unfavorable	Fair

Source: Bieniawski, 1984.

TABLE 6.9 Assessment of Joint Orientation Effect on Stability of Dam Foundation

	Dip 10°–30°			
	Dip direction			
Dip 0°–10°	***Upstream***	***Downstream***	**Dip 30°–60°**	**Dip 60°–90°**
Very favorable	Unfavorable	Fair	Favorable	Very unfavorable

TABLE 6.10 Adjustment for Joint Orientation

Joint orientation assessment for	Very favorable	Favorable	Fair	Unfavorable	Very unfavorable
Tunnels	0	−2	−5	−10	−12
Raft foundation	0	−2	−7	−15	−25
Slopes*	0	−5	−25	−50	−60

**It is recommended to use slope mass rating (SMR; Chapter 18).*

Source: Bieniawski, 1979.

ESTIMATION OF RMR

RMR should be determined as an algebraic sum of ratings for all of the parameters given in Tables 6.1 to 6.5 and Table 6.10 after adjustments for orientation of discontinuities given in Tables 6.8 and 6.9. The sum of the ratings for the four parameters (Tables 6.2 to 6.5) is called the "rock condition rating," which discounts the effect of the compressive strength of intact rock material and orientation of joints (Goel, Jethwa, & Paithankar, 1996). Heavy blasting creates new fractures. Experience suggests that 10 points should be added to get RMR for undisturbed rock masses in situations where tunnel boring machines (TBMs) or road headers are used for tunnel excavation and 3 to 5 points can be added depending upon the quality of the controlled blasting. Solving Eq. (6.7) gives RMR_{TBM}.

On the basis of RMR values for a given engineering structure, the rock mass is sorted into five classes: very good (RMR 100–81), good (80–61), fair (60–41), poor (40–21), and very poor (<20) as shown in Table 6.11.

With wider tunnels and caverns, the RMR obtained may be somewhat less than obtained from drifts, because in drifts intrusions of weaker rocks and joint sets having lower joint condition ratings may be missed. A separate RMR should be obtained for tunnels of different orientations after taking into account the orientation of the tunnel axis with respect to the critical joint set (Table 6.8).

The classification may be used for estimating many useful parameters such as the unsupported span, the stand-up time, the bridge action period, and the support pressure for an underground opening as shown in the following section. It may also be used for selecting a method of excavation and the permanent support system. Cohesion, angle of internal friction, modulus of deformation of the rock mass, and allowable bearing pressure for foundations may also be estimated to analyze the stability of rock slopes. Back analysis of rock slopes in distress is a more reliable approach for assessment of shear strength parameters. It also recommends cut slope angle along hill roads and rail lines. Correlations suggested in the next section should

TABLE 6.11 Design Parameters and Engineering Properties of Rock Mass

S. No.	Parameter/ properties of rock mass	RMR (rock class)				
		100–81 (I)	*80–61 (II)*	*60–41 (III)*	*40–21 (IV)*	*<20 (V)*
1	Classification of rock mass	Very good	Good	Fair	Poor	Very poor
2	Average stand-up time	20 years for 15 m span	1 year for 10 m span	1 week for 5 m span	10 hours for 2.5 m span	30 minutes for 1 m span
3	Cohesion of rock mass (MPa)*	>0.4	0.3–0.4	0.2–0.3	0.1–0.2	<0.1
4	Angle of internal friction of rock mass	>45°	35–45°	25–35°	15–25°	<15°
5	Allowable bearing pressure (T/m^2)	600–440	440–280	280–135	135–45	45–30
6	Safe cut slope (°) (Waltham, 2002)	>70	65	55	45	<40

During earthquake loading, the above values of allowable bearing pressure may be increased by 50% in view of rheological behavior of rock masses (see Chapter 20).

**These values are applicable to slopes only in saturated and weathered rock mass.*

Source: Bieniawski, 1993.

be used for feasibility studies and preliminary designs only. In situ tests supported with numerical modeling could be essential, particularly for a large opening such as a cavern.

APPLICATIONS OF RMR

The following engineering properties of rock masses may be obtained using RMR. If the rock mass rating lies within a given range, the value of engineering properties can be interpolated between the recommended range of properties.

Average Stand-up Time for an Arched Roof

The stand-up time depends upon an effective (unsupported) span of the opening, which is defined as the width of the opening or the distance between the tunnel face and the last support (whichever is smaller). For arched openings the stand-up time would be significantly higher than for a flat roof. Controlled blasting further increases the stand-up time as damage to the rock mass is decreased. For tunnels with an arched roof the stand-up time is related to the rock mass class in Table 6.11 (Figure 6.1). Do not unnecessarily delay supporting the roof in a rock mass with high stand-up time as this may lead to deterioration in the rock mass, which ultimately reduces the stand-up time. Lauffer (1988) observed that the stand-up time improves by one class of RMR value in excavations by TBM.

Cohesion and Angle of Internal Friction

Assuming that a rock mass behaves as a Coulomb material, its shear strength depends upon cohesion and angle of internal friction. RMR is used to estimate the cohesion and angle of internal friction (Table 6.11). Usually the strength parameters are different for peak failure and residual failure conditions. In Table 6.11, only peak failure values are given. These values are applicable to slopes only in saturated and weathered rock masses. Cohesion is small under low normal stresses due to rotation of rock blocks. The angle of internal friction of even highly weathered rock masses (RMR << 25) is generally more than

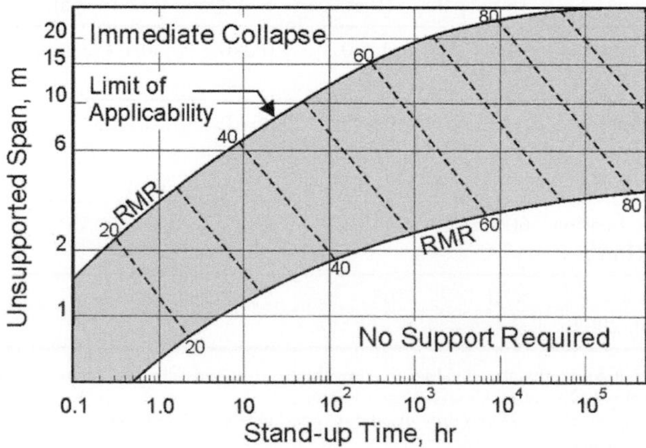

FIGURE 6.1 Stand-up time versus unsupported span for various rock mass classes according to RMR. *(From Bieniawski, 1984)*

14 degrees (Chapter 13, in the section Dynamic Strength of Rock Mass). Further RMR along the failure surface may be much less than on the slope in distress. However, the cohesion is one order of magnitude higher in tunnels because joints are relatively discontinuous, tight, and widely spaced. Joints may have smaller lengths than those near rock slopes. See the section Shear Strength of Rock Masses in this chapter and Chapter 16.

Modulus of Deformation

The following correlations are suggested for determining the modulus of deformation of rock masses. Modulus of deformation (E_d) is obtained from the loading cycle of the uniaxial jacking test, whereas the elastic modulus of rock mass (E_e) is found from the unloading cycle.

Modulus Reduction Factor

Figure 6.2 illustrates the correlation between RMR and the modulus reduction factor (MRF), which is defined as a ratio of the modulus of deformation of a rock mass to the elastic modulus of the rock material obtained from the core. Thus, the modulus of deformation of a rock mass (E_d) can be determined as a product of MRF corresponding to a given RMR (Figure 6.2) and the elastic modulus of the rock material (E_r) from the following equation (Singh, 1979):

$$E_d = E_r \cdot \text{MRF} \tag{6.1}$$

There is an approximate correlation between the modulus of deformation and RMR suggested by Bieniawski (1978) for hard rock masses ($q_c > 100$ MPa).

$$E_d = 2\,\text{RMR} - 100, \text{GPa (applicable for RMR} > 50) \tag{6.2}$$

Serafim and Pereira (1983) suggested the following correlation:

$$E_d = 10^{(\text{RMR}-10)/40}, \text{GPa (applicable for RMR} < 50 \text{ also)} \tag{6.3}$$

These correlations are shown in Figure 6.3. Here q_c means average uniaxial crushing strength of the intact rock material in MPa.

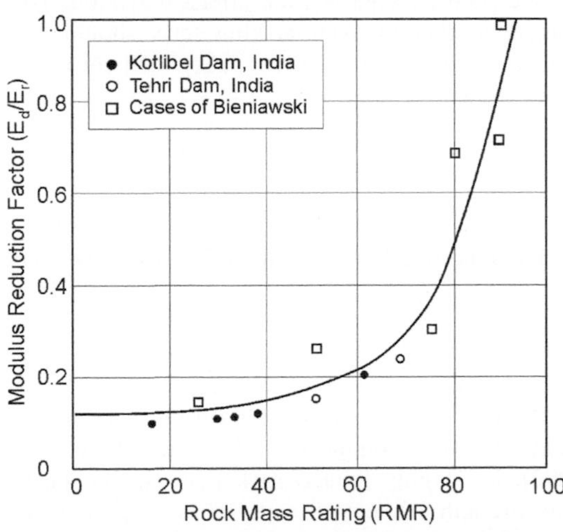

FIGURE 6.2 Relationship between rock mass rating (RMR) and modulus reduction factor. *(From Singh, 1979)*

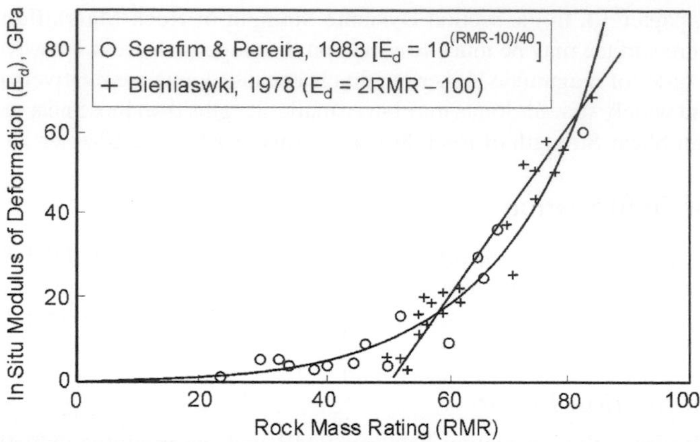

FIGURE 6.3 Correlation between modulus of deformation of rock masses and RMR. *(From Bieniawski, 1984)*

The modulus of deformation of a dry and weak rock mass (q_c < 100 MPa) around underground openings located at depths exceeding 50 m is dependent upon confining pressure due to overburden and may be determined by the following correlation (Verman, 1993):

$$E_d = 0.3\, H^{\alpha} \cdot 10^{(RMR-20)/38}, \text{GPa} \tag{6.4a}$$

where α = 0.16 to 0.30 (higher for poor rocks) and H = depth of location under consideration below ground surface in meters ≥ 50 m.

Read, Richards, and Perrin (1999) suggested the following correlation:

$$E_d = 0.1(RMR/10)^{0.3}, \text{GPa} \tag{6.4b}$$

Table 8.14 summarizes various correlations for the assessment of the modulus of deformation.

The modulus of deformation of poor rock masses with water sensitive minerals decreases significantly after saturation and with passage of time after excavation. To design dam foundations, it is recommended that uniaxial jacking tests should be conducted very carefully soon after the excavation of drifts, particularly for poor rock masses in saturated conditions (Mehrotra, 1992).

Allowable Bearing Pressure

Allowable bearing pressure for a 12 mm foundation settlement is also related to RMR and may be estimated using Table 6.11 (Mehrotra, 1992). Chapter 20 discusses this subject in greater detail.

Shear Strength of Rock Masses

Table 16.1 summarizes the non-linear shear strength equations for various rock mass ratings, degree of saturation, and rock types. The recommended criterion is based on 43 block shear tests by Mehrotra (1992). For highly jointed rock masses, the shear strength (τ) may not be governed by the strength of the rock material as suggested by

Hoek and Brown (1980). The results show that saturation significantly affects the shear strength of rock mass (see Figure 16.1).

For hard and massive rock masses (RMR > 60), the shear strength is proportional to the UCS. It follows that block shear tests on saturated rock blocks should be conducted for design of concrete dams and stability of abutments (see Table 20.8).

Estimation of Support Pressure

In 1983, Unal, on the basis of his studies in coal mines, proposed the following correlation for estimation of support pressure using RMR for openings with a flat roof:

$$p_v = \left[\frac{100 - RMR}{100}\right] \cdot \gamma \cdot B \tag{6.5}$$

where p_v = support pressure; γ = unit weight of rock; and B = tunnel width.

Goel and Jethwa (1991) evaluated Eq. (6.5) for application to rock tunnels with arched roofs by comparing the measured support pressures with estimates from Eq. (6.5) and found that Eq. (6.5) cannot be used with rock tunnels. They found that the estimated support pressures were unsafe for all sizes of tunnels under squeezing ground conditions. Further, the estimates for non-squeezing ground conditions were unsafe for small tunnels (diameter up to 6 m) and over-safe for large tunnels (diameter > 9 m), which implies that the size effect is overemphasized for arched openings. This observation is logical since bending moments in a flat roof increase geometrically with the opening size, unlike in an arched roof.

Subsequently, using the measured support pressure values from 30 instrumented Indian tunnels, Goel and Jethwa (1991) proposed Eq. (6.6) for estimating the short-term support pressure for arched underground openings in both squeezing and non-squeezing ground conditions in tunneling by conventional blasting methods using steel rib supports (but not in the rock burst condition).

$$p_v = \frac{7.5\, B^{0.1} \cdot H^{0.5} - RMR}{20\, RMR}, \text{ MPa} \tag{6.6}$$

where B = span of opening in meters; H = overburden or tunnel depth in meters (Eq. 6.6 applicable for H = 50 to 600 m); p_v = short-term roof support pressure in MPa; and RMR = actual (disturbed) post-excavation rock mass rating just before supporting.

Bieniawski (1984) provided guidelines for a selection of tunnel supports (Table 6.12). This is applicable to tunnels excavated with conventional drilling and blasting methods. These guidelines depend upon factors such as depth below surface (to take care of overburden pressure or the in situ stress), tunnel size, and shape and method of excavation. The support measures in Table 6.12 are for permanent supports.

The interrelation between RMR and Q is presented in Chapter 9, in the section Interrelation Between Q and RMR. Figure 13.3 offers criteria for various types of rock bursts according to RMR and σ_1/q_c values.

PRECAUTIONS

It must be ensured that double-accounting for a parameter is not done in the analysis of rock structures or in estimating the rating of a rock mass. For example, if pore water pressure is considered in the analysis of rock structures, it should not be accounted for in RMR. Similarly, if orientation of joint sets is considered in stability analysis of rock

TABLE 6.12 Guidelines for Excavation and Support of Rock Tunnels in Accordance with the Rock Mass Rating System

		Supports		
Rock mass class	Excavation	*Rock bolts (20 mm diameter, fully grouted)*	*Conventional shotcrete*	*Steel sets*
Very good rock RMR = 81–100	Full face; 3 m advance	Generally, no support required except for occasional spot bolting		
Good rock RMR = 61–80	Full face; 1.0–1.5 m advance; complete support 20 m from face	Locally, bolts in crown 3 m long, spaced 2.5 m, with occasional wire mesh	50 mm in crown where required	None
Fair rock RMR = 41–60	Heading and bench; 1.5–3 m advance in heading; commence support after each blast; complete support 10 m from face	Systematic bolts 4 m long, spaced 1.5–2 m in crown and walls with wire mesh in crown	50–100 mm in crown and 30 mm in sides	None
Poor rock RMR = 21–40	Top heading and bench; 1.0–1.5 m advance in top heading; install support concurrently with excavation 10 m from face	Systematic bolts 4–5 m long, spaced 1–1.5 m in crown and wall with wire mesh	100–150 mm in crown and 100 mm in sides	Light to medium ribs spaced 1.5 m where required
Very poor rock RMR <20	Multiple drifts; 0.5–1.5 m advance in top heading; install support concurrently with excavation; shotcrete as soon as possible after blasting	Systematic bolts 5–6 m long, spaced 1–1.5 m in crown and walls with wire mesh; bolt invert	150–200 mm in crown, 150 mm in sides, and 50 mm on face	Medium to heavy ribs spaced 0.75 m with steel lagging and forepoling if required; close invert

Shape: Horseshoe; width: 10 m; vertical stress <25 MPa; construction: drilling and blasting.

Source: Bieniawski, 1984.

slopes, the same should not be accounted for in RMR. The following example illustrates how to obtain strength parameters.

It is cautioned that the RMR system is found to be unreliable in very poor rock masses, so care should be exercised when applying the RMR system to such rock masses. The Q-system is more reliable for tunneling in weak rock masses.

Example 6.1

The rock mass parameters are listed in Table 6.13 for rock slopes of about 100 m in height and a slope angle of 80 degrees along a dam reservoir in the upper Himalayas in gneiss rocks in a highly seismic zone. The height of the concrete dam is 60 m, and the joints are oriented favorably. The geological strength index (GSI) is about 45. Because of this, the strength parameters for a circular wedge analysis should be used.

The factor of safety of a slope was calculated as 1.0 for existing static conditions for a completely dry condition. However, this slope was seen to be stable geologically (with SMR = 79). GSI gave a very high value of cohesion of rock mass (D = 0) that yielded a factor of safety of 2.2, which is too high.

TABLE 6.13 Rock Parameters for RMR_{basic}

S. No.	Parameters	Value	Rating
1	Point load lump strength (I_L) = 1.6 MPa $q_c = 15\ I_L = 15 \times 1.6 = 24$ MPa	$q_c \cong 24$ MPa	4
2	Rock quality designation (RQD) in % $J_v = 13$, RQD = 115 – 3.3, $J_v = 72$	72	13
3	Spacing of discontinuities	0.2–0.3 m	10
4	Condition of discontinuities	—	22
	(i) Discontinuity length	0.5–0.7 m	—
	(ii) Aperture	0	—
	(iii) Roughness	Slightly rough	—
	(iv) Infilling thickness	0	—
	(v) Degree of weathering	Moderately	—
5	Groundwater	Always for completely dry condition	15
	Total RMR_{basic}	—	64
	Cohesion	—	300 kPa
	Angle of internal friction	—	35°

ROCK MASS EXCAVABILITY INDEX FOR TBM

Bieniawski (2007) analyzed over 500 case histories to develop the rock mass excavability (RME) index to estimate the performance of double-shield and open-type TBMs. Excavability is defined as the rate of excavation expressed in machine performance in meters per day.

Bieniawski et al. (2006) found that the parameters with stronger influence on the average rate of advance (ARA), expressed in m/day, are abrasivity (or drillability), discontinuity spacing, and stand-up time. In addition, it was decided to include the two basic rock parameters—UCS of the rock material and groundwater inflow—because in some cases these two factors strongly influence the TBM advance. Once these five parameters were selected, a weighted distribution was performed. These weights have been statistically analyzed, minimizing the error in the ARA prediction and resulting in the ratings shown in Table 6.14. Thus, the RME index is based on the five input parameters listed in the table together with the ratings associated with each.

Out of the five parameters listed in Table 6.14, three parameters—uniaxial crushing strength, discontinuities in the front of the tunnel, and groundwater inflow—can be easily obtained by an experienced engineering geologist. For stand-up time for TBM excavated

TABLE 6.14 Input Parameters for the RME Index

Uniaxial compressive strength of intact rock (0–25 points)

q_c (MPa)	<5	5–30	30–90	90–180	>180
Rating	4	14	25	14	0

Drillability — Drilling rate index (01–15 points)

DRI	<80	80–65	65–50	50–40	<40
Rating	15	10	7	3	0

Discontinuities in front of the tunnel face (0–30 points)

	Homogeneity		*Number of joints per meter*					*Orientation with respect to tunnel axis*		
	Homogeneous	Mixed	0–4	4–8	8–15	15–30	>30	Perpendicular	Oblique	Parallel
Rating	10	0	2	7	15	10	0	5	3	0

Stand-up time for TBM excavated tunnels (0–25 points)

Hours	<5	5–24	24–96	96–192	>192
Rating	0	2	10	15	25

Groundwater inflow (0–5 points)

Liter/sec	>100	70–100	30–70	10–30	<10
Rating	0	1	2	4	5

Source: Bieniawski, 2007.

tunnels, it is required that RMR be estimated. Figure 6.1 shows the RMR chart for estimation of the stand-up time for tunnels. Since this chart was originally developed for drill and blast tunnels, the following correlation is available between the $RMR_{D\&B}$ and RMR_{TBM} based on the work by Alber (2000).

$$RMR_{TBM} = 0.8 \times RMR_{D\&B} + 20 \tag{6.7}$$

Construction by TBM generally results in higher RMR values than for the same tunnel section excavated by drilling and blasting because of the favorable circular shape and less damage to the surrounding rock mass by machine boring.

The RME index is obtained by summation of the five input parameters in Table 6.14, which tabulates the ratings appropriate for the ranges listed. Using the RME index in Eq. (6.8), the "theoretical" average rate of advance (ARA_T) in m/day of TBM can be estimated (Bieniawski et al., 2006).

$$ARA_T = 0.422\ RME - 11.61 \tag{6.8}$$

Subsequently, to get the "real" average rate of advance (ARA_R) of TBM from ARA_T, Bieniawski (2007) suggested three adjustment factors:

1. Influence of the TBM crew (F_E): The TBM crew who handles the tunneling machine every day has an important influence on the performance achieved. The adjustment factor of the TBM crew is listed in Table 6.15.
2. Influence of the excavated length (F_A): As tunnel excavation increases, the TBM performance is increased because of the adaptation of the machine. The quantitative effect of this adjustment adaptation factor (F_A) is given in Table 6.16.
3. Influence of tunnel diameter (F_D): Equation (6.8) was derived for tunnels with diameters close to 10 m. Taking into account the influence of different tunnel diameters, D (in meters), on the advance rate of TBM, a coefficient (F_D) is proposed as seen in Eq. (6.9) (Bieniawski, 2007).

$$F_D = -0.007D^3 + 0.1637D^2 - 1.2859D + 4.5158 \tag{6.9}$$

Therefore, for $D = 10$ m, $F_D = 1.0$, whereas for $D = 8$ m, $F_D = 1.2$, but for $D = 12$ m, $F_D = 0.5$, that is, one-half of the coefficient for $D = 10$ m.

Combining the effect of the three adjustment factors, the ARA_R can be estimated from Eq. (6.10).

$$ARA_R = ARA_T \cdot F_E \cdot F_A \cdot F_D \tag{6.10}$$

TABLE 6.15 Adjustment Factor for the Influence of TBM Crew (F_E) on TBM Advance Rate

Effectiveness of the crew handling TBM and terrain	Crew factor (F_E)
Less than efficient	0.88
Efficient	1.0
Very efficient	1.15

TABLE 6.16 Adjustment Factor for the Influence of Excavated Length (F_A) on TBM Advance Rate

Tunnel length excavated (km)	Adaptation factor (F_A)
0.5	0.68
1.0	0.80
2.0	0.90
4.0	1.00
6.0	1.08
8.0	1.12
10.0	1.16
12.0	1.20

Further, Bieniawski (2007) evaluated Eq. (6.10) and found that this equation gives reliable results for double-shield TBM in rock with strength less than 45 MPa and open type TBM in rock with strength more than 45 MPa. Another method of estimating the advance rate of TBM is presented in Chapter 14 based on Q_{TBM}.

TUNNEL ALIGNMENT

The following checklist may be followed for an economical, trouble-free alignment of a long tunnel.

1. Does the tunnel pass through young mountains?
2. Is there an intra-thrust zone?
3. Are there active and inactive fault/thrust zones?
4. Where are the thick shear zones?
5. Is rock cover excessive?
6. Is pillar width between tunnels adequate?
7. Are there thermic zones of ground temperature that are too high?
8. What is the least rock cover or shallow tunnel beneath the gullies/river/ocean?
9. Are there water-charged rock masses?
10. Are there swelling rocks?
11. Are joints oriented unfavorably or is the strike parallel to the tunnel axis (Table 6.8)? Is the tunnel along an anticline (favorable) or syncline (unfavorable)?
12. Mark expected tunneling conditions and corresponding methods of excavation along all alignments according to Chapter 7.
13. In which reaches, open/single-shield/double-shield, should TBMs be used in very long tunnels?
14. In which reaches are conventional drill and blast methods recommended?
15. Is it likely that a landslide-dam will be formed and lake water will enter the tailrace tunnel and powerhouse cavern, and so forth?

16. What are the expected costs of tunneling for different alignments along with their periods of completion?
17. What is the possible surveying error, especially in the hilly terrain?

Without a list to follow, "*mega chaos is self-organizing.*"

REFERENCES

Alber, M. (2000). Advance rates for hard rock TBMs and their effects on project economics. *Tunnelling and Underground Space Technology*, *15*(1), 55–64.

Bieniawski, Z. T. (1973). Engineering classification of jointed rock masses. *Transactions of the South African Institution of Civil Engineers*, *15*(12), 335–344.

Bieniawski, Z. T. (1978). Determining rock mass deformability, experience from case histories. *International Journal of Rock Mechanics and Mining Sciences—Geomechanics Abstracts*, *15*, 237–247.

Bieniawski, Z. T. (1979). The geomechanics classification in rock engineering applications. In *Proceedings of the 4th Congress of the International Society for Rock Mechanics* (Vol. 2, pp. 41–48). ISRM Montreux, September 2–8.

Bieniawski, Z. T. (1984). *Rock mechanics design in mining and tunnelling* (p. 272). Rotterdam: A. A. Balkema.

Bieniawski, Z. T. (1993). In J. A. Hudson (Ed.), *Classification of rock masses for engineering: The RMR system and future trends, comprehensive rock engineering* (Vol. 3, pp. 553–574). New York: Pergamon Press.

Bieniawski, Z. T., Celada, B., & Galera, J. M. (2007). Predicting TBM excavability. In *Tunnels and Tunnelling International*, September. p. 25.

Bieniawski, Z. T., Caleda, B., Galera, J. M., & Alvares, M. H. (2006). Rock mass excavability (RME) index. In *ITA World Tunnel Congress* (Paper no. PITA06-254), April, Seoul.

Edelbro, C. (2003). Rock mass strength—A review. In *Technical Review* (p. 132). Lulea University of Technology.

Goel, R. K., & Jethwa, J. L. (1991). Prediction of support pressure using RMR classification. In *Proceedings of the Indian Geotechnical Conference* (pp. 203–205). Surat, India.

Goel, R. K., Jethwa, J. L., & Paithankar, A. G. (1996). Correlation between Barton's Q and Bieniawski's RMR—A new approach. *International Journal of Rock Mechanics and Mining Sciences—Geomechanics Abstracts*, *33*(2), 179–181.

Hoek, E., & Brown, E. T. (1980). *Underground excavations in rocks. Institution of Mining and Metallurgy* (p. 527). London: Maney Publishing.

ISO 14689-1. (2003). (E). *Geotechnical investigation and testing—Identification and classification of rock—Part 1: Identification and description* (pp. 1–16). Geneva: International Organization for Standardization.

ISRM. (1978). Description of discontinuities in a rock mass. *International Journal of Rock Mechanics and Mining Sciences—Geomechanics Abstracts*, *15*, 319–368.

Kaiser, P. K., MacKay, C., & Gale, A. D. (1986). Evaluation of rock classifications at B.C. Rail Tumbler Ridge Tunnels. *Rock mechanics and rock engineering* (Vol. 19, pp. 205–234). New York: Springer Verlag.

Lauffer, H. (1988). Zur Gebirgsklassifizierung bei Frasvortrieben. *Felsbau*, *6*(3), 137–149.

Mehrotra, V. K. (1992). *Estimation of engineering properties of rock mass* (p. 267). Ph.D. Thesis. Uttarakhand, India: IIT Roorkee.

Read, S. A. L., Richards, L. R., & Perrin, N. D. (1999). Applicability of the Hoek-Brown Failure Criterion to New Zealand Greywacke Rocks. In *Proceedings of the 9th International Society for Mechanics Congress* (Vol. 2, pp. 655–660). Paris.

Serafim, J. L., & Pereira, J. P. (1983). Considerations of the geomechanics classification of Bieniawski. In *International Symposium of Engineering and Geological Underground Construction* (pp. II.33–II.42). Lisbon: LNEC.

Singh, B. (1979). Geological and geophysical investigation in rocks for engineering projects. In *International Symposium of In Situ Testing of Soils and Performance of Structures* (pp. 486–492).

Unal, E. (1983). *Design guidelines and roof control standards for coal mine roofs* (p. 355). Ph.D. Thesis. University Park: Pennsylvania State University.

Verman, M. K. (1993). *Rock mass-tunnel support interaction analysis* (p. 258). Ph.D. Thesis. Uttarakhand, India: IIT Roorkee.

Waltham, T. (2002). *Foundations of engineering geology* (2nd ed., p. 92). London: Spon Press.

Chapter 7

Tunneling Hazards

The most incomprehensible fact about nature is that it is comprehensible.

Albert Einstein

INTRODUCTION

The knowledge of potential tunneling hazards plays an important role in the selection of excavation method and designing a support system for underground openings. The tunneling media could be stable/competent (and/or non-squeezing) or squeezing/failing depending upon the in situ stress and the rock mass strength. A weak overstressed rock mass would experience squeezing ground condition (Dube & Singh, 1986), whereas a hard and massive overstressed rock mass may experience rock burst condition. When the rock mass is not overstressed, the ground condition is called "stable" or "competent" (non-squeezing).

There are two possible situations when tunneling in competent ground conditions: (1) no supports are required, or a self-supporting condition and (2) supports are required for stability, or a non-squeezing condition. The squeezing ground condition has been divided into four classes on the basis of tunnel closures by Hoek (2001) as minor, severe, very severe, and extreme squeezing ground conditions (Table 7.1 and Figure 26.4).

Tunneling through the squeezing ground condition is a very slow and hazardous process because the rock mass around the opening loses its inherent strength under the influence of in situ stresses. This may result in development of high support pressure and tunnel closures. Tunneling under the non-squeezing ground condition, on the other hand, is comparatively safe and easy because the inherent strength of the rock mass is maintained. Therefore, the first important step is to assess whether a tunnel would experience a squeezing ground condition or a non-squeezing ground condition. This decision controls the selection of the realignment, excavation method, and the support system. For example, a large tunnel could possibly be excavated full face with light supports under the non-squeezing ground condition. It may have to be excavated by a heading and benching method with a flexible support system under the squeezing ground condition.

Non-squeezing ground conditions are common in most tunneling projects. Squeezing conditions are common in the lower Himalayas in India, the Alps, and other young mountains where the rock masses are weak, highly jointed, faulted, folded, and tectonically disturbed, and the overburden is high.

TABLE 7.1 Classification of Ground Conditions for Tunneling

S. No.	Ground condition class	Subclass	Rock behavior
1	Competent self-supporting	—	Massive rock mass requires no support for tunnel stability
2	Incompetent non-squeezing	—	Jointed rock mass requires support for tunnel stability; tunnel walls are stable and do not close
3	Raveling	—	Chunks or flakes of rock mass begin to drop out of the arch or walls after the rock mass is excavated
4	Squeezing	Minor squeezing ($u_a/a = 1–2.5\%$) Severe squeezing ($u_a/a = 2.5–5\%$) Very severe squeezing ($u_a/a = 5–10\%$) Extreme squeezing ($u_a/a > 10\%$) (Hoek, 2001)	Rock mass squeezes plastically into the tunnel both from the roof and the walls, and the phenomenon is time dependent; rate of squeezing depends upon the degree of overstress; may occur at shallow depths in weak rock masses like shales, clay, etc.; hard or strong rock masses under high cover may experience slabbing/popping/rock burst
5	Swelling	—	Rock mass absorbs water, increases in volume, and expands slowly into the tunnel (e.g., in montmorillonite clay)
6	Running	—	Granular material becomes unstable within steep shear zones
7	Flowing/sudden flooding	—	A mixture of soil-like material and water flows into the tunnel; the material can flow from invert as well as from the face crown and wall and can flow for large distances, completely filling the tunnel and burying machines in some cases; the discharge may be 10–100 L/sec which can cause sudden flood; a chimney may be formed along thick shear zones and weak zones
8	Rock burst	—	A violent failure in hard (brittle) and massive rock masses of Class II* type when subjected to high stress

u_a = radial tunnel closure; a = tunnel radius; u_a/a = normalized tunnel closure in percentage.
**Uniaxial compressive strength (UCS) test on Class II type rock shows reversal of strain after peak failure (Figure 3.2).*

Source: Singh and Goel, 1999.

TUNNELING CONDITIONS

Various conditions encountered during tunneling are summarized in Table 7.1. Table 7.2 outlines the method of excavation, the type of support, and precautions for various ground conditions. As per the guidelines of the Austrian Society for Geomechanics, various conditions of ground behavior type (BT) and description of potential failure modes during excavation of the unsupported rock mass have been summarized in Table 7.3 (Solak, 2009). Table 7.4 summarizes different conditions for tunnel collapse caused by unforeseen geological conditions and inadequacy of design models or support systems (Vlasov, Makovski, & Merkin, 2001).

The Commission on Squeezing Rocks in Tunnels of the International Society for Rock Mechanics (ISRM) has published *Definitions of Squeezing* as reproduced here (Barla, 1995):

Squeezing of rock is the time-dependent large deformation, which occurs around a tunnel and other underground openings, and is essentially associated with creep caused by (stress) exceeding shear strength (limiting shear stress). Deformation may terminate during construction or continue over a long time period.

This definition is complemented by the following additional statements:

- Squeezing can occur in both rock and soil as long as the particular combination of induced stresses and material properties pushes some zones around the tunnel beyond the limiting shear stress at which creep starts.
- The magnitude of the tunnel convergence associated with squeezing, the rate of deformation, and the extent of the yielding zone around the tunnel depend on the geological conditions, the in situ stresses relative to rock mass strength, the groundwater flow and pore pressure, and the rock mass properties.
- Squeezing of rock masses can occur as squeezing of intact rock, as squeezing of infilled rock discontinuities, and/or along bedding and foliation surfaces, joints, and faults.
- Squeezing is synonymous with overstressing and does not comprise deformations caused by loosening as might occur at the roof or at the walls of tunnels in jointed rock masses. Rock bursting phenomena do not occur during squeezing.
- Time-dependent displacements around tunnels of similar magnitudes as in squeezing ground conditions may also occur in rocks susceptible to swelling. While swelling always implies volume increase due to penetration of the air and moisture into the rock, squeezing does not, except for rocks exhibiting dilatant behavior. However, it is recognized that in some cases squeezing may be associated with swelling.
- Squeezing is closely related to the excavation, support techniques, and sequence adopted in tunneling. If the support installation is delayed, the rock mass moves into the tunnel and a stress redistribution takes place around it. Conversely, if the rock deformations are constrained, squeezing will lead to long-term load build-up on rock support.

The ground pressure developing far behind the tunnel face in a heavily squeezing ground depends on the amount of support resistance during the yielding phase. The higher the yield pressure of the support, the lower the final load. A targeted reduction in ground pressure can be achieved by installing a support that accommodates a larger deformation (which is a well-known principle), as well as by selecting a support that yields at a higher pressure.

TABLE 7.2 Method of Excavation, Type of Support, and Precautions to Be Adopted for Different Ground Conditions

S. No.	Ground conditions	Excavation method	Type of support	Precautions
1	Self-supporting/ competent	TBM or full face drill and controlled blast	No support or spot bolting with a thin layer of shotcrete to prevent widening of joints	Look out for localized wedge/shear zone; past experience discourages use of TBM if geological conditions change frequently
2	Non-squeezing/ incompetent	Full face drill and controlled blast by boomers	Flexible support; shotcrete and pre-tensioned rock bolt supports of required capacity; steel fiber reinforced shotcrete (SFRS) may or may not be required	First layer of shotcrete should be applied after some delay but within the stand-up time to release the strain energy of rock mass
3	Raveling	Heading and bench; drill and blast manually	Steel support with struts/pre-tensioned rock bolts with SFRS	Expect heavy loads including side pressure
4	Minor squeezing	Heading and bench; drill and blast	Full column grouted rock anchors and SFRS; floor to be shotcreted to complete a support ring	Install support after each blast; circular shape is ideal; side pressure is expected; do not have a long heading, which delays completion of support ring
5	Severe squeezing	Heading and bench; drill and blast	Flexible support; full-column grouted highly ductile rock anchors and SFRS; floor bolting to avoid floor heaving and to develop a reinforced rock frame; in case of steel ribs, these should be installed and embedded in shotcrete to withstand high support pressure	Install support after each blast; increase the tunnel diameter to absorb desirable closure; circular shape is ideal; side pressure is expected; instrumentation is essential

6	Very severe squeezing and extreme squeezing	Heading and bench in small tunnels and multiple drift method in large tunnels; use forepoling if stand-up time is low	Very flexible support; full-column grouted highly ductile rock anchors and thick SFRS; yielding steel ribs with struts when shotcrete fails repeatedly; steel ribs may be used to supplement shotcrete to withstand high support pressure; close ring by erecting invert support; encase steel ribs in shotcrete, floor bolting to avoid floor heaving; sometimes steel ribs with loose backfill are also used to release the strain energy in a controlled manner (tunnel closure of more than 4% will not be permitted)	Increase the tunnel diameter to absorb desirable closure; provide invert support as early as possible to mobilize full support capacity; long-term instrumentation is essential; circular shape is ideal
7	Swelling	Full face or heading and bench; drill and blast	Full-column grouted rock anchors with SFRS shall be used all around the tunnel; increase 30% thickness of shotcrete due to weak bond of the shotcrete with rock mass; erect invert strut; the first layer of shotcrete is sprayed immediately to prevent ingress of moisture into rock mass	Increase the tunnel diameter to absorb the expected closure; prevent exposure of swelling minerals to moisture, monitor tunnel closure
8	Running and flowing	Multiple drift with forepoles; grouting of the ground is essential; shield tunneling may be used in soil conditions; realign the tunnel	Full-column grouted rock anchors and SFRS; concrete lining up to face, steel liner in exceptional cases with shield tunneling; use probe hole to discharge groundwater; face should also be grouted, bolted, and shotcreted	Progress is very slow; trained crew should be deployed; in reach of sudden flooding, the tunnel is realigned by-passing the same, if ground is not groutable; monitor rate of flow of seepage
9	Rock burst	Full face drill and blast	Fiber reinforced shotcrete with full-column resin anchors immediately after excavation	Micro-seismic monitoring is essential

TABLE 7.3 General Categories of Ground Behavior Types

S. No.	Behavior type	Description of potential failure modes/ mechanisms during excavation of the unsupported rock mass
1	Stable	Stable rock mass with the potential of small local gravity-induced falling or sliding of blocks
2	Discontinuity controlled block failure	Deep reaching, discontinuity controlled; gravity-induced falling and sliding of blocks; occasional local shear failure
3	Shallow stress-induced failure	Shallow stress-induced brittle and shear failures in combination with discontinuity and gravity controlled failure of the rock mass
4	Deep-seated stress-induced failure	Deep-seated stress-induced brittle and shear failures in combination with large displacements
5	Rock burst	Sudden and violent failure of the rock mass caused by highly stressed brittle rocks and the rapid release of accumulated strain energy
6	Buckling failure	Buckling of rocks with a narrowly spaced discontinuity set; frequently associated with shear failure
7	Shear failure under low confining pressure	Potential for excessive overbreak and progressive shear failure with the development chimney type failure; caused mainly by a deficiency of side pressure
8	Raveling ground	Flow of cohesionless dry or moist intensely fractured rocks or soil
9	Flowing ground	Flow of intensely fractured rocks or soil with high water content
10	Swelling	Time-dependent volume increase of the rock mass caused by physicochemical reaction of rock and water in combination with stress relief, leading to inward movement of the tunnel perimeter
11	Frequently changing behavior	Rapid variations of stresses and deformations, caused by heterogeneous rock mass conditions or block-in-matrix rock situation of a tectonic melange (brittle fault zone)

Source: Solak, 2009.

Furthermore, high yield pressure reduces the risk of violating the clearance profile and increases the safety level of roof instabilities (loosening) during the deformation phase (Cantieni & Anagnostou, 2009).

A comparison between squeezing and swelling phenomena by Jethwa (1981) is given in Table 7.5. Figure 7.1 shows how radial displacements significantly vary with time within the broken zone. The radial displacement, however, tends to converge at the interface boundary of the elastic and the broken zones. Figure 7.2 shows that a compaction zone is formed within this large broken zone so that the rate of tunnel wall closure is arrested.

TABLE 7.4 Quality Aspects Related to Tunnel Collapses

S. No.		Type	Phenomenon	Cause	Remedial measures
1	Ground collapse	Ground collapse near the portal	During the excavation of the upper half section of the portal the tunnel collapsed and the surrounding ground slid to the river side	Ground collapse was caused by the increase of pore water pressure due to rain for five consecutive days	• Installation of anchors to prevent landslides • Construction of counterweight embankment, which can also prevent landslide • Installation of pipe roofs to strengthen the loosened crown
2		Landslide near the portal	Cracks appeared in the ground surface during the excavation of the side drifts of the portal, and the slope near the portal gradually collapsed	Excavation of the toe of the slope composed of strata disturbed the stability of soil, and excavation of the side drifts loosened the natural ground, which led to landslide	• Caisson type pile foundations were constructed to prevent unsymmetrical ground pressure • Vertical reinforcement bars were driven into the ground to increase its strength
3		Collapse of the crown of cutting face	10 to 30 m^3 of soil collapsed and supports settled during excavation of the upper half section	The ground loosened and collapsed due to the presence of heavily jointed fractured rock mass at the crown of the cutting face, and the vibration caused by the blasting for the lower half section (hard rock)	• Roof bolts were driven into the ground to stabilize the tunnel crown • To strengthen the ground near the portal and talus, chemical injection and installation of vertical reinforcement bars were conducted

Continued

TABLE 7.4 Quality Aspects Related to Tunnel Collapses—Cont'd

S. No.		Type	Phenomenon	Cause	Remedial measures
4		Collapse of fault fracture zone	After completion of blasting and mucking, flaking of sprayed concrete occurred behind the cutting face, following which 40 to 50 m^3 of soil collapsed along a 7 m section from the cutting face; later it extended to 13 m from the cutting face and the volume of collapsed soil reached 900 m^3	The fault fracture zone above the collapsed cutting face loosened due to blasting, and excessive concentrated loads were imposed on supports, causing the shear failure and collapse of the sprayed concrete	• Reinforcement of supports behind the collapsed location (additional sprayed concrete, additional rock bolts) • Addition of the number of the measurement section • Hardening of the collapsed muck by chemical injection • Air milk injection into the voids above the collapsed portions • Use of supports with a higher strength
5	Distortion of supports	Distortion of tunnel supports	During excavation by the full face tunneling method, steel supports considerably settled and foot protection concrete cracked	Bearing capacity of the ground at the bottom of supports decreased due to prolonged immersion by groundwater	• Permanent foot protection concrete was placed to decrease the concentrated load • An invert with drainage was placed
6		Distortion of lining concrete due to unsymmetrical ground pressure	During the excavation of the upper half section, horizontal cracks ranging in width from 0.1 to 0.4 mm appeared in the arch portion of the mountainside concrete lining, while subsidence reached the ground surface on the valley side	Landslide was caused due to the steep topography with asymmetric pressure and the ground with lower strength, leading to the oblique load on the lining concrete	• Earth anchors were driven into the mountainside ground to withstand the oblique load • Ground around the tunnel was strengthened by chemical injection; subsidence location was filled

7		Distortion of tunnel supports due to swelling pressure	Hexagonal cracks appeared in the sprayed concrete and the bearing plates for rock bolts were distorted due to the sudden inward movement of the side walls of the tunnel	Large swelling pressure was generated by swelling clay minerals in mudstone	• Sprayed concrete and face support bolts on the cutting face were provided to prevent weathering • A temporary invert was placed in the upper half section by spraying concrete
8		Heaving of a tunnel in service	Heaving occurred in the pavement surface six months after the commencement of service, causing cracks and faulting in the pavement. Heaving reached as large as 25 cm	A fault fracture zone containing swelling clay minerals, which was subjected to hydrothermal alteration, existed in the distorted section; plastic ground pressure caused by this fracture zone concentrated on the base course of the weak tunnel section without invert	• To restrict the plastic ground pressure, rock bolts and sprayed concrete were applied to the soft sandy soil beneath the base course • Reinforced invert concrete was placed
9	Adverse effects on the surrounding environment	Adverse effects of vibration due to blasting on the adjacent existing tunnel	During the construction of a new tunnel, which runs parallel to the side wall of the existing portal, cracks appeared in the lining (made of bricks) of the existing tunnel	The voids behind the existing tunnel loosened and the lining was distorted due to the vibration of the blasting for construction of the new tunnel	• Steel supports and temporary concrete lining were provided to protect the existing tunnel • Backfill grouting was carried out • Excavation was carried out by the non-blasting rock breaking method and the limit for chemical agent was set to mitigate the vibration

Continued

TABLE 7.4 Quality Aspects Related to Tunnel Collapses—Cont'd

S. No.	Type	Phenomenon	Cause	Remedial measures
10	Ground settlement due to the excavation for dual-tunnel directly beneath residential area	Considerable distortion of supports occurred in the embankment section; although additional bolts were driven into the ground and additional sprayed concrete was provided, ground surface settlement exceeded 100 mm	Since the soil characteristics in the embankment section were worse than expected, the ground settlement was considerably increased by the construction of tunnels following the dual-tunnel	• Pipe roofs were driven from inside the tunnel to reduce ground surface settlement

Summary of different conditions for tunnel collapses caused by geological unforeseen conditions and inadequacy of design, models, or support systems.

Source: Vlasov et al., 2001.

TABLE 7.5 Comparison between Squeezing and Swelling Phenomena

Parameter	Squeezing	Swelling
1. Cause	Small volumetric expansion of weak and soft ground upon stress-induced shear failure; compaction zone can form within broken zone	Volumetric expansion due to ingress of moisture in ground containing swelling minerals
2. Closure		
Rate of closure	Very high initial rate, up to several centimeters per day for the first 1–2 weeks of excavation	1. High rate for several weeks till moisture penetrates deep into the ground
	Reduces with time	2. Decreases with time as moisture penetrates into the ground deeply with difficulty
Period	May continue for years in exceptional cases	3. May continue for years if the moist ground is scooped out to expose fresh ground
3. Extent	The affected zone can be several tunnel diameters thick	The affected zone is several meters thick; post-construction saturation may increase swelling zone significantly
4. Failure	The rock blocks are crushed in the broken zone	The rock blocks are not crushed during swelling; poor rocks are pulverized due to swelling

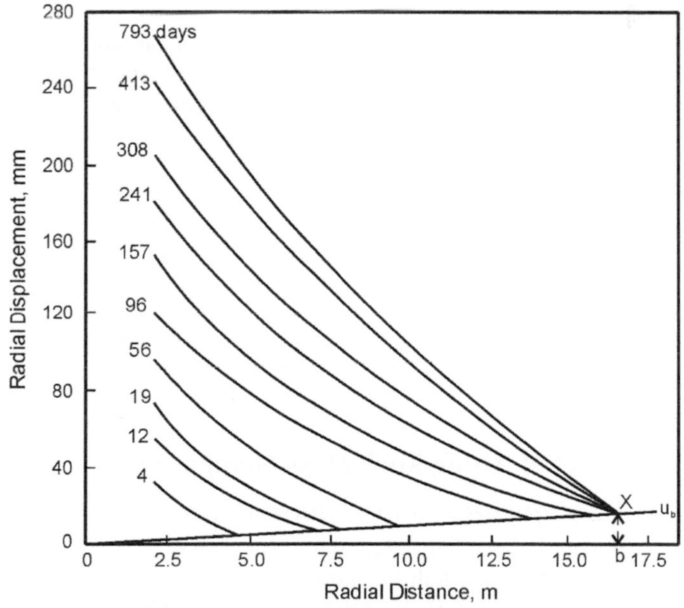

FIGURE 7.1 Observed variation of radial displacement with radial distance within slates/phyllites of the Giri Tunnel, India. *(From Jethwa, 1981)*

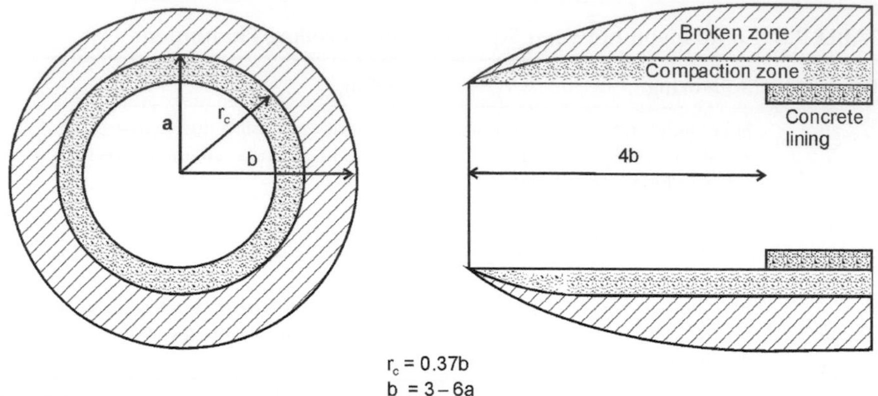

FIGURE 7.2 Compaction zone within broken zone in the squeezing ground condition. *(From Jethwa, 1981)*

Various approaches for estimating the ground conditions for tunneling on the basis of Q and rock mass number, N (Q with SRF = 1), are illustrated in the following sections (Chapters 8 and 9 describe Q and N, respectively, in greater detail).

EMPIRICAL APPROACH FOR PREDICTING GROUND CONDITIONS

Singh et al. Criterion

Singh, Jethwa, Dube, and Singh (1992) suggested an empirical approach based on case histories in the Himalayas and by collecting data on Barton, Lien, and Lunde (1974) rock mass quality (Q) and overburden (H). It implies that a squeezing ground condition would be encountered if

$$H >> 350\ Q^{1/3}\ \text{meters} \tag{7.1}$$

and a non-squeezing ground condition would be encountered if

$$H << 350\ Q^{1/3}\ \text{meters} \tag{7.2}$$

For computing Q, the SRF rating of 2.5 should be used in Eqs. (7.1) and (7.2).

It is suggested that future efforts should be made to account for the ratio of horizontal to vertical in situ stresses.

Criterion of Goel et al. Using Rock Mass Number (N)

Prediction of Non-Squeezing and Squeezing Ground Conditions

To avoid the uncertainty in obtaining appropriate SRF ratings in the rock mass quality (Q) of Barton et al. (1974), Goel, Jethwa, and Paithankar (1995) suggested rock mass number (N), defined as in Eq. (7.3), for proposing the criteria of estimating ground conditions for tunneling.

$$N = [Q]_{SRF=1} \tag{7.3}$$

Equation (7.3) suggests that N is Q with an SRF of 1.

Other parameters considered are the tunnel depth (H) in meters to account for stress condition or SRF indirectly, and tunnel width (B) to take care of the strength reduction of the rock mass with size. The values of three parameters — the rock mass number (N), the tunnel depth (H), and the tunnel diameter or width (B) — were collected covering a wide variety of ground conditions varying from highly jointed and fractured rock masses to massive rock masses.

All the data points were plotted on a log-log graph (Figure 7.3) between rock mass number (N) and $HB^{0.1}$. Figure 7.3 shows zones of tunneling conditions/hazards based upon the values of $HB^{0.1}$ and N. Here H is the overburden in meters, B is the width of the tunnel or cavern in meters, and N is the rock mass number (Chapter 9). It should be noted that B should be more than the size of self-supporting tunnels (Eqs. 7.7 and 7.9).

In Figure 7.3, a clear line (AB) demarcating the squeezing and non-squeezing cases is obtained. The equation of this line is

$$H = (275\ N^{0.33}) \cdot B^{-0.1} \text{ meters} \tag{7.4}$$

where H = tunnel depth or overburden in meters and B = tunnel span or diameter in meters. The points lying above line AB (Eq. 7.4) represent squeezing ground conditions, whereas those below this line represent the non-squeezing ground condition. This can be explained as follows.

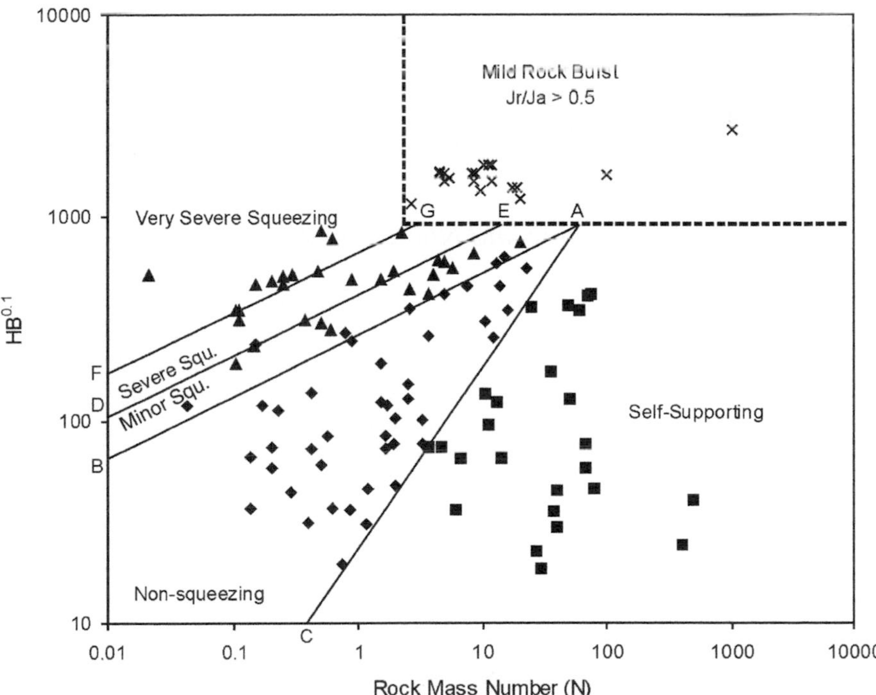

FIGURE 7.3 Plot between rock mass number (N) and $HB^{0.1}$ for predicting ground conditions.

For a Squeezing Ground Condition

$$H >> (275\ N^{0.33}) \cdot B^{-0.1} \text{ meters} \tag{7.5}$$

$$\frac{J_r}{J_a} \leq \frac{1}{2}$$

For a Non-Squeezing Ground Condition

$$H << (275\ N^{0.33}) \cdot B^{-0.1} \text{ meters} \tag{7.6}$$

How is a stress- and strength-related rock mass condition estimated using the Q or N? The rock strength is related to the Q or N and stress is related to tunnel depth (H) as given above. Chapter 8 also presents the correlation between the rock mass strength and Q.

The use of Eq. (7.4) is explained with the help of the following example.

Example 7.1

In a hydroelectric project in India a tunnel was driven through metabasics with a rock mass number (N) of 20, tunnel depth (H) of 635 m, tunnel diameter (B) of 5.8 m, and $J_r/J_a \approx 0.35$.

Using Eq. (7.4), the calculated value of H is 620 m for squeezing; however, the actual depth is 635 m. This satisfies the squeezing ground condition represented by inequality (7.5). To avoid the squeezing ground condition, the designers could either realign the tunnel to reduce the cover or make it pass through a rock mass having a higher N value.

Equation (7.4) also explains why a drift cannot represent the ground condition in the main tunnel, because a drift would normally be smaller in size and not experience as much squeezing as the larger main tunnel.

Prediction of Self-Supporting and Non-Squeezing Ground Conditions

As presented in Chapter 6, Bieniawski (1973) neglected the effect of in situ stress/tunnel depth (H) while obtaining the span of an unsupported or self-supporting tunnel using RMR. Barton et al. (1974) proposed Eq. (8.12) for the unsupported span, but did not give adequate weightage to tunnel depth in the stress reduction factor (SRF; Chapter 8).

Goel et al. (1995) developed an additional criterion to estimate the self-supporting tunneling condition. In Figure 7.3, demarcation line CA was obtained to separate the self-supporting condition from the non-squeezing condition. The equation of this line is obtained as follows:

$$H = 23.4\ N^{0.88}\ B_s^{-0.1} \text{ meters,} \tag{7.7}$$

where B_s = unsupported span or span of self-supporting tunnel in meters.

Equation (7.7) suggests that for a self-supporting tunnel condition

$$H << 23.4\ N^{0.88}\ B_s^{-0.1} \text{ meters,} \tag{7.8}$$

$$B_s = 2\ ESR\ Q^{0.4} \text{ meters (after Barton et al., 1974)} \tag{7.9}$$

Prediction of Degree of Squeezing

Degree of Squeezing and Its Effect on Tunneling

The degree of squeezing can be represented by tunnel closure (Singh, Jethwa, & Dube, 1995) as follows:

Mild or minor squeezing	Closure 1–3% of tunnel diameter
Moderate or severe squeezing	Closure 3–5% of tunnel diameter
High or very severe squeezing	Closure >5% of tunnel diameter

On the basis of the previous limits of closures, out of 29 squeezing cases, 14 cases denote mild or minor squeezing, 6 cases represent moderate or severe squeezing, and 9 cases pertain to high or very severe squeezing ground conditions.

It may be added here that tangential strain ε_θ is equal to the ratio of tunnel closure and diameter. If it exceeds the failure strain ε_f of the rock mass, squeezing will occur. Mild squeezing may not begin even if closure is 1% and less than ε_f in most cases (see the section Critical Strain on Rock Mass in Chapter 13).

Considering the previously mentioned limits of closure, it is possible to draw two more demarcation lines, DE and FG, in the squeezing zone in Figure 7.3. The equation of line DE separating cases of mild from moderate squeezing ground conditions is obtained as:

$$H = (450\ N^{0.33}) \cdot B^{-0.1}\ \text{meters} \tag{7.10}$$

Similarly, the equation of line FG (Figure 7.3) separating the moderate and high squeezing conditions is obtained as:

$$H = (630\ N^{0.33}) \cdot B^{-0.1}\ \text{meters} \tag{7.11}$$

All of the equations obtained from Figure 7.3 for predicting ground conditions are summarized in Table 7.6. The squeezing ground condition has not been encountered in tunnels where J_r/J_a was found to be more than 0.5.

It is important to know in advance, if possible, the location of rock burst or squeezing conditions, because the support systems are different in each condition. Kumar (2002) classified modes of failures according to values of joint roughness number (J_r) and joint alteration number (J_a), as shown in Figure 7.4. It is observed that mild rock burst

TABLE 7.6 Prediction of Ground Condition Using N

S. No.	Ground conditions	Correlations for predicting ground condition
1	Self-supporting	$H < 23.4\ N^{0.88} \cdot B^{-0.1}$ and $1000\ B^{-0.1}$ and $B < 2\ Q^{0.4}$ m
2	Non-squeezing	$23.4\ N^{0.88} \cdot B^{-0.1} < H < 275 N^{0.33} \cdot B^{-0.1}$
3	Mild squeezing	$275\ N^{0.33} \cdot B^{-0.1} < H < 450 N^{0.33} \cdot B^{-0.1}$ and $J_r/J_a < 0.5$
4	Moderate squeezing	$450\ N^{0.33} \cdot B^{-0.1} < H < 630 N^{0.33} \cdot B^{-0.1}$ and $J_r/J_a < 0.5$
5	High squeezing	$H > 630 N^{0.33} \cdot B^{-0.1}$ and $J_r/J_a < 0.25$
6	Mild rock burst	$H \cdot B^{0.1} > 1000$ m and $J_r/J_a > 0.5$ and $N > 1.0$

Source: Goel, 1994.

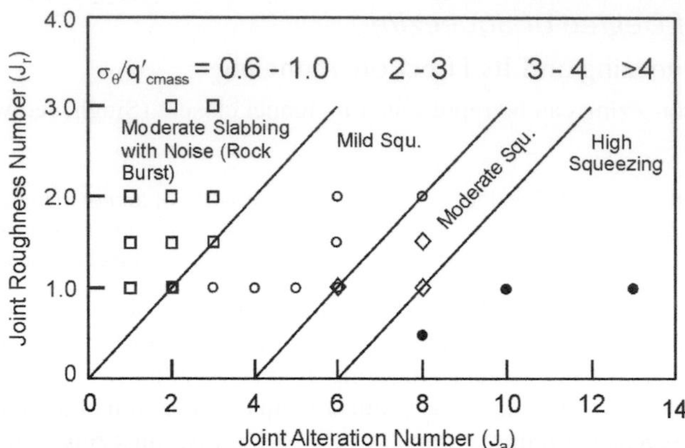

FIGURE 7.4 Prediction of ground condition. *(From Kumar, 2002)*

occurred only where J_r/J_a exceeds 0.5. This observation confirmed the study of Singh and Goel (2002). If J_r/J_a was significantly less than 0.5, then a squeezing phenomenon was encountered in many tunnels under high overburden in the Himalayas. Thus, a semi-empirical criterion for mild rock burst in the tunnels is suggested as follows:

$$\frac{\sigma_\theta}{q'_{cmass}} = 0.60 - 1.0 \tag{7.12}$$

and

$$J_r/J_a > 0.50 \tag{7.13}$$

where q'_{cmass} = biaxial strength of rock mass (Eq. 7.14) and σ_θ = maximum tangential stress at tunnel periphery. Predictions should be made on the basis of Figures 7.3 and 7.4.

Rock Burst

The upper right corner zone in Figure 7.3 is marked by dotted lines. Spalling and mild to moderate rock burst cases in tunnels from Indian hydroelectric and mining projects are in this region, which indicates a probable zone of rock burst condition. The inter-block shear strength parameter (J_r/J_a) of Barton et al. (1974) is found to be more than 0.5 for all tunneling cases encountering the mild to moderate rock burst condition.

Criterion of Bhasin and Grimstad

Using the results of Eq. (7.1), Bhasin and Grimstad (1996) developed a monogram (Figure 7.5) between rock mass strength, in situ stress, and rock behavior in tunnels with rock mass quality (Q) for estimating the ground conditions.

THEORETICAL/ANALYTICAL APPROACH

Theoretically, the squeezing conditions around a tunnel opening are encountered if (Eq. 13.20)

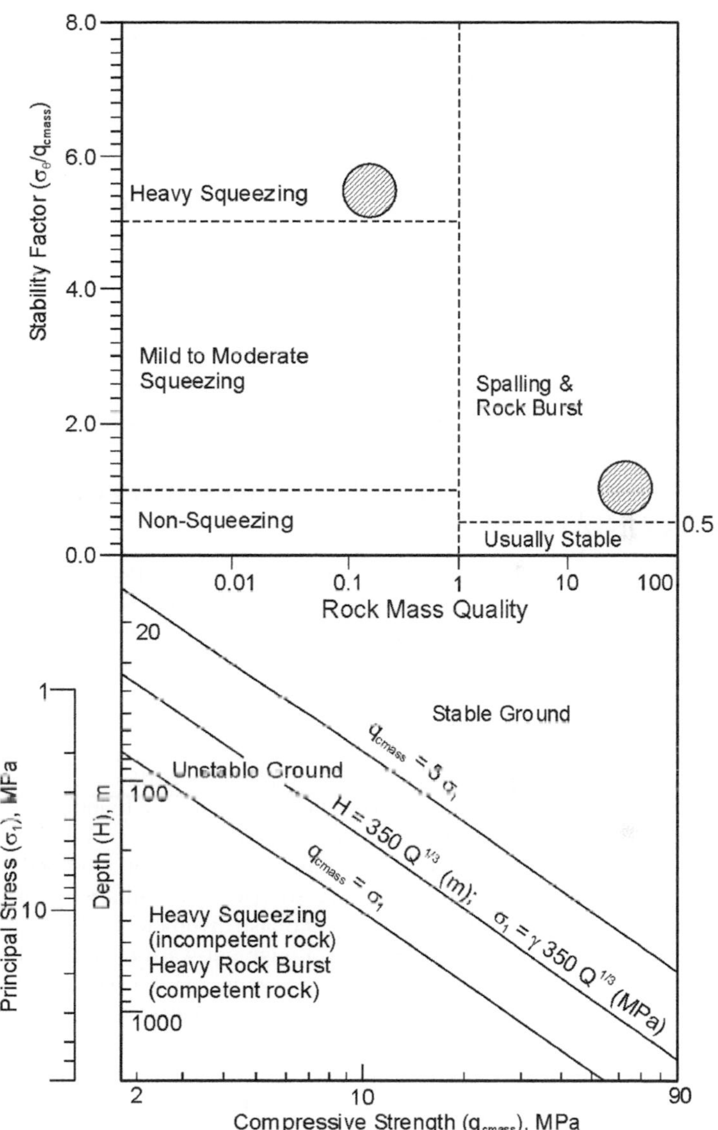

FIGURE 7.5 Monogram for prediction of tunnel stability. *(From Bhasin and Grimstad, 1996)*

$$\sigma_\theta > \text{strength} = q_{cmass} + P_O\ A/2 = q'_{cmass} \tag{7.14}$$

where σ_θ is the tangential stress and q_{cmass} is the uniaxial compressive strength (UCS) of the rock mass, Po is in situ stress along the tunnel axis, and A is the rock parameter proportional to friction (Chapter 13). Practically, Eq. (7.14) can be written as follows for a circular tunnel under a hydrostatic stress field:

$$2\,P > q_{cmass} + P \cdot A/2 \tag{7.15}$$

where P is the magnitude of the overburden pressure. Therefore, it may be noted that squeezing may not occur in hard rocks with a high value of parameter A.

Use of Eq. (7.14) for predicting the squeezing ground condition poses practical difficulties as the measurement of the in situ stress and determination of the in situ compressive strength of a rock mass are both time-consuming and expensive.

ISRM classifies squeezing rock/ground condition as follows:

Degree of squeezing	σ_θ/q_{cmass} (ISRM)	$q_{cmass}/(\gamma \cdot H)$ (Barla, 1995)
No squeezing	<1.0	>1.0
Mild squeezing	1.0–2.0	0.4–1.0
Moderate squeezing	2.0–4.0	0.2–0.4
High squeezing	>4.0	<0.2

This approach may be used reliably depending upon the values of σ_θ and q_{cmass}.

EFFECT OF THICKNESS OF WEAK BAND ON SQUEEZING GROUND CONDITION

From the 29 km tunnel of the Nathpa-Jhakri project in Himachal Pradesh (H.P.), India, it is suggested that squeezing does not take place if the thickness of the band of weak rock mass is less than approximately $2 \cdot Q^{0.4}$ meters. However, more project data is needed for a better correlation.

SUDDEN FLOODING OF TUNNELS

The inclined beds of impervious rocks (shale, phyllite, schist, etc.) and pervious rocks (crushed quartzites, sandstone, limestone, fault, etc.) may be found along a tunnel alignment. The heavy rains/snow charge the beds of pervious rocks with water like an aquifer. While tunneling through the impervious bed into a pervious bed, seepage water may suddenly gush out. The authors have studied four similar case histories at the Chhibro-Khodri, Maneri Bhali, BSL, and Dulhasti hydroelectric projects in the Himalayas where sudden flood accompanied by a huge out-wash of sand and boulders occurred ahead of the tunnel face where several shear zones existed. This flooding problem becomes dangerous where the pervious rock mass is squeezing ground due to the excessive overburden. In two projects in the Himalayas, the machines and tunnel boring machines (TBMs) are partly buried (Kadkade, 2007).

Seepage should be monitored near the portal regularly. The discharge of water should be plotted along the chainage of the face of the tunnel. If the peak discharge is found to increase with tunneling, it is very likely that sudden flooding of the tunnel may take place with further tunneling. Consult the international experts before tackling such situations.

CHIMNEY FORMATION

There may be local thick shear zones dipping toward a tunnel face. The soil/gouge may fall down rapidly unless it is carefully supported immediately after excavation. There are chances of formation of a high cavity/chimney along the thick shear zone. The chimney may be very high in water-charged rock mass. This cavity should be completely backfilled by lean concrete.

TABLE 7.7 Properties of Various Gases That May Be Present in a Tunnel

Gas	Density	Color	Odor	Source	Physiological effect on workers
Oxygen (O_2)	1.11	None	None	Air is normally 20.93% O_2	At least 20% is required to sustain normal health; workers become dizzy if concentration drops to 15%; some workers may die at 12.5%; most will faint at a concentration of 9%; and death will occur at 6% or less
Nitrogen (N_2)	0.97	Yellow	None	Air is normally 78.10% N_2	Nitrogen has no ill effect on persons except to dilute air and decrease O_2%
Carbon dioxide (CO_2)	1.50	None	None	Air is normally 0.03% CO_2; CO_2 is produced by decaying timber and fires, and is present in diesel exhaust	CO_2 acts as a respiratory stimulant and may increase effects of other harmful contaminants; at 5% CO_2, breathing is laborious; a concentration of 10% can be endured for only a few minutes
Carbon monoxide (CO)	0.97	None	None	Present in diesel exhaust and blast fumes	CO is absorbed into the blood rather than O_2. In time, very small concentrations will produce symptoms of poisoning. A concentration slightly greater than 0.01% will cause a headache or possibly nausea. A concentration of 0.2% is fatal
Methane (CH_4)	0.55	None	None	Present in certain rock formations containing carbonaceous materials	Has no ill effect on persons except to dilute air and decrease O_2%; it is dangerous because of its explosive properties; methane is explosive in the concentration range of 5.5 to 14.8%, being most explosive at a concentration of 9.5%
Hydrogen sulfide (H_2S)	1.19	None	Rotten eggs	Present in certain rock formations and sometimes in blast fumes	Extremely poisonous — 0.06% will cause serious problems in a few minutes

Continued

TABLE 7.7 Properties of Various Gases That May Be Present in a Tunnel—Cont'd

Gas	Density	Color	Odor	Source	Physiological effect on workers
Sulfur dioxide (SO_2)	2.26	None	Burning sulfur	Present in diesel exhaust and blast fumes	Strongly irritating to mucous membranes at low concentrations; can be kept below objectionable levels by limiting fuel sulfur content to 0.5%
Oxides of nitrogen	Approx. 1.5	Yellow-brown	Stings nose	Present in diesel exhaust and blast fumes	NO_2 is most toxic; all oxides of nitrogen cause severe irritation of the respiratory tract at high concentrations; acute effects may be followed by death in a few days to several weeks owing to permanent lung damage

Source: Mathews, 1996.

ENVIRONMENTAL HAZARDS DUE TO TOXIC OR EXPLOSIVE GASES AND GEOTHERMAL GRADIENT

There are serious environmental hazards due to toxic or explosive gases while tunneling in the argillaceous rocks. Sometimes methane gas is emitted by blasted shales. Improper ventilation also increases concentration of toxic gases like carbon monoxide, carbon dioxide, hydrogen sulfide, and sulfur dioxide, so additional ventilation capacity is required. If there is methane gas emission, permissible electrical equipment may be used. Attention should be given to the physical properties of the gases, as some gases tend to collect either in high or low pockets in a tunnel complex. Table 7.7 summarizes the properties of the previously mentioned gases found in tunnels (Mathews, 1996). Monitoring of gases and oxygen should be carried out near the face of a tunnel where blast fumes and gas emissions are heaviest. Oxygen must be maintained at a level of 20% or greater. Dust inside the tunnel should also be controlled to reduce health hazards; therefore, the wet drilling method is recommended for both blast holes and bolt holes.

As rock engineers go deeper and deeper, they will have to face high temperatures. The temperature may increase at a rate of about 30°C per kilometer. This is in addition to the average ground temperature, which is equal to the average temperature in a year. The temperature inside a 1400 m deep NJPC tunnel in the Himalayas, in India, was more than 45°C. The efficiency of workers in such a high temperature was reduced drastically. They worked for two to three hours, frequently bathing in buckets of ice-filled water. If possible, cool fresh air should be used for ventilation to maintain a working temperature of around 30°C at the tunnel face.

CONCLUDING REMARKS

Rock has extraordinary geological occurrence (EGO) problems. Enormous time and money are lost due to unforeseen tunneling hazards, particularly in the Himalayas and other young mountain chains. Generally, if a shear zone or a weak zone is not seen within 200 m in the lower Himalayas, it means that it has been missed. Thus, geological uncertainties may be managed by adopting a strategy of tunnel construction that copes with most tunneling conditions. *A hazard foreseen is a hazard controlled.* Therefore, it is desirable to use safe and effective tunneling methodology based on detailed engineering geological explorations before and during the tunnel construction. The modern trend of insuring the tunneling machine and the losses due to delays because of unexpected geological and geohydrological conditions takes care of the contractor's interests.

REFERENCES

Barla, G. (1995). Squeezing rocks in tunnels. *ISRM News Journal*, *2*(3 & 4), 44–49.

Barton, N., Lien, R., & Lunde, J. (1974). Engineering classification of rock masses for the design of tunnel support. In *Rock mechanics* (Vol. 6, pp. 189–236). New York: Springer-Verlag.

Bhasin, R., & Grimstad, E. (1996). The use of stress-strength relationship in the assessment of tunnel stability. In *Proceedings of the Recent Advances in Tunnelling Technology* (pp. 183–196). New Delhi, India: CSMRS.

Bieniawski, Z. T. (1973). Engineering classification of jointed rock masses. *Transactions of the South African Institution of Civil Engineers*, *15*(12), 335–344.

Cantieni, L., & Anagnostou, G. (2009). The interaction between yielding supports and squeezing ground. *Tunnelling and Underground Space Technology*, *24*, 309–322.

Dube, A. K., & Singh, B. (1986). Study of squeezing pressure phenomena in a tunnel—Part I and II. *Tunnelling and Underground Space Technology*, *1*(1), 35–39 (Part I), 41–48 (Part II).

Goel, R. K. (1994). *Correlations for predicting support pressures and closures in tunnels* (p. 308). Ph.D. Thesis. Nagpur, India: Nagpur University308.

Goel, R. K., Jethwa, J. L., & Paithankar, A. G. (1995). Indian experiences with Q and RMR systems. *Tunnelling and Underground Space Technology*, *10*(1), 97–109.

Hoek, E. (2001). Big tunnels in hard rock, the 36th Karl Terzaghi Lecture. *Journal of Geotechnical and Geo-environmental Engineering*, *127*(9), 726–740.

Jethwa, J. L. (1981). *Evaluation of rock pressures in tunnels through squeezing ground in lower Himalayas* (p. 272). Ph.D. Thesis. Uttarakhand, India: Department of Civil Engineering, IIT Roorkee272.

Kadkade, D. G. (2007). Case history of head race tunnel of Dulhasti Hydroelectric Project. *Journal of Rock Mechanics and Tunnelling Technology*, *13*(1), 41–54.

Kumar, N. (2002). *Rock mass characterization and evaluation of supports for tunnels in Himalayas* (p. 295). Ph.D. Thesis. Uttarakhand, India: WRDM, IIT Roorkee295.

Mathews, A. A. (1996). Material handling and construction plant. In J. O. Bickel, T. R. Kuesel & E. H. King (Eds.), *Tunnel engineering handbook* (2nd ed., pp. 231–267). New York: Chapman & Hall and New Delhi: CBS Publishers.

Singh, B., & Goel, R. K. (1999). *Rock mass classification — A practical approach in civil engineering* (Chap. 7, p. 267). Amsterdam: Elsevier Science Ltd.

Singh, B., & Goel, R. K. (2002). *Software for engineering control of landslide and tunnelling hazards* (p. 344). Rotterdam: A. A. Balkema (Swets & Zeitlinger).

Singh, B., Jethwa, J. L., & Dube, A. K. (1995). A classification system for support pressure in tunnels and caverns. *Journal of Rock Mechanics and Tunnelling Technology*, *1*(1), 13–24.

Singh, B., Jethwa, J. L., Dube, A. K., & Singh, B. (1992). Correlation between observed support pressure and rock mass quality. *Tunnelling and Underground Space Technology*, *7*(1), 59–74.

Solak, T. (2009). Ground behavior evaluation for tunnels in blocky rock masses. *Tunnelling and Underground Space Technology*, *24*, 323–330.

Vlasov, S. N., Makovski, L. V., & Merkin, V. E. (2004). ITA/AITES accredited material on quality in tunnelling, ITA-AITES Working Group 16, Final Report. In C. Q. Oggeri & G. Ova (Eds.), *Accidents in transportation and subway tunnels* (Vol. 19, pp. 239–272). *Tunnelling and Underground Space Technology*. Moscow: Elex KM.

Chapter 8

Rock Mass Quality Q-System

Genius is 99 per cent perspiration and 1 per cent inspiration.

Bernard Shaw

THE Q-SYSTEM

Barton, Lien, and Lunde (1974) at the Norwegian Geotechnical Institute (NGI) originally proposed the Q-system of rock mass classification on the basis of approximately 200 case histories of tunnels and caverns. They defined the rock mass quality (Q) by the following causative factors:

$$Q = [RQD/J_n][J_r/J_a][J_w/SRF] \quad (8.1a)$$

where RQD = Deere's Rock Quality Designation ≥ 10,

$$= 115 - 3.3\, J_v \leq 100 \quad (8.1b)$$

J_n = joint set number, J_r = joint roughness number for critically oriented joint set, J_a = joint alteration number for critically oriented joint set, J_w = joint water reduction factor, SRF = stress reduction factor to consider in situ stresses and according to the observed tunneling conditions, and J_v = volumetric joint count per m^3 (see Chapter 4 for details).

For various rock conditions, the ratings (numerical value) of these six parameters are assigned. The six parameters given in Eq. (8.1a) are defined in the next section. The goal of the Q-system is to characterize the rock mass and preliminary empirical design of the support system for tunnels and caverns (see the section Design on Supports later in this chapter). There are 1260 case records to prove the efficacy of this design approach; it is the best classification system for tunnel supports (Kumar, 2002).

Rock Quality Designation

RQD is discussed in Chapter 6 and in more detail in Chapter 4. The RQD value in percentage is also the rating of RQD for the Q-system. In a poor rock mass where RQD is less than 10%, a minimum value of 10 should be used to evaluate Q (Table 8.1). If the rock cores are unavailable, the RQD can be estimated by the volumetric joint count (J_v) from Eq. (8.1b). The RQD estimated from J_v is usually conservative. The J_v is the sum of frequencies of all joint sets per meter in a pit of 1 m × 1 m × 1 m.

TABLE 8.1 Rock Quality Designation

	Condition	RQD
A	Very poor	0–25
B	Poor	25–50
C	Fair	50–75
D	Good	75–90
E	Excellent	90–100

Where RQD is reported or measured as ≤ 10 (including 0), a nominal value of 10 is used to evaluate Q in Eq. (8.1a). RQD intervals of 5, such as 100, 95, 90, etc., are sufficiently accurate.

Source: Barton et al., 1974.

Joint Set Number (J_n)

The parameter J_n, representing the number of joint sets, is often affected by foliations, schistocity, slaty cleavages or beddings, and so forth. If strongly developed, these parallel discontinuities should be counted as a complete joint set. If there are few joints visible or only occasional breaks in rock core due to these features, then they should be counted as "a random joint set" while evaluating J_n from Table 8.2. Rating of J_n is approximately equal to square of the number of joint sets.

TABLE 8.2 Joint Set Number (J_n)

	Condition	J_n
A	Massive, no or few joints	0.5–1.0
B	One joint set	2
C	One joint set plus random	3
D	Two joint sets	4
E	Two joint sets plus random	6
F	Three joint sets	9
G	Three joint sets plus random	12
H	Four or more joint sets, random, heavily jointed, "sugar cube," etc.	15
J	Crushed rock, earth-like	20

For intersections use ($3.0 \cdot J_n$). For portals use ($2.0 \cdot J_n$).

Source: Barton et al., 1974.

Joint Roughness Number and Joint Alteration Number (J_r and J_a)

The parameters J_r and J_a, given in Tables 8.3 and 8.4, respectively, represent roughness and degree of alteration of joint walls or filling materials. The parameters J_r and J_a should be obtained for the weakest critical joint set or clay-filled discontinuity in a given zone. If the joint set or the discontinuity with the minimum value of (J_r/J_a) is favorably oriented for stability, then a second less favorably oriented joint set or discontinuity may be of greater significance, and its value (J_r/J_a) should be used when evaluating Q from Eq. (8.1a). Refer to Tables 6.8 & 6.9 for the critical orientation of the joint sets.

Joint Water Reduction Factor (J_w)

The parameter J_w (Table 8.5) is a measure of water pressure, which has an adverse effect on the shear strength of joints. This is due to reduction in the effective normal stress across joints. Adding water may cause softening and possible wash-out in the case of clay-filled joints. The value of J_w should correspond to the future groundwater condition where seepage erosion or leaching of chemicals can alter permeability of rock mass significantly. For a water-carrying tunnel excavated through a dry rock mass, select class B for the J_w rating (Table 8.5).

TABLE 8.3 Joint Roughness Number (J_r)

	Condition	J_r
	(a) Rock wall contact and *(b) Rock wall contact before 10 cm shear*	
A	Discontinuous joint	4.0
B	Rough or irregular, undulating	3.0
C	Smooth, undulating	2.0
D	Slickensided, undulating	1.5
E	Rough or irregular, planar	1.5
F	Smooth, planar	1.0
G	Slickensided, planar	0.5
	(c) No rock wall contact when sheared	
H	Zone containing clay minerals thick enough to prevent rock wall contact	1.0
J	Sandy, gravelly, or crushed zone thick enough to prevent rock wall contact	

Descriptions refer to small-scale features and intermediate-scale features, in that order. Add 1.0 if the mean spacing of the relevant joint set is greater than 3m. $J_r = 0.5$ can be used for planar, slickensided joints having lineation, provided the lineations are favorably oriented. J_r and J_a classification is applied to the joint set or discontinuity that is least favorable for stability both from the point of view of orientation and shear resistance, τ.

Source: Barton, 2002.

TABLE 8.4 Joint Alteration Number (J_a)

	Condition	ϕ_r approx. (degree)	J_a
	(a) Rock wall contact (no mineral filling, only coating)		
A	Tightly healed, hard, non-softening, impermeable filling, i.e., quartz or epidote		0.75
B	Unaltered joint walls, surface staining only	25–35	1.0
C	Slightly altered joint walls; non-softening mineral coatings, sandy particles, clay-free disintegrated rock, etc.	25–30	2.0
D	Silty or sandy clay coatings, small clay fraction (non-softening)	20–25	3.0
E	Softening or low friction clay mineral coatings, i.e., kaolinite and mica; also chlorite, talc, gypsum, and graphite, etc., and small quantities of swelling clays (discontinuous coatings, 1–2 mm or less in thickness)	8–16	4.0
	(b) Rock wall contact before 10 cm shear (thin mineral fillings)		
F	Sandy particles, clay-free disintegrated rock, etc.	25–30	4.0
G	Strongly over-consolidated, non-softening clay mineral fillings (continuous, <5 mm in thickness)	16–24	6.0
H	Medium or low over-consolidation, softening, clay mineral fillings (continuous, <5 mm in thickness)	12–16	8.0
J	Swelling clay fillings, i.e., montmorillonite (continuous, <5 mm in thickness); value of J_a depends on percent of swelling clay-size particles, and access to water, etc.	6–12	8–12
	(c) No rock wall contact when sheared (thick mineral fillings)		
K, L, M	Zones or bands of disintegrated or crushed rock and clay (see G, H, J for description of clay condition)	6–24	6, 8, or 8–12
N	Zones or bands of silty or sandy clay, small clay fraction (non-softening)	—	5.0
O, P, R	Thick, continuous zones or bands of clay (see G, H, J for description of clay condition)	6–24	10, 13, or 13–20

Source: Barton, 2002.

Stress Reduction Factor

The stress reduction factor (SRF) parameter (Table 8.6) is a measure of (1) loosening pressure during an excavation through shear zones and clay-bearing rock masses, (2) rock stress q_c/σ_1 in a competent rock mass where q_c is the uniaxial compressive strength (UCS) of rock material and σ_1 is the major principal stress before

TABLE 8.5 Joint Water Reduction Factor (J_w)

	Condition	Approx. water pressure (MPa)	J_w
A	Dry excavation or minor inflow, i.e., 5 lt./min locally	<0.1	1
B	Medium inflow or pressure, occasional outwash of joint fillings	0.1–0.25	0.66
C	Large inflow or high pressure in competent rock with unfilled joints	0.25–1.0	0.5
D	Large inflow or high pressure, considerable outwash of joint fillings	0.25–1.0	0.33
E	Exceptionally high inflow or water pressure at blasting, decaying with time	>1.0	0.2–0.1
F	Exceptionally high inflow or water pressure continuing without noticeable decay	>1.0	0.1–0.05

Factors C to F are crude estimates. Modify Jw if drainage measures are installed.
Special problems caused by ice formation are not considered.
For general characterization of rock masses distant from excavation influences, the use of J_w = 1.0, 0.66, 0.5, 0.33, etc., as depth increases from, say, 0–5, 5–25, 25–250 to >250 m is recommended, assuming that RQD/J_n is low enough (e.g., 0.5–25) for good hydraulic conductivity. This will help to adjust Q for some of the effective stress and water softening effects in combination with appropriate characterization values of SRF. Correlations with depth-dependent static modulus of deformation and seismic velocity will then follow the practice used when these were developed.

Source: Barton, 2002.

TABLE 8.6 Stress Reduction Factor

	Conditions	SRF
	(a) Weakness zones intersecting excavation, which may cause loosening of rock mass when tunnel is excavated	
A	Multiple occurrences of weakness zones containing clay or chemically disintegrated rock, very loose surrounding rock (any depth)	10.0
B	Single-weakness zones containing clay or chemically disintegrated rock (depth of excavation ≤50 m)	5.0
C	Single-weakness zones containing clay or chemically disintegrated rock (depth of excavation >50 m)	2.5
D	Multiple-shear zones in competent rock (clay-free), loose surrounding rock (any depth)	7.5
E	Single-shear zones in competent rock (clay-free) (depth of excavation ≤50 m)	5.0
F	Single-shear zones in competent rock (clay-free) (depth of excavation >50 m)	2.5

Continued

TABLE 8.6 Stress Reduction Factor—Cont'd

	Conditions				SRF
	(a) Weakness zones intersecting excavation, which may cause loosening of rock mass when tunnel is excavated				
G	Loose, open joints, heavily jointed or "sugar cube," etc. (any depth)				5.0
	(b) Competent rock, rock stress problems				
		qc/σ1	σ_θ/qc	SRF (old)	SRF (new)
H	Low stress, near surface, open joints	>200	<0.01	2.5	2.5
J	Medium stress, favorable stress condition	200–10	0.01–0.3	1.0	1.0
K	High stress, very tight structure; usually favorable to stability, may be unfavorable to wall stability	10–5	0.3–0.4	0.5–2.0	0.5–2.0
L	Moderate slabbing after >1 hour in massive rock	5–3	0.5–0.65	5–9	5–50
M	Slabbing and rock burst after a few minutes in massive rock	3–2	0.65–1.0	9–15	50–200
N	Heavy rock burst (strain-burst) and immediate dynamic deformations in massive rock	<2	>1	15–20	200–400
	(c) Squeezing rock; plastic flow of incompetent rock under the influence of high rock pressures				
O	Mild squeezing rock pressure			1–5	5–10
P	Heavy squeezing rock pressure			>5	10–20
	(d) Swelling rock; chemical swelling activity depending on presence of water				
Q	Mild swelling rock pressure			5–10	
R	Heavy swelling rock pressure			10–15	

Reduce these SRF values by 25–50% if the relevant shear zones only influence but do not intersect the excavation. This will also be relevant for characterization.
For strongly anisotropic virgin stress field (if measured): when $5 \leq \sigma_1/\sigma_3 \leq 10$, reduce q_c to 0.75 q_c; when $\sigma_1/\sigma_3 > 10$, reduce q_c to 0.50 q_c (where q_c is unconfined compressive strength), σ_1 and σ_3 are major and minor principal stresses, and σ_θ is the maximum tangential stress (estimated from elastic theory).
Few case records available where depth of crown below surface is less than span width; suggest SRF increase from 2.5 to 5 for such cases (see H).
Cases L, M, and N are usually most relevant for support design of deep tunnel excavation in hard massive rock masses, with RQD/J_n ratios from about 50–200.
For general characterization of rock masses distant from excavation influences, the use of SRF = 5, 2.5, 1.0, and 0.5 is recommended as depth increases from, say, 0–5, 5–25, 25–250, >250 m. This will help to adjust Q for some of the effective stress effects, in combination with appropriate characterization values of J_w. Correlations with depth-dependent static modulus of deformation and seismic velocity will then follow the practice used when these were developed.
Cases of squeezing rock may occur for depth $H > 350Q^{1/3}$ (Singh & Goel, 2006). Rock mass compressive strength can be estimated from $q_{cmass} \approx 7\gamma (Q)^{1/3}$ (MPa); γ is the rock density in t/m^3, and q_{cmass} = rock mass compressive strength.

Source: Barton, 2002.

excavation, and (3) squeezing or swelling pressures in incompetent rock masses. SRF can also be regarded as a total stress parameter. Ratings for SRF are given in Table 8.6. For competent rock masses (Category B of SRF), new ratings of SRF are listed in Table 8.6 as proposed by Grimstad and Barton (1993). The SRF should be classified according to the observed behavior of rocks and by sound engineering judgment. However, it may be difficult to predict the tunneling conditions in advance in complex geological situations. For predicting the ground conditions, the modified Q-value (N-value, i.e., Q with SRF = 1) discussed in Chapter 9 and Figure 7.3 can be used.

1. SRF should be reduced where micro-folding occurs and its axis is nearly parallel to the strike of walls of caverns or tunnels. The accumulated high stresses may be released locally during excavation (leading to failure of rock bolts in weak rocks).
2. In jointed rocks under high overburden (H > 1000 m), rock burst may not occur due to strength enhancement by intermediate stress (σ_2) along the axis of the underground opening (cases L, M, and N in Table 8.6). SRF should be selected according to the *observed* rock burst condition and not the *expected* rock burst condition (cases L, M, and N in Table 8.6).
3. It would be better if in situ stresses are measured at the tunneling projects, and the maximum tangential stress (σ_0) is obtained to determine SRF accurately.

Ratings of all the six parameters are given in Tables 8.1 to 8.6. The ratings of these parameters obtained for a given rock mass are substituted in Eq. (8.1a) to solve for rock mass quality (Q).

As seen from Eq. (8.1a), the rock mass quality (Q) may be considered a function of only three parameters, which are approximate measures of

a. Block size (RQD/J_n):	It represents overall structure of rock mass (Table 4.5)
b. Inter - block shear strength (J_r/J_a):	It has been found that $\tan^{-1}(J_r/J_a)$ is a fair approximation of the actual peak sliding angle of friction along the clay-coated joints (Table 8.7). This has been later modified by Barton (2008) as given in Eq. (8.16).
c. Active stress (J_w/SRF):	It is an empirical factor describing the active effective stress

The first quotient (RQD/J_n) represents the rock mass structure and is a measure of block size or the size of the wedge formed by the presence of different joint sets (see Table 4.5). In a given rock mass, the rating of parameter J_n could increase with the tunnel size in situations where additional joint sets are encountered. Hence, it is not advisable to use a Q-value obtained from a small drift to estimate the support pressure for a large tunnel or a cavern. It would be more appropriate to obtain J_n from drill core observations or a borehole camera.

The second quotient (J_r/J_a) represents the roughness and frictional characteristics of joint walls or filling materials. It should be noted that the value of J_r/J_a is collected for the critical joint set, that is, the joint set most unfavorable for the stability of a key rock block in the roof.

The third quotient (J_w/SRF) is an empirical factor describing an "active stress condition." SRF is a measure of (1) loosening pressure during an excavation through

TABLE 8.7 Estimation of Angle of Internal Friction from the Parameters J_r and J_a

Description	J_r	$\tan^{-1}(J_r/J_a)$				
(a) Rock wall contact		*(Thin coatings)*				
		J_a = 0.75	*1.0*	*2.0*	*3.0*	*4.0*
A. Discontinuous joints	4.0	79°	76°	63°	53°	45°
B. Rough, undulating	3.0	76°	72°	56°	45°	37°
C. Smooth, undulating	2.0	69°	63°	45°	34°	27°
D. Slickensided, undulating	1.5	63°	56°	37°	27°	21°
E. Rough, planar	1.5	63°	56°	37°	27°	21°
F. Smooth, planar	1.0	53°	45°	27°	18°	14°
G. Slickensided, planar	0.5	34°	27°	14°	9.5°	7.1°
(b) Rock wall contact when sheared		*(Thin filling)*				
	J_r	*J_a = 4.0*	*6*	*8*	*12*	
A. Discontinuous joints	4.0	45°	34°	27°	18°	
B. Rough, undulating	3.0	37°	27°	21°	14°	
C. Smooth, undulating	2.0	27°	18°	14°	9.5°	
D. Slickensided, undulating	1.5	21°	14°	11°	7.1°	
E. Rough, planar	1.5	21°	14°	11°	7.1°	
F. Smooth, planar	1.0	14°	9.5°	7.1°	4.7°	
G. Slickensided, planar	0.5	7°	4.7°	3.6°	2.4°	
(c) No rock wall contact when sheared		*(Thick filling)*				
	J_r	*J_a = 5*	*6*	*8*	*12*	
Nominal roughness of discontinuity rock walls	1.0	11.3°	9.5°	7.1°	4.8°	
	J_r	*J_a = 13*	*16*	*20*	—	
	1.0	4.4°	3.6°	2.9°	—	

Source: Barton, 2002.

shear zones and clay-bearing rocks; (2) rock stress in competent rocks; and (3) squeezing pressure in plastic incompetent rocks, which can be regarded as a total stress parameter. The water reduction factor Jw is a measure of water pressure, which has an adverse effect on the shear strength of joints due to reduction in effective normal stress. Adding water causes softening and possible outwash in clay-filled joints.

JOINT ORIENTATION AND THE Q-SYSTEM

Barton et al. (1974) stated that joint orientation was not as important a parameter as expected, because the orientation of many types of excavation can be, and normally are, adjusted to avoid the maximum effect of unfavorably oriented major joints. Barton et al. (1974) also stated that the parameters J_n, J_r, and J_a appear to play a more important role than the joint orientation, because the number of joint sets determines the degree of freedom for block movement (if any); the frictional and dilatational characteristics (J_r) can counterbalance the down-dip gravitational component of weight of wedge formed by the unfavorably oriented joints. If joint orientation had been included the classification system would be less general, and its essential simplicity lost.

However, it is still suggested to collect the rating for J_r/J_a for the most critical joint set. The critical joint set or "very unfavorable joint set" with respect to tunnel axis can be obtained from Table 6.8.

UPDATING THE Q-SYSTEM

The Q-system (originally created in 1974) has been updated on several occasions during the last few years, and it is now based on 1260 case records where the installed rock support has been correlated to the observed Q-values. The original parameters of the Q-system have not been changed, but some of the ratings for the SRF have been altered by Grimstad and Barton (1993). The new SRF ratings for competent rocks are shown in Table 8.6. These rates were created because a hard massive rock under high stress requires far more support than those recommended by the Q-value with old SRF ratings as proposed by Barton et al. (1974). In the original Q-system, this problem was addressed in a supplementary note with instructions on how to support spalling or rock burst zones with closely spaced end-anchored rock bolts and triangular steel plates. Tunnels under high stresses in hard rocks suggest less bolting, but extensive use of steel fiber reinforced shotcrete (SFRS), an unknown product when the Q-system was first developed in 1974. The updating of the Q-system has shown that in the most extreme case of high stress and hard massive (unjointed) rocks, the maximum SRF value has to be increased from 20 to 400 to give a Q-value that correlates with the modern rock supports shown in Figure 8.5. With moderately jointed rocks, the SRF needs to be significantly reduced according to the observed tunneling conditions (Kumar, 2002).

Also, overburden height (H) should be considered in addition to SRF in Table 8.6 when obtaining the support pressure of squeezing ground conditions (see the section Correlation by Singh et al. (1992)). It is our feeling that old values of SRF should not be changed when assessing the Q-value of jointed rocks.

COLLECTION OF FIELD DATA

The length of core or rock exposures used for evaluating the first four parameters (RQD, J_n, J_r, and J_a) depends on the uniformity of the rock mass. If there is little variation, a core or wall length of 5–10 m should be sufficient. However, a closely jointed shear zone a few meters wide with alternate sound rock is necessary to evaluate these parameters separately if it is considered that the closely jointed shear zones are wide enough to justify special treatment (i.e., additional shotcrete) compared to only systematic bolting in the remainder of the excavation. If, on the other hand, the shear zones are less than 0.5 m in width and occur frequently, then an overall reduced value of Q for the entire tunnel reach may be most appropriate since increased support is likely to be applied uniformly along the entire length of such variable zones. In such cases a core or wall length of 10–50 m may be needed to obtain an overall picture of the reduced rock mass quality.

1. Values of the rock mass quality (Q) should be obtained separately for the roof, the floor, and two walls, particularly when the geological description of the rock mass is not uniform around the periphery of an underground opening.
2. With power tunnels the value of J_w for calculation of ultimate support pressures should be reduced assuming that seepage water pressure in Table 8.5 is equal to the internal water pressure after commissioning the hydroelectric projects.

Suggestions for Beginners

Beginners may find it difficult to select a single rating for a particular parameter. They may opt for a range of ratings or two ratings or values for tension-free judgment. Subsequently, a geometrical mean can be obtained from the minimum and maximum values for a representative value of the parameter. According to the authors, this not only reduces the bias but also generates confidence among users. For the purpose of eliminating the bias of an individual, the ratings for different parameters should be given a range in preference to a single value.

To overcome the problem of selecting a representative rating of various parameters, NGI has proposed a geotechnical chart (Figure 8.1). The main body of the geotechnical chart consists of rectangular graduated areas for making numerous individual observations of joints and jointing characteristics in the form of a histogram. NGI proposed that efforts should be made to estimate approximate percentages of the various qualities of each observed parameter — 10% poorest, 60% most typical, 30% best or maximum value — since the weighted average from all of the histograms masks the extreme values. For example, the values of Q parameters collected at a location are shown in Table 8.8.

Using the weighted average value of each parameter, a more realistic Q can be obtained from Eq. (8.1a). The weighted average value is obtained using the percentage weightage mentioned previously and as shown next for RQD.

A weighted average for RQD in Table 8.8 is obtained as

$$(10 \times 25 + 60 \times 65 + 30 \times 85)/100 = 67$$

Similarly, weighted averages can be obtained for other parameters like the joint alteration number (J_a), joint roughness number (J_r), and so forth, as proposed by NGI.

CLASSIFICATION OF THE ROCK MASS

The rock mass quality (Q) is a very sensitive index and its value varies from 0.001 to 1000. Use of the Q-system is specifically recommended for tunnels and caverns with an arched roof. On the basis of the Q-value, the rock masses are classified into nine categories (Table 8.9). Rock mass quality varies from Q_{min} to Q_{max}, so the average rock mass quality of $(Q_{max} \times Q_{min})^{1/2}$ may be assumed in the design calculations.

The Q-values will be higher where a tunnel boring machine (TBM) or a road header is used to smooth the surface of excavation. The Q-value, on the other hand, in the tunnel blast method will be lower because of high overbreaks and the development of new fractures. To minimize the negative effect of blasting on Q, use a controlled blasting technique. The blasting effects are better in the rock masses having a Q-value between 1 and 30.

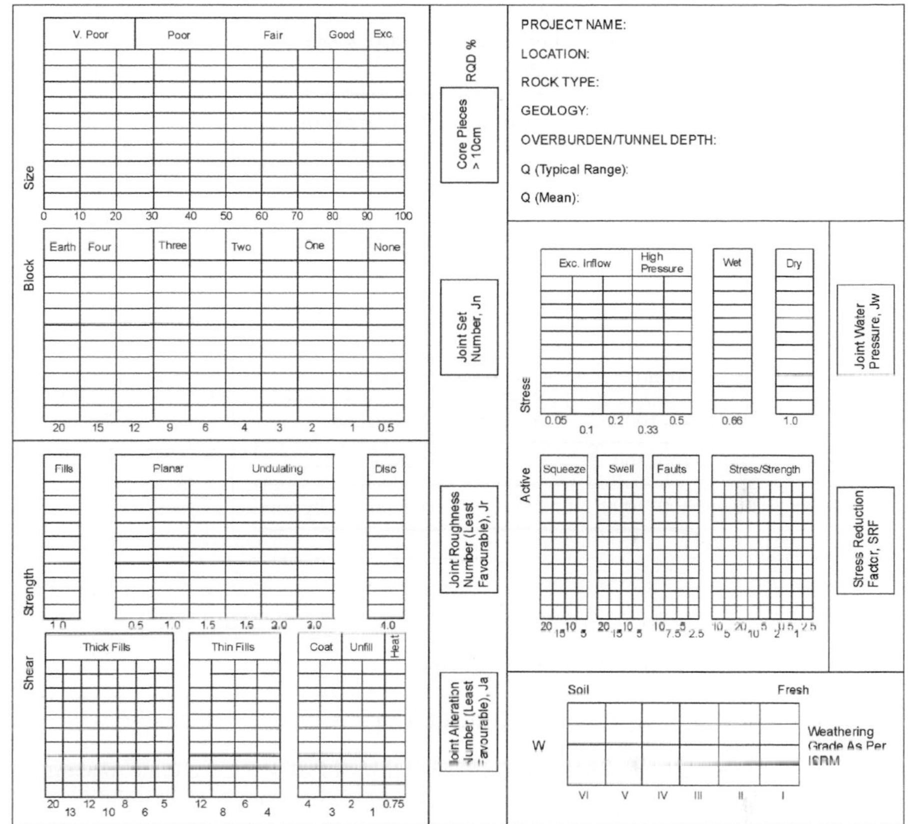

FIGURE 8.1 Data sheet for recording Q parameters. *(From Barton, 1993)*

TABLE 8.8 Weighted Average Method of Obtaining Q-Value

Parameter of Q	Poorest value (10%)	Most typical value (60%)	Maximum value (30%)	Weighted average
RQD	25	65	85	67
J_n	12	9	—	9.42
J_r	1.5	3	4	2.05
J_a	4	2	1	1.9
J_w	0.66	1	1	0.966
SRF	7.5	5	2.5	4.5

Source: Barton, 1993.

TABLE 8.9 Classification of Rock Mass Based on Q-Values

Q	Group	Classification
0.001–0.01		Exceptionally poor
0.01–0.1	3	Extremely poor
0.1–1		Very poor
1–4	2	Poor
4–10		Fair
10–40		Good
40–100	1	Very good
100–400		Extremely good
400–1000		Exceptionally good

ESTIMATION OF SUPPORT PRESSURE

Using the Approach of Barton et al. (1974)

Barton et al. (1974, 1975) plotted support capacities of 200 underground openings against the rock mass quality (Q) and found the following empirical correlation for ultimate support pressure (Figure 8.2):

$$p_v = (0.2/J_r)Q^{-1/3} \tag{8.2}$$

$$p_h = (0.2/J_r)Q_w^{-1/3} \tag{8.3}$$

where p_v = ultimate roof support pressure in MPa, p_h = ultimate wall support pressure in MPa, and Q_w = wall factor.

Figure 8.2 shows the correlation for Eq. (8.2). The center line of the shaded band should be used when assessing the support pressure in the roof.

Dilatant joints or J_r values play a dominant role in the stability of underground openings. Consequently, support capacities may be independent of the opening size, unlike what Terzaghi (1946) thought and Table 5.2 illustrated.

The wall factor (Q_w) is obtained after multiplying Q by a factor that depends on the magnitude of Q as given in this table.

Range of Q	Wall factor Q_w
>10	5.0 Q
0.1–10	2.5 Q
<0.1	1.0 Q

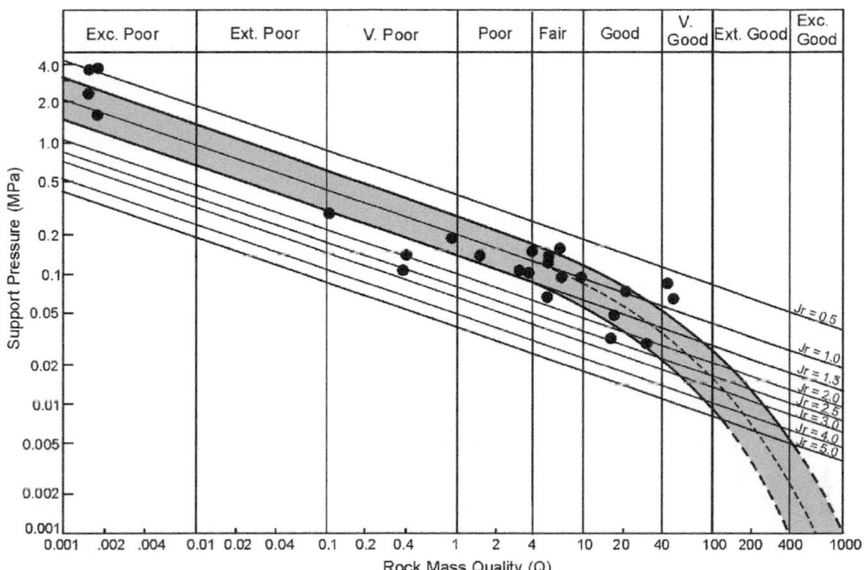

FIGURE 8.2 Correlation between support pressure and rock mass quality Q. *(From Barton et al., 1974)*

Barton et al. (1974) further suggested that if the number of joint sets is less than three, Eqs. (8.2) and (8.3) are expressed as Eqs. (8.4a) and (8.4b), respectively,

$$p_v = \frac{0.2 \cdot J_n^{1/2}}{3 \cdot J_r} \cdot Q^{-1/3}, \text{ MPa} \tag{8.4a}$$

$$p_h = \frac{0.2 \cdot J_n^{1/2}}{3 \cdot J_r} \cdot Q_W^{-1/3}, \text{ MPa} \tag{8.4b}$$

They felt that the short-term support pressure can be obtained after substituting 5Q in place of Q in Eq. (8.2). Thus, the ultimate support pressure is obtained as 1.7 times the short-term support pressure.

The Q-value in dynamic condition is half of the Q-value in static conditions ($Q_{dyn} = Q_{static}/2$; Barton, 2008). According to Bhasin (personal communication), the support capacity as calculated from UDEC increased by 10 to 40% of static capacity in the shallow tunnels in seismic regions.

Bhasin and Grimstad (1996) suggested the following correlation for predicting support pressure in tunnels through poor rock masses (say, $Q < 4$):

$$p_v = \frac{40 \text{ B}}{J_r} \cdot Q^{-1/3}, \text{ kPa} \tag{8.5}$$

where B is diameter or span of the tunnel in meters. Equation (8.5) shows that the support pressure increases with tunnel size B in poor rock masses.

The Q referred to in Eq. (8.5) is actually the post-excavation quality of a rock mass, because in tunnels the geology of the rock mass is usually studied after blasting and an on-the-spot decision is made for support density.

Correlation by Singh et al. (1992)

Vertical or Roof Support Pressure

The observed roof support pressure is related to the short-term rock mass quality (Q_i) for 30 instrumented tunnels from the following empirical correlation

$$p_v = \frac{0.2}{J_r} \cdot Q_i^{-0.33} \cdot f \cdot f' \cdot f'', \text{ MPa} \tag{8.6}$$

$$f = 1 + (H - 320)/800 \geq 1 \tag{8.7}$$

where $Q_i = 5Q$ = short-term rock mass quality soon after the underground excavation; p_v = short-term roof support pressure in MPa; f = correction factor for overburden (Eq. 8.7); f′ = correction factor for tunnel closure (Table 8.10) obtained from Figure 8.3 for squeezing ground condition ($H > 350\ Q^{1/3}$ and $J_r/J_a < 1/2$) and = 1 in non-squeezing ground; f″ = correction factor for the time after excavation (Eq. 8.8) and support erection; and H = overburden above crown or tunnel depth below ground level in meters.

TABLE 8.10 Correction Factor f′ for Tunnel Closure

S. No.	Rock condition	Support system	Tunnel closure (ua/a), %	Correction factor, f′
1	Non-squeezing ($H < 350\ Q^{0.33}$)	—	<1	1.0
2	Squeezing ($H > 350\ Q^{0.33}$, $J_r/J_a < 0.5$)	Very stiff	<2%	>1.8
3	-do-	Stiff	2–4%	0.85
4	-do-	Flexible	4–6%	0.70
5	-do-	Very flexible	6–8%	1.15
6	-do-	Extremely flexible	>8%	1.8

Tunnel closure depends significantly on method of excavation. In extreme squeezing ground conditions, heading and benching method may lead to tunnel closure >8%.

Tunnel closures more than 4% of tunnel span should not be allowed, otherwise support pressures are likely to build up rapidly due to failure of rock arch. In such cases, additional rock anchors should be installed immediately to arrest the tunnel closure within a limiting value of 4% of width.

Steel ribs with struts may not absorb more than 2% tunnel closure. Thus, SFRS is suggested as an immediate support at the face to be supplemented with steel arches behind the face in situations where excessive closures are encountered.

The minimum spacing between the parallel tunnels is 5B center to center in squeezing ground, where B is the width of a tunnel.

Source: Singh et al., 1992.

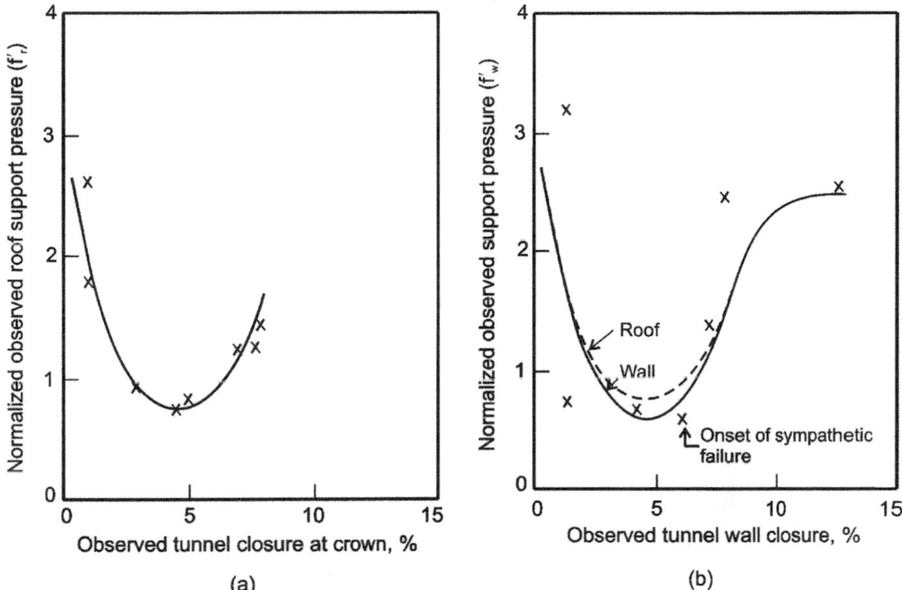

FIGURE 8.3 Correction factor for (a) roof closure and (b) wall closure under squeezing ground conditions. *(From Singh et al., 1992)*

In developing Eq. (8.6), the correction factors have been applied in steps. First, the correction factor for tunnel depth is applied, next comes the correction for tunnel closure, and finally there is the correction for time after support erection (Singh et al., 1992). Grimstad and Barton (1993) agreed on the overburden correction factor from Eq. (8.7).

Values of correction factors for tunnel closure (f′) can be obtained from Table 8.10 based on the design value of tunnel closure. Table 8.10 is derived from Figures 8.3a and b between normalized tunnel closure (u_a/a) and the correction factor for tunnel closure f′ defined in Eq. (8.6). Figures 8.3a and b represent normalized observed ground response (reaction) curves for tunnel roofs and walls, respectively, in squeezing ground. The closure must be controlled to be less than 4% of tunnel width, otherwise the support pressure is likely to jump as shown in Figures 8.3a and b.

Palmstrom and Broch (2006) raised a very interesting question: What value of SRF should be used in the criterion of squeezing grounds ($H > 350\ Q^{1/3}$ m)? The SRF in S. No. 2 in Table 8.10 is equal to 2.5 as in situ rock mass was in peak failure condition (Singh et al., 1992; Kumar, 2002). Therefore the Q-value should be corrected for SRF = 2.5 when predicting minimum depth of overburden (H) for squeezing ground conditions. Thus the Q-value should also be corrected in Eqs. (13.9) and (13.12b) for the rock mass strength as SRF will be about 2.5 at the time of peak failure. Palmstrom and Broch (2006) raised another question: Which rock types demonstrate squeezing in the Himalayas? The squeezing conditions were encountered in tunnels in schist, phyllites, slates, shales, clay stones, sandstones, metabasics, fault gouge, and weak rocks only where H exceeds 350 $Q^{1/3}$ m and J_r/J_a was less than 0.5. Otherwise rock burst occurred.

The correction factor f″ for time was found as

$$f'' = \log(9.5\, t^{0.25}) \tag{8.8}$$

where t is time in months after support installation. Goel et al. (1995b) verified correction factors f and f′ for the Maneri-Uttarkashi tunnel (H = 700 to 900 m). Kumar (2002) confirmed all three correction factors from a study of the behavior of a 27 km long NJPC tunnel in the Himalayas in India (H < 1400 m). Incorporating these correction factors, Singh et al. (1992) proposed the following correlation for ultimate tunnel support pressure, p_{ult}, after about 100 years ($f'' = 5^{1/3} = 1.7$):

$$p_{ult} = \frac{0.2}{J_r} \cdot Q^{-1/3} \cdot f \cdot f', \text{MPa} \tag{8.9}$$

Dube (1979) and Jethwa (1981) observed concentric broken zones in nine tunnels in squeezing grounds. Singh et al. (1992) also studied the effect of tunnel size (2–22 m) on support pressures. They inferred no significant effect of size on observed support pressure. This aspect is further discussed in Chapter 9.

Horizontal or Wall Support Pressure

To estimate the wall support pressure, Eq. (8.6) can be used with short-term wall rock mass quality Q_{wi} in place of Q_i. The short-term wall rock quality Q_{wi} for short-term wall support pressure is obtained after multiplying Q_i by a factor that depends on the magnitude of Q as given next:

(i) For $Q > 10$; $Q_{wi} = 5.0 \cdot Q_i = 25\,Q$,
(ii) For $0.1 < Q < 10$; $Q_{wi} = 2.5 \cdot Q_i = 12.5\,Q$, and
(iii) For $Q < 0.1$; $Q_{wi} = 1.0 \cdot Q_i = 5\,Q$

The observed short-term wall support pressure is generally insignificant in non-squeezing rock conditions. Therefore, it is recommended that these may be neglected in tunnels in rock masses of good quality from group 1 in Table 8.9 (Q > 10).

Although the wall support pressure would be negligible under non-squeezing ground conditions, high wall support pressure is common with poor ground or squeezing ground conditions. Therefore, invert struts with steel ribs are used when the estimated wall support pressure requires using a wall support in exceptionally poor rock conditions and highly squeezing ground conditions. In different conditions the New Austrian Tunneling Method (NATM) or the Norwegian Method of Tunneling (NMT) is a better choice.

Ultimate Support Pressure in Special Conditions

Long-term monitoring at the Chhibro cavern (with a steel rib support system in the roof and a prestressed rock anchor in the wall) of the Yamuna hydroelectric project in India has enabled researchers to study the support pressure trend with time and with saturation. The study, based on 10 years of monitoring, shows that the ultimate support pressure—at the roof for water-charged rock masses with erodible joint fillings—may rise up to 6 times the short-term support pressure (Mitra, 1990). No time-dependent effect was noticed in the walls of the cavern except near the thick plastic shear zone. The monitoring also suggests that for tunnels/caverns located near faults/shear zones/thrusts (with plastic gouge) in seismic areas, the ultimate support pressure might be about 25% more due to accumulated strains in the rock mass along the fault.

Extrapolating the support pressure values for 100 years, a study by Singh et al. (1992) showed that the ultimate support pressure would be about 1.75 times the short-term support pressure under non-squeezing ground conditions, whereas in squeezing ground conditions, Jethwa (1981) estimated that the ultimate support pressure would be 2 to 3 times the short-term support pressure.

Evaluation of Barton et al. (1974) and Singh et al. (1992) Approaches

Support pressures estimated from Eqs. (8.2) and (8.3) for various test sections have been compared with the measured values. The estimates are reasonable (correlation coefficient r = 0.81) for tunnel sections through non-squeezing ground conditions. In squeezing ground conditions, the estimated support pressures never exceeded 0.7 MPa, whereas the measured values were as high as 1.2 MPa for larger tunnels. Therefore, it is thought that the Q-system may be unsafe for larger tunnels (diameter > 9 m) under highly squeezing ground conditions (Goel et al., 1995a).

The estimated support pressures from Eq. (8.6) are also compared with the measured values for non-squeezing and squeezing ground conditions. It has been found that the correlation of Singh et al. (1992) provides reasonable estimates of support pressures.

Limitations of the Q-System

Kaiser, Mackay, and Gale (1986) opined that SRF is probably the most contentious parameter. They concluded that it may be appropriate to neglect the SRF during rock mass classification and to assess the detrimental effects of high stresses separately. However, they have not given an alternate approach to assess high stress effect. Keeping this problem in mind, Goel et al. (1995a) proposed rock mass number N, that is, stress-free Q and incorporated stress-effect in the form of tunnel depth H, to suggest a new set of empirical correlations for estimating support pressures. This aspect is discussed in Chapter 9.

ESTIMATION OF DEFORMATION OR CLOSURE

Barton (2008) plotted the tunnel roof and wall deformations with Q on a log-log scale (Figure 8.4) to develop equations for predicting the deformation or closure in underground openings. He has also introduced the "competence factor"—ratio of stress to strength—directly in Eqs. (8.10) and (8.11).

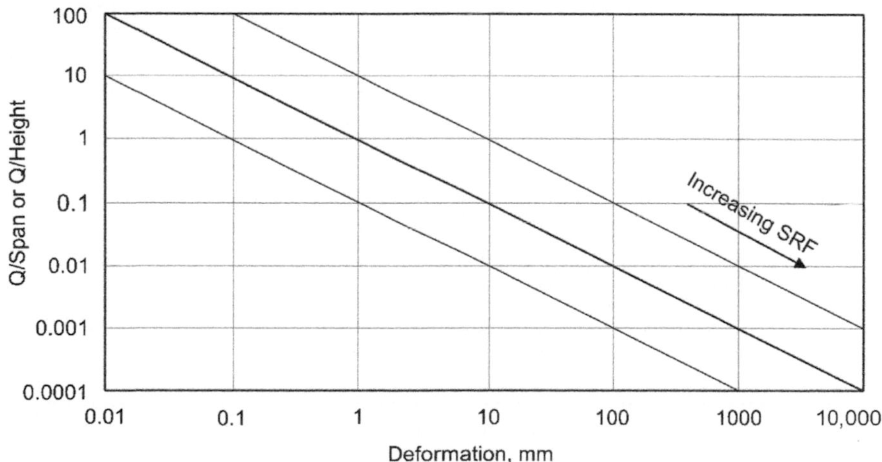

FIGURE 8.4 Deformation vs. Q/Span or Q/Height. *(From Barton, 2008)*

$$\Delta_v = \frac{\text{Span}}{100\ Q}\sqrt{\frac{\sigma_v}{q_c}} \tag{8.10}$$

$$\Delta_h = \frac{\text{Height}}{100\ Q}\sqrt{\frac{\sigma_h}{q_c}} \tag{8.11}$$

where Δ_v and Δ_h = roof and wall deformations, respectively; σ_v and σ_h = in situ vertical and horizontal stresses, respectively, in MPa; and q_c = UCS of intact rock material in MPa.

UNSUPPORTED SPAN

Barton et al. (1974) proposed the following equation for estimating equivalent dimension ($D_{e'}$) of a self-supporting or an unsupported tunnel

$$D_{e'} = 2.0(Q^{0.4}),\ \text{meters} \tag{8.12}$$

if $H < 23.4\ N^{0.88}\ B_s{}^{-0.1}$ meters where $D_{e'} = $ equivalent dimension and $\frac{\text{span, diameter, or height in meters } (B_s)}{\text{ESR}}$; Q = rock mass quality; and ESR = excavation support ratio (Table 8.11).

In equivalent dimension, the span or diameter is used for analyzing the roof support and the height of the wall for wall support. The excavation support ratio (ESR) appropriate to a variety of underground excavations is listed in Table 8.11.

General requirements for permanently unsupported openings are:

(a) $J_n < 9, J_r > 1.0, J_a < 1.0, J_w = 1.0, SRF < 2.5$

Further, conditional requirements for permanently unsupported openings are given next.

(b) If $RQD < 40$, need $J_n < 2$

TABLE 8.11 Values of Excavation Support Ratio

	Type of excavation	ESR
A	Temporary mine openings, etc.	2–5
B	Permanent mine openings, water tunnels for hydro power (excluding high pressure penstocks), pilot tunnels, drifts and headings for large openings, surge chambers	1.6–2.0
C	Storage caverns, water treatment plants, minor road and railway tunnels, access tunnels	1.2–1.3
D	Power stations, major road and railway tunnels, civil defense chambers, portals, intersections	0.9–1.1
E	Underground nuclear power stations, railway stations, sports and public facilities, factories, major gas pipeline tunnels	0.5–0.8

ESR should be increased by 1.5 times, Q by 5, and Q_w by 5, for temporary supports.

Source: Barton, 2008.

(c) If $J_n = 9$, need $J_r > 1.5$ and RQD > 90
(d) If $J_r = 1.0$, need $J_w < 4$
(e) If SRF > 1, need $J_r > 1.5$
(f) If span > 10 m, need $J_n < 9$
(g) If span > 20 m, need $J_n < 4$ and SRF < 1

DESIGN OF SUPPORTS

The Q-value is related to tunnel support requirements with the equivalent dimensions of the excavation. The relationship between Q and the equivalent dimension of an excavation determines the appropriate support measures, as depicted in Figure 8.5. The bolt and anchor length, l_b and l_a, respectively, are determined in terms of excavation width B or height H in meters for roofs and walls, respectively, using Eqs. (8.13) and (8.14a,b) proposed by Barton et al. (1974).

$$l_b = 2 + (0.15\ B \text{ or } H/ESR),\ m \tag{8.13}$$

$$In\ Roof \quad l_a = 0.40\ B/ESR,\ m \tag{8.14a}$$

$$In\ Walls \quad l_a = 0.35\ H/ESR,\ m \tag{8.14b}$$

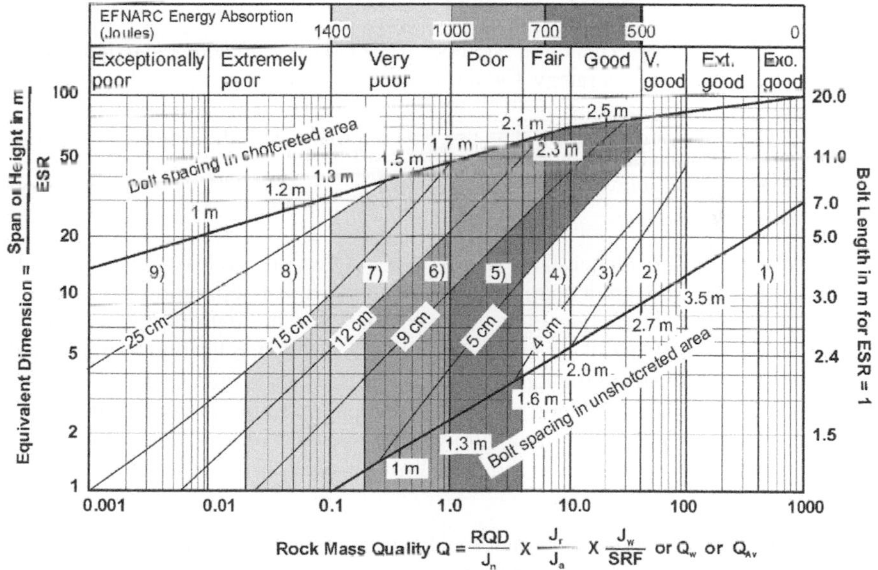

REINFORCEMENT CATEGORIES

1) Unsupported
2) Spot bolting, sb
3) Systematic bolting, B
4) Systematic bolting (and unreinforced shotcrete, 4 to 10 cm, B(+S)
5) Fiber reinforced shotcrete and bolting, 5 to 9cm, S(fr)+B
6) Fiber reinforced shotcrete and bolting, 9 to 12 cm, S(fr)+B
7) Fiber reinforced shotcrete and bolting, 12 to 15 cm, S(fr)+B
8) Fiber reinforced shotcrete > 15 cm, reinforced ribs of shotcrete and bolting, S(fr), RRS+B
9) Cast concrete lining, CCA

FIGURE 8.5 Grimstad and Barton (1993) chart for the design of support including the required energy absorption capacity of SFRS suggested by Papworth (2002).

The problem with the Norwegian design approach is that, although the thickness of SFRS is given, there is no toughness requirement indicated. With the wide range in performance for different fibers and the fiber content in SFRS, the SFRS generically expressed in the Grimstad and Barton (1993) chart could range in toughness from 400 to 1400 J of energy absorption based on the EFNARC panel test for 25 mm deflection. The energy absorption is the area below the load-deflection curve of the SFRS in the panel test. Given the structural requirements of the SFRS, only the thickness of SFRS is not satisfactory. Hence, as suggested by Papworth (2002), the energy absorption capacity of SFRS is also included in the original design chart of Grimstad and Barton (1993; Figure 8.5).

Figure 8.5 is recommended for tunneling in poor rock conditions (see the section Experiences in Poor Rock Condition later in this chapter) provided that more steel fibers are added in shotcrete so that energy absorption or toughness is increased as shown in the top row of this figure. If fly-ash is used as the admixture in shotcrete/SFRS, Kadkade (2009) suggested using fly-ash obtained from an electrostatic precipitator. Figure 8.5 does not give the capacity of rock bolts, so TM software may be used to design the support system (Singh & Goel, 2006).

A high percentage of rebound loss of shotcrete mix along with steel fiber is a very important factor to keep in mind while designing the fiber content and shotcrete thickness.

NEW AUSTRIAN TUNNELING METHOD

The name "New Austrian Tunneling Method" (NATM) is a misnomer as it is not a method of tunneling but a strategy for tunneling that has a considerable uniformity and sequence.

The NATM is based on the "build as you go" approach with the following caution:

Not too stiff, Nor too flexible
Not too early, Nor too late

The NATM accomplishes tunnel stabilization by controlled stress release. The surrounding rock is transformed from a complex load system to a self-supporting structure together with the installed support elements, provided that the detrimental loosening, resulting in a substantial loss of strength, is avoided. The self-stabilization by controlled stress release is achieved by introducing the so-called "semi-rigid lining," that is, systematic rock bolting with the application of a shotcrete lining. This offers a certain degree of immediate support and the flexibility to allow stress release through radial deformation. The development of shear stresses in shotcrete lining in an arched roof is thus reduced to a minimum (Singh & Goel, 2006).

1. NATM is based on the principle that the capacity of the rock mass should be taken to support itself by carefully controlling the forces in the redistribution process, which takes place in the surrounding rock mass when a cavity is made. This is also called "tunneling with rock support." The main feature of this method is that the rock mass in the immediate vicinity of the tunnel excavation is made to act as a load-bearing member together with the supporting system. The outer rock mass ring is activated by means of systematic rock bolting together with shotcrete. The main carrying members of the NATM are the shotcrete and the systematically anchored rock arch.
2. The installation of systematic rock bolting with shotcrete lining allows limited deformations but prevents loosening of the rock mass. In the initial stage it requires

very small forces to prevent rock mass from moving in, but once movement has started, large forces are required. Therefore, NATM advocates installation of supports within stand-up time to prevent movements. Where deformation rates are large, slotted shotcrete lining (i.e., shotcrete sprayed in longitudinal sections separated by expansion joints) helps the problem. In non-squeezing ground conditions, the stresses in the shotcrete may be reduced significantly if the spray of the shotcrete is slightly delayed; however, the delay should be within the stand-up time. A safe practice is to spray a sealing shotcrete layer immediately.

3. In static consideration a tunnel should be treated as a thick wall tube consisting of a bearing ring of rock arch and supporting lining. Since a tube can act as a tube only if it is closed, the closing of the ring becomes of paramount importance, especially where the foundation rock is incapable of withstanding high support pressure in squeezing ground conditions.
4. Due to stress-redistributions when a cavity is excavated, a full face heading is considered most favorable. Drivage in different stages complicates the stress-redistribution phenomenon and destroys the rock mass. When full face tunneling is not possible, as in the Chhibro-Khodri Tunnel and many more tunnels of India due to very little stand-up time and the associated chances of rock falls and cavities, engineers changed to a heading and benching method and struggled to achieve the targeted drivage rates in the absence of shotcrete support.
5. How should the capacity of a rock to support itself be used? This is accomplished by providing an initial shotcrete layer followed by systematic rock bolting, spraying additional shotcrete, and using steel ribs, if necessary. With the Loktak Tunnel, NATM without steel arches in high squeezing grounds would have required several layers of shotcrete that could not be accommodated without compromising the available finished bore. The spacing of steel arches (with invert struts) is adjusted to suit the squeezing ground condition. The behavior of the protective support and the surrounding rock during the stress redistribution process has to be monitored and controlled, if necessary, by different measurements.
6. Shotcrete in a water-charged rock mass should be applied in small patches leaving gaps for effective drainage.

Thus, the basic principles of NATM are summarized as

- Mobilization of rock mass strength
- Shotcrete protection to preserve the load-carrying capacity of the rock mass
- Monitoring the deformation of the excavated rock mass
- Providing flexible but active supports
- Closing of invert to form a load-bearing support ring to control deformation of the rock mass

The NATM appears most suitable for soft ground that can be machine or manually excavated, where jointing and overbreak are not dominant, where a smooth profile can often be formed by smooth blasting, and where a complete load-bearing ring can (and often should) be established. Monitoring plays a significant role in deciding the timing and the extent of secondary support.

Despite the comments by an experienced NATM pioneer that "it is not usually necessary to provide support in hard rocks," Norwegian tunnels require more than 50,000 m^3 of fiber reinforced shotcrete and more than 100,000 rock bolts each year (*World Tunnelling*, 1992). Two major tunneling nations, Norway and Austria, have long traditions in using shotcrete and rock bolts for tunnel supports, yet there are significant differences in philosophy and areas of application between the two.

NORWEGIAN METHOD OF TUNNELING

NMT appears most suitable for good rock masses even where jointing and overbreak are dominant, and where the drill and blasting method or hard rock TBMs are the most common methods of excavation. Bolting is the dominant form of rock support since it mobilizes the strength of the surrounding rock mass the best. Potentially unstable rock masses with clay-filled joints and discontinuities increasingly need shotcrete and SFRS [S(fr)] to supplement systematic bolting (B). It is understood in NMT that [B + S(fr)] are the two most versatile tunnel support methods, because they can be applied to any profile as a temporary or as a permanent support just by changing thickness and bolt spacing. A thick, load-bearing ring (reinforced rib in shotcrete (RRS)) can be formed as needed, and matches an uneven profile better than lattice girders or steel sets. These support requirements based on the Q-system are shown in Figure 8.5. The essential features of the NMT are summarized in Table 8.12 (*World Tunnelling*, 1992).

ROCK MASS CHARACTERIZATION

The chaos theory appears to be applicable at the micro-level only in nature and mostly near the surface. Further, chaos is self-organizing. For engineering use, the overall (weighted average) behavior is all that is needed. Since there is perfect harmony in nature at the macro-level, the overall behavior should also be harmonious. Hence, in civil engineering the chaos theory seems to find only limited applications. In civil engineering practice, simple continuum characterization is more popular for large stable structures. Thus, when behavior of jointed rock masses is discussed, the civil engineer is really talking about the most probable continuum behavior of rock masses.

For caverns, empirical design should be checked by software such as UDEC/3DEC, FLAC, or FEM. To be used, they require the knowledge of deformation and strength characteristics of rock mass and joints.

To develop correlations between Q and other engineering/geophysical parameters, Barton (2008) suggested using the term $Q_c = Q(q_c/100)$.

Cohesion and Angle of Internal Friction

Barton (2008) suggested the following correlations to obtain the cohesive strength (c_p) and angle of internal friction or frictional strength (ϕ_p) of the rock mass.

$$c_p = \frac{RQD}{J_n} \times \frac{1}{SRF} \times \frac{q_c}{100} \text{ MPa} \tag{8.15}$$

$$1.12.4\phi_p = \tan^{-1}\left(\frac{J_r}{J_a} \times J_w\right), \text{ degrees} \tag{8.16}$$

Barton (2008) further recommended that the cohesive strength (c_p) represents the component of the rock mass requiring shotcrete or mesh or concrete support. Similarly, the angle of internal friction or frictional strength (ϕ_p) represents the component of the rock mass requiring the bolting. He further suggested that the rock masses with low c_p values require more shotcrete, whereas rock masses with low ϕ_p values require more rock bolts.

TABLE 8.12 Essential Features of NMT

S. No.	Features
1.	*Areas of usual application*
	Jointed rock, harder end of scale (q_c = 3 to 300 MPa)
	Clay-bearing zones, stress slabbing (Q is 0.001 to 10)
2.	*Usual methods of excavation*
	Drill and blast hard rock, TBM, hand excavation in clay zones
3.	*Temporary support and permanent support may be any of the following*
	• CCA, S(fr) + RRS + B, B + S(fr), B + S, B, S(fr), S, sb, (NONE) • Temporary support forms part of permanent support • Mesh reinforcement not used • Dry process shotcrete not used • Steel sets or lattice girder not used, RRS used in clay zones • Contractor chooses temporary support • Owner/consultant chooses permanent support • Final concrete lining less frequently used, i.e., B + S(fr) is usually the final support
4.	*Rock mass characterization for*
	• Predicting rock mass quality • Predicting support needs • Updating of both during tunneling (monitoring in critical cases only)
5.	*The NMT gives low costs and*
	• Rapid advance rates in drill and blast tunnels • Improved safety • Improved environment

CCA = cast concrete arches; S(fr) = steel fiber reinforced shotcrete; RRS = reinforced steel ribs in shotcrete; B = systematic bolting; S = conventional shotcrete; sb = spot bolting; NONE = no support needed.

Source: *World Tunnelling*, 1992.

Modulus of Deformation of Rock Mass

In India a large number of hydroelectric power projects have been completed recently and several projects are still under construction. These projects have generated a bulk of instrumentation data that have been analyzed by Mitra (1990), Mehrotra (1992), Verman (1993), Goel (1994), and Singh (1997). These new data and their analyses led to a revision of the existing empirical relations and formulation of new correlations subsequently described in this chapter.

Modulus of deformation varies considerably; it occurs more in the horizontal direction than in the vertical direction. However, a mean value of modulus of deformation can be obtained by using the following relation (Barton, 2008).

$$E_d = 10\left(\frac{Q \cdot q_c}{100}\right)^{1/3} \quad \text{GPa} < E_r \text{ [for } Q = 0.1 \text{ to } 100 \text{ and } q_c = 10 - 200 \text{ MPa]} \tag{8.17}$$

This relation agrees with the correlations of Bieniawski (1978) and Serafim and Pereira (1983). The value of UCS of rock material (q_c) can be chosen from Table 8.13 when test results are not available.

Analysis of the field data gives the following correlation for the modulus of deformation (E_d) of weak and nearly dry rock masses with a coefficient of correlation of 0.85 (Singh, 1997):

$$E_d = H^{0.2} \cdot Q^{0.36}, \text{ GPa} \tag{8.18}$$

where Q is the rock mass quality at the time of uniaxial jacking test and H is the overburden above the tunnel in meters >50 m. Mehrotra (1992) found a significant effect from saturation on E_d of water sensitive (argillaceous) rocks. It is thus seen that the modulus of deformation of weak rock masses is pressure dependent. This correlation is suggested for static analysis of underground openings and concrete dams. Further, the test data of 30 uniaxial jacking tests suggested the following correlation for elastic modulus E_e during the unloading cycle (Singh, 1997).

$$E_e = 1.5\, Q^{0.6}\, E_r^{\,0.14}, \text{ GPa} \tag{8.19}$$

where E_r = modulus of elasticity of rock material in GPa and Q = rock mass quality at the time of uniaxial jacking test in drift.

Equation 8.19 is valid for both dry and saturated rock masses. It is suggested for dynamic analysis of concrete dams subjected to impulsive seismic loads due to a high intensity earthquake at a nearby epicenter (active fault). Other correlations are summarized in Table 8.14. The average value of E_d from various correlations may be assumed for stress analysis rejecting its values that are too high and too low.

Special Anisotropy of Rock Mass

Jointed rock masses have very low shear modulus due to very low shear stiffness of joints. The shear modulus of a jointed rock mass has been back analyzed by Singh (1973) as follows:

$$G \approx E_d/10, \text{ GPa} \tag{8.20}$$

The axis of anisotropy is naturally along the weakest joint or a bedding plane. Low shear modulus changes stress distribution drastically in the foundations. Kumar (1988) studied its effect on lined tunnels and found it to be significant.

Another feature of special anisotropy of the rock mass with critically oriented joint sets is that its lateral strain ratio ($\varepsilon_x/\varepsilon_z$) may be as high as 2.79 along the dip direction; its lateral strain ratio in the transverse direction is much lower (Singh & Singh, 2008; Samadhiya, Viladkar, & Al-obaydi, 2008):

$$\text{Lateral strain ratio} = \varepsilon_x/\varepsilon_z = 0.6 - 2.79 \tag{8.21}$$

The degree of anisotropy decreases with increasing confining stress and disappears at σ_3 equal to UCS (q_c). Grouting can reduce the degree of anisotropy even at a shallow depth.

The distinct element method (3DEC; Itasca, 2000) appears to automatically simulate this special kind of rock mass anisotropy (in strength, low shear modulus, modulus of

TABLE 8.13 Average Uniaxial Compressive Strength (q_c) of a Variety of Rocks, Measured on 50 mm Diameter Samples

Type of rock	q_c, MPa	Type of rock	q_c, MPa	Type of rock	q_c, MPa	Type of rock	q_c, MPa
Andesite (I)	150	Granite (I)	160	Marble (M)	<100>	Shale (S, M)	95
Amphibolite (M)	<160>	Granitic Gneiss (M)	100	Micagneiss (M)	90	Siltstone (S, M)	<80>
Augen Gneiss (M)	160	Granodiorite (I)	160	Micaquartzite (M)	85	Slate (M)	<190>
Basalt (I)	160	Granulite (M)	<90>	Micaschist (M)	<80>	Syenite (I)	150
Clay Schist (S, M)	55	Gneiss (M)	130	Phyllite (M)	<50>	Tuff (S)	<25>
Diorite (I)	140	Greenschist (M)	<75>	Quartzite (M)	<190>	Ultrabasic (I)	160
Dolerite (I)	200	Greenstone (M)	110	Quartzitic Phy. (M)	100	Clay (hard)	0.7
Dolomite (S)	<100>	Greywacke (M)	80	Rhyolite (I)	85	Clay (stiff)	0.2
Gabbro (I)	240	Limestone (S)	90	Sandstone (S, M)	<100>	Clay (soft)	0.03
				Serpentine (M)	135	Silt, sand (approx.)	0.0005

(I)=Igneous; (M)=Metamorphic; (S)=Sedimentary; < > = Large Variation

Source: Palmstrom, 2000.

TABLE 8.14 Empirical Correlations for Overall Modulus of Deformation of Rock Mass in the Non-Squeezing Ground Condition (GSI & RMR << 100)

Authors	Expression for E_d (GPa)	Conditions	Recommended for
Bieniawski (1978)	$E_d = 2\ RMR - 100$	$q_c > 100$ MPa and $RMR > 50$	Dams
Serafim & Pereira (1983)	$E_d = 10^{(RMR-10)/40}$	$q_c \geq 100MPa$	Dams
Nicholson & Bieniawski (1990)	$E_d/E_r = 0.0028\ RMR^2 + 0.9\ e^{(RMR/22.82)}$	—	
Verman (1993)	$E_d = 0.3\ H\alpha.\ 10^{(RMR-20)/38}$	$\alpha = 0.16$ to 0.30 (higher for poor rocks) $q_c \leq 100$ MPa; $H \geq 50$ m; $J_w = 1$ Coeff. of correlation = 0.91	Tunnels
Mitri et al. (1994)	$E_d/E_r = 0.5[1\text{-}\cos(\pi\ RMR/100)]$	—	
Singh (1997)	$E_d = Q^{0.36}\ H^{0.2}$ $E_e = 1.5Q^{0.6}\ E_r^{0.14}$	$Q < 10$; $J_w = 1$ Coeff. of correlation for $E_e = 0.96$; $J_w \leq 1$	Dams and slopes Dams
Hoek et al. (2002)	$E_d = \left(1 - \frac{D}{2}\right)\sqrt{\frac{q_c}{100}} \cdot 10^{((GSI-10)/40)}$ $E_d = \left(1 - \frac{D}{2}\right) \cdot 10^{((GSI-10)/40)}$	$q_c \leq 100$ MPa D = disturbance factor (Table 26.4) $q_c \geq 100$ MPa	
Adachi & Yoshida (2002)	$E_d = 10^{(0.0431R-0.8853)}$	For weak rocks, R = In situ average Schmidt hammer rebound number	
Barton (2008)	$E_d = 10[Q \cdot q_c/100]^{1/3} < E_r$	$Q = 0.1 - 100$ $q_c = 10 - 200$ MPa	Tunnels
Zhang & Einstein (2004)	$\frac{E_d}{E_r} = 10^{0.0186\ RQD-1.91}$	For $0 \leq RQD \leq 100$	Preliminary analysis
Hoek & Diederichs (2006)	$E_d = \left[0.02 + \frac{1 - D/2}{1 + \exp((60 + 15\ D - GSI)/11)}\right]$		Tunnels, caverns, and dam foundations

The above correlations are expected to provide a mean value of modulus of deformation.

deformation, high lateral ratio, and permeability and post-peak characteristics of work softening and pre-stressing due to σ_2) and is recommended for Q-values between 0.1 and 100 where $H < 350\,Q^{1/3}$ meters (i.e., in the case of non-squeezing blocky rock mass).

Q-Wave versus P-Wave Velocity

A correlation between seismic P-wave velocity and rock mass quality Q has been proposed by Barton (2002) on the basis of approximately 2000 measurements for a rough estimation of Q ahead of the tunnel face using seismic P-wave velocity:

$$Q = \frac{100}{q_c} 10^{[(V_p - 3500)/1000]}, \text{ for } 500\text{ m} > H > 25\text{ m} \tag{8.22}$$

$$\frac{V_s}{V_p} = 0.15 \text{ to } 0.66 \tag{8.23}$$

where V_p is P-wave velocity in meters per second and q_c is the UCS of rock material in MPa. V_s is the shear wave velocity of rock masses.

For good and fair quality granites and gneisses, an even better fit is obtained using the relation $Q = (V_p - 3600)/50$ (Barton, 1991). Figure 8.6 illustrates the approximate values of rock mass quality before underground excavation for a known P-wave velocity for different values of depth of overburden (H). It should be noted that P-wave velocity increases rapidly with the depth of overburden. Figure 8.6 also suggests the following correlation between mean static modulus of deformation in roof (in GPa) and support pressure (in MPa).

$$p_{roof} = f \cdot f'/E_{d(mean)}, \text{ MPa} \tag{8.24}$$

The advantage of this correlation is that cross-hole seismic tomography may be used in a more direct and accurate manner for specifying expected rock qualities and potential rock support needs in tender documents. In the future it may be possible to assess

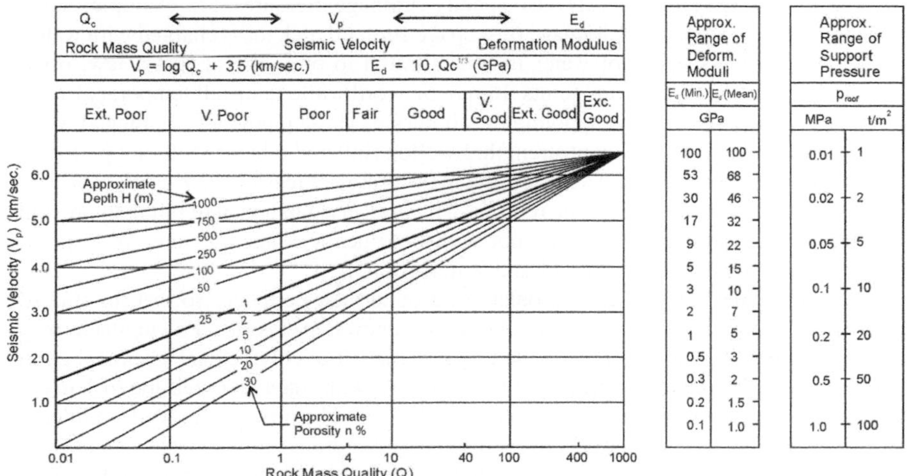

FIGURE 8.6 An integration of V_p, Q, q_c, depth, porosity, and static deformation modulus (E_d), which was developed stage by stage by trial and error using field data. *(From Barton, 2002)*

Q-values at great depths along tunnels by the seismic refraction survey on the ground level before it is excavated. The Q-values after squeezing or rock burst or seepage erosion may be significantly less than Q-values before tunneling, that is, during seismic survey.

Improvement in Q by Grouting

According to Barton (2002), the in situ permeability (k) of rock mass near the surface is of the order of (for Q = 0.01 to 100, H < 25 m and 1 lugeon = 1.0×10^{-5} cm/sec)

$$k \approx \frac{1}{Q_c} = \frac{100}{Q \cdot q_c}, \text{ lugeons} \tag{8.25}$$

This is a surprisingly simple correlation, yet it is true for Q between 0.01 and 100. Rock mass quality may be improved significantly by grouting rock masses with cement grout, which would be proportional to the decrease in the maximum value of permeability of a grouted rock mass in any direction. Thus, the required capacity of support systems for underground openings may be reduced substantially. The long grout holes will drain off any water in the rock masses effectively, reducing construction problems in the water-charged rock masses (flowing ground condition).

Grouting of the rock mass with permeability above 1 lugeon is feasible at sites with cement particles with a maximum size of 100–150 μm. Micro-fine and ultra-fine cements with maximum size particles of 15–30 μm may be used in fair rock masses with physical apertures of about 0.05–0.10 μm. The rule of thumb is that the maximum size particle should be more than three or four times the physical aperture of joints (Barton, 2002). The assumption is that the grout will follow the path of least resistance, which is predominantly the most permeable and least normally stressed joint set. Thus, the least J_r/J_a value will also be improved. With his contributions, Barton (2002) proved why construction engineers often grouted weak rock masses (see Example 23.1).

DRAINAGE MEASURES

The drainage system should be fully designed before the construction of the tunnel and cavern. The NATM and the NMT also specify drainage measures. For example, radial gaps are left unshotcreted for drainage of seepage for hard rock mass charged with water.

Very often the seepage of water is concentrated to only one or just a few, often tubular, openings in fissures and joints. It may be worthwhile to install temporary drainage pipes in such areas before applying the shotcrete. These pipes can be plugged when the shotcrete has gained sufficient strength. Swellex (inflated tubular) bolts are preferred in water-charged rock masses. Cement grouted bolts are not feasible here as grout will be washed out. Resin grout may not be reliable. The seals used in concrete lining to prevent seepage in the road/rail tunnels may not withstand heavy water pressure. Waterproof lining makes a tunnel moistureproof and waterproof.

Pressure tunnels are generally grouted all around their periphery so that the ring of grouted rock mass is able to withstand heavy groundwater pressure. Polyurethane may be used to grout the water-charged jointed rock masses. The polyurethane chemical grout swells up to 26 times its size when it comes in contact with water and seals the rock mass.

Deep drainage holes in water-charged tunnels should be provided in walls to release seepage pressure and discharge the water into side drains of adequate capacity in the transportation (railroad) tunnels. Discharge from side drains and selected drainage holes should be monitored even after completion of the tunnel to ascertain the clogging of drainage holes.

EXPERIENCES IN POOR ROCK CONDITIONS

SFRS has proved very successful in the 6.5 km long tunnel for the Uri Hydel project and desilting underground chambers of NJPC in the Himalayas in India. The main advantage is that a smaller thickness of SFRS is needed. No weld mesh is required to reinforce the shotcrete. Provided that the shotcrete is graded and sprayed properly, there is less rebound because of steel fibers. This method is now economical, safer, and faster than the conventional shotcrete. The contour blasting technique is adopted to excavate the tunnel where SFRS is to be used. The selection of the right ingredients and tight quality control over application are keys to the success of SFRS.

Grimstad and Barton (1993) proposed a chart (Figure 8.5) for designing the support system for poor rock conditions. Later Barton (2002) improved this chart slightly for good rock conditions. It gives the thickness of SFRS, spacing, and length of rock bolt corresponding to the rock mass quality (Q) and width or height of the tunnel or cavern. The spacing of rock bolts in the shotcreted area is indicated on the upper left side of Figure 8.5 corresponding to the Q-value. The values at the bottom right side of the figure are the spacing of rock bolts for unshotcreted areas. Q_{av} should be used in place of shear zones. Q_w needs to be used for supports in walls. Palmstrom and Broch (2006) considered the chart suitable for blocky mass ($0.1 < Q < 40$ and $3 \text{ m} < B < 40 \text{ m}$). The Q-system has much wider applications. Example 8.1 illustrates the use of Figure 8.5. Example 9.3 shows how Q may be estimated from N-value due to a realistic guess of tunneling hazards where SRF is difficult to judge on site in advance of tunneling.

In squeezing and swelling grounds, the supports (steel ribs) fail invariably, but a state of equilibrium is reached eventually. The suggested construction approach is to remove the damaged supports and replace them by stiffer and stronger steel ribs embedded in SFRS. The SFRS layers are sprayed until the rate of tunnel wall closure is reduced to 2 to 3 mm per month. With rock-burst-prone rocks, the failed supports are replaced by the ductile SFRS and resin anchor support system to arrest propagation of fractures in the brittle rocks. The segmented reinforced-concrete lining is recommended within intra-thrust zones with active faults (Singh & Goel, 2006).

Using mesh (weld mesh, etc.) has been unsatisfactory when there were overbreaks in a tunnel after blasting. Soon after the weld mesh was spread between bolts and shotcrete, it started rebounding the shotcrete and could not penetrate inside the mesh and fill the gap between the mesh and the overbreak. Consequently, gaps were left above the shotcrete; the sound of the hammer struck above the mesh indicated hollow areas. Loosely fitted welded wire mesh vibrates as a result of blast vibrations, causing subsequent loosening of the shotcrete.

Because mesh-reinforced shotcrete has been unsatisfactory in handling overbreak situations, it is recommended that mesh with plain shotcrete should not be used where there is an uneven tunnel surface due to high overbreaks. In such cases, the thickness of shotcrete should be increased sufficiently (e.g., by 10 mm).

CONCLUDING REMARKS

Some doubts have been expressed as to whether or not in situ stress and water pressure should be considered in rock mass classification. This is questioned because they are external and internal boundary conditions of a rock structure that are taken into account in all software packages. The real response of rock masses is often highly coupled or interacting.

The SRF depends upon the height of overburden. Hence, it is an external boundary condition. However, high overburden pressure causes damage to the rock mass structure, which needs to be considered in a rock mass classification system. It is worth seeing the time-dependent squeezing and rock burst phenomenon in deep tunnels as it helps to develop the total concept of rock mass quality.

The seepage water pressure in rock joints, on the other hand, represents an internal boundary condition. The high water pressure softens weak argillaceous rock masses due to seepage erosion and long-term weathering of rock joints, particularly with coating of soft material such as clay. So the joint water reduction factor (J_w) also needs to be considered for both rock mass classification and rock mass characterization.

The classification of rock mass does not mean that the correlation should be obtained with the rock mass classification rating only. Correlations with rock mass classification rating and other important parameters such as height of overburden, UCS, modulus of elasticity of rock material, size of opening, and so forth, should also be used. The objective is to improve the coefficient of correlation significantly so it is practical and simple to understand.

There is worldwide appreciation of the utility of the (post-excavation) rock mass quality Q-system for empirical design of support systems for tunnels and caverns. The classification approach is really an amazing civil engineering application. Recently, the Q-system has been successfully extended to rock mass characterization (Barton, 2002).

The following list includes remarks about the use of shotcrete and SFRS.

1. In a poor rock mass, the support capacity of the rock bolts (or anchors) is small in comparison to that of shotcrete and SFRS, which is generally the main element of the long-term support system for resisting heavy support pressures in tunnels in weak rock masses.
2. The untensioned full-column grouted bolts (called anchors) are more effective than pre-tensioned rock bolts in supporting weak rock masses.
3. The thickness of SFRS is about half of the thickness of plain shotcrete without reinforcement.
4. SFRS has been used successfully in mild and moderate squeezing ground conditions and tectonically disturbed rock masses with thin shear zones.
5. The NMT is based on the philosophy of NATM to form a load-bearing ring all around a tunnel. NMT offers site-specific design tables for plain shotcrete and a design chart for SFRS. By following their philosophy, the tunnel engineer benefits from the extensive experience of the past NATM and the modern NMT.
6. Quality control in tunnel construction by experts should be made mandatory.

Example 8.1

In a major hydroelectric project in dry quartzitic phyllite, the rock mass quality is found to be in the range of 6 to 10. The joint roughness number (J_r) is 1.5 and the joint alteration number (J_a) is 1.0 for critically oriented joints in the underground machine hall. The width of the cavern is 25 m, its height is 50 m, and the roof is arched. The overburden is 450 m, $J_w = 1.0$, and SRF = 2.5. Suggested design of the support system is as follows.

The average rock mass quality is $(6 \times 10)^{1/2} = 8$ (approximately). The overburden above the crown is less than $350\ (8)^{1/3} = 700$ m, hence the rock mass is non-squeezing. The correction factor for overburden $f = 1 + (450 - 320)/800 = 1.16$. The correction for tunnel closure $f' = 1.0$. Short-term support pressure in the roof from Eq. 8.6 is ($f'' = 1$)

$$(0.2/1.5)\ (5 \times 8)^{-1/3}\ 1.16 = 0.045 \text{ MPa}$$

Short-term wall support pressure is

$$= (0.2/1.5)\ (5 \times 2.5 \times 8)^{-1/3}\ 1.16 = 0.033 \text{ MPa (practically negligible)}$$

Ultimate support pressure in the roof from Eq. (8.9) is given by

$$p_{roof} = (0.2/1.5)\ (8)^{-1/3}\ 1.16 = 0.077 \text{ MPa}$$

Ultimate wall support pressure (see the section Estimation of Support Pressure in this chapter) is given by

$$p_{wall} = (0.2/1.5)\ (2.5 \times 8)^{-1/3}\ 1.16 = 0.057 \text{ MPa}$$

The modulus of deformation of the rock mass is given by Eq. (8.18):

$$E_d = (8)^{0.36}\ (450)^{0.2} = 7.0 \text{ GPa}$$

The excavation support ratio is 1.0 for important structures. Figure 8.5 gives the following support system in the roof:

Bolt length = 6 m
Bolt spacing = 2.2 m
Thickness of SFRS = 90 mm

Figure 8.5 is also useful in recommending the following wall support system of the cavern ($Q_w = 2.5 \times 8 = 20$, ESR = 1, height = 50 m)

Bolt length = 11 m
Bolt spacing = 2.5 m
Thickness of SFRS = 70 mm

Example 8.2

A 2.4 m wide and D-shaped new canal tunnel is tangentially joining an existing 2.4 m wide and lined canal tunnel. The rock mass quality (Q) is likely to vary from 0.4 to 4.0 in gneiss with $J_r = 1.0$, $J_a = 2.0$, SRF = 10, average fracture spacing = 0.5 m, safe-bearing capacity = 200 T/m^2 (2 MPa), and under overburden of 390 m. The tunnel is in a highly seismic zone. The maximum width of the opening at the intersection of tunnels is 6.3 m. The bolt capacity is 10 T and the bolt length is limited to 2.0 m due to the small size of the new tunnel. The UCS of SFRS is found to be 15 MPa after only 7 days. Groundwater can seep into the opening near shear zones. The project authorities can close the existing canal tunnel for only one month. Design a safe support system near the intersection of the tunnels.

The minimum effective rock mass quality near the intersection is $0.4/3 = 0.13$, considering three times the joint set number. Mild squeezing is likely to occur as $H < 350Q^{1/3}$ m and $J_r/J_a < 0.5$ here. The vertical support pressure is estimated by correlation (Eq. 8.5) of Bhasin and Grimstad (1996) as follows ($Q < 4$):

$$p_V = \frac{40\ B}{J_r} \cdot Q^{-1/3} = \frac{40 \times 6.3\ (0.13)^{-1/3}}{1.0} = 497\ \text{KPa}$$

$$= 0.5\ \text{MPa (static)}$$

$$= 0.5 \times 1.25\ \text{MPa (dynamic)}\ [Q_{dynamic} = Q/2]$$

Total support pressure = 0.62 + 0.05 (seepage pressure) = 0.67 MPa.

The high value of support pressure is justified as the pillar between the tunnels is too small near their intersection. The wall support pressure may be of the same order as the vertical support pressure in the squeezing ground.

Figure 8.5 suggests the following design parameters for Q = 0.13 and B = 6.3 m:

Bolt length = 2.5 m
Bolt spacing = 1.4 m
SFRS thickness = 12 cm

The shear strength of the SFRS may be 0.20 × 15 = 3.0 MPa. The rational method (Chapter 12) gives the revised design details as follows:

Bolt length = 2.0 m
Bolt spacing = 1.0 m ≤ half-bolt length
Size of base plate of resin bolt = $(10/200)^{1/2} = 0.25$ m
Capacity of steel ribs = 125 T
Steel rib spacing = 0.6 m
SFRS thickness = 25 cm

$$\text{Support capacity of steel ribs} = \frac{P_{rib}}{S_{rib}\ B} = \frac{125}{0.6 \times 6.3} = 0.33\ \text{MPa}$$

$$\text{Support capacity of SFRS} = \frac{2t_{sc}\ q_{sc}}{0.6\ B} = \frac{2 \times 0.25 \times 3}{0.6 \times 6.3} = 0.4\ \text{MPa}$$

Total capacity > 0.33 + 0.40 = 0.73 MPa > 0.67 MPa (hence safe)

The support capacity of bolts is considered negligible. The revised support system needs to be installed in the length of new tunnel equal to 3 B, that is, 3 × 2.4 = 7.2 m. The steel ribs should be provided with the invert struts to withstand high wall support pressures and should be embedded in SFRS all around including the bottom. The steel ribs that buckle during squeezing should be replaced one by one and shotcreted again. Smooth blasting is recommended near the old tunnel to cause minimum damage to its concrete lining. Finally, there will be a concrete lining 15 cm thick for smooth flow of water in the proposed new canal tunnel. The rock mass should be grouted up to a depth of 2 m beyond the concrete lining to reduce seepage loss of water.

REFERENCES

Adachi, T., & Yoshida, N. (2002). In situ investigation on mechanical characteristics of weak rocks. In V. M. Sharma & K. R. Saxena (Eds.), *In situ characterisation of rocks* (Chap. 4, p. 358). New Delhi: Oxford & IBH Publishing Co. Pvt. Ltd. and The Netherlands: A.A. Balkema.

Barton, N. (1991). Geotechnical design. In *World Tunnelling* (pp. 410–416). November.

Barton, N. (1993). Application of Q-system and index tests to estimate shear strength and deformability of rock masses. In *Workshop on Norwegian Method of Tunnelling* (pp. 66–84). New Delhi, India.

Barton, N. (2002). Some new Q-value correlations to assist in site characterisation and tunnel design. *International Journal of Rock Mechanics and Mining Sciences, 39*, 185–216.

Barton, N. (2008). *Training course on rock engineering* (p. 502). New Delhi, India: Organized by ISRMTT & CSMRS. December 10–12.

Barton, N., Lien, R., & Lunde, J. (1974). *Engineering classification of rock masses for the design of tunnel support* (NGI Publication No. 106, p. 48). Oslo: Norwegian Geotechnical Institute.

Barton, N., Lien, R., & Lunde, J. (1975). Estimation of support requirements for underground excavations (pp. 163–177). *XVIth Symposium on Rock Mechanics*, Minneapolis: University of Minnesota.

Bhasin, R., & Grimstad, E. (1996). The use of stress-strength relationship in the assessment of tunnel stability. In *Proceedings of the Recent Advances in Tunnelling Technology* (pp. 183–196), New Delhi, India: CSMRS.

Bieniawski, Z. T. (1978). Determining rock mass deformability, experience from case histories. *International Journal of Rock Mechanics and Mining Sciences—Geomechanics Abstracts, 15*, 237–247.

Dube, A. K. (1979). *Geomechanical evaluation of tunnel stability under failing rock conditions in a Himalayan tunnel* (p. 212). Ph.D. Thesis. Uttarakhand, India: IIT Roorkee.

Goel, R. K. (1994). *Correlations for predicting support pressures and closures in tunnels* (p. 308). Ph.D. Thesis. Uttarakhand, India: Nagpur University.

Goel, R. K., Jethwa, J. L., & Paithankar, A. G. (1995a). Indian experiences with Q and RMR systems. *Tunnelling and Underground Space Technology, 10*(1), 97–109.

Goel, R. K., Jethwa, J. L., & Paithankar, A. G. (1995b). Tunnelling through the young Himalayas—A case history of the Maneri-Uttarkashi power tunnel. *Engineering Geology, 39*, 31–44.

Grimstad, E., & Barton, N. (1993). Updating of the Q-system for NMT. In *Proceedings of the International Symposium on Sprayed Concrete—Modern Use of Wet Mix Sprayed Concrete for Underground Support*, Oslo: Fagernes, Norwegian Concrete Association.

Hoek, E., Carranza-Torres, C., & Corkum, B. (2002). Hoek-Brown Failure Criterion—2002. In *North American Rock Mechanics Symposium* (5th ed., Vol. 1, pp. 267–273). 17th Tunnel Association of Canada, NARMS-TAC Conference, Toronto.

Hoek, E., & Diederichs, M. S. (2006). Empirical estimation of rock mass modulus. *International Journal of Rock Mechanics and Mining Sciences, 43*, 203–215.

Itasca. (2000). *Universal Distinct Element Code*. Minneapolis, MN.

Jethwa, J. L. (1981). *Evaluation of rock pressures in tunnels through squeezing ground in Lower Himalayas* (p. 272). Ph.D. Thesis. Uttarakhand, India: Department of Civil Engineering, University of Roorkee.

Kadkade, D. G. (2009). *Remarks during Indorock 2009*. New Delhi, November 12–13.

Kaiser, P. K., Mackay, C., & Gale, A. D. (1986). Evaluation of rock classification at B.C. Rail Tumbler Ridge Tunnels. *Rock Mechanics and Rock Engineering, 19*, 205–234.

Kumar, N. (2002). *Rock mass characterisation and evaluation of supports for tunnels in Himalaya* (p. 289). Ph.D. Thesis. Uttarakhand, India: WRDM, IIT Roorkee.

Kumar, P. (1988). *Development and application of infinite elements for analysis of openings in rock mass* (p. 192). Ph.D. Thesis. Uttarakhand, India: IIT Roorkee.

Mehrotra, V. K. (1992). *Estimation of engineering properties of rock mass* (p. 267). Ph.D. Thesis. Uttarakhand, India: University of Roorkee.

Mitra, S. (1990). *Studies on long-term behaviour of underground powerhouse cavities in soft rocks*. Ph.D. Thesis. Uttarakhand, India: IIT Roorkee.

Mitri, H. S., Edrissi, R., & Henning, J. (1994). *Finite Element Modelling of Cable Bolted Stopes in Hard Rock Underground Mines* (pp. 14–17). SME Annual Meeting.

Nicholson, G. A., & Bieniawski, Z. T. (1990). A non-linear deformation modulus based on rock mass classification. *Geotechnical and Geological Engineering*, *8*, 181–202.

Palmstrom, A. (2000). Recent developments in rock support estimates by the RMI. *Journal of Rock Mechanics and Tunnelling Technology*, *2*(1), 1–24.

Palmstrom, A., & Broch, E. (2006). Use and misuse of rock mass classification systems with particular reference to the Q-System. *Tunnelling and Underground Space Technology*, *21*, 575–593.

Papworth, F. (2002). *Design Guidelines for the use of Fibre-Reinforced Shotcrete in Ground Support, Shotcrete* (pp. 16–21). *www.shortcrete.org/pdf_files/0402Papworth.pdf.*

Samadhiya, N. K., Viladkar, M. N., & Al-obaydi, M. A. (2008). Numerical implementation of anisotropic continuum model for rock masses, Technical Note. *International Journal of Geomechanics*, ASCE, *8*(2), 157–161.

Serafim, J. L., & Pereira, J. P. (1983). Considerations of the geomechanics classification of Bieniawski. *International Symposium of Engineering and Geological Underground Construction* Vol 1, (pp. II.33–II.42). Lisbon: LNEC.

Singh, B. (1973). Continuum characterization of jointed rock mass, Part II: Significance of low shear modulus. *International Journal of Rock Mechanics and Mining Sciences—Geomechanics Abstracts*, *10*, 337–349.

Singh, B., & Goel, R. K. (2006). J. A. Hudson (Ed.), *Tunnelling in weak rocks* (p. 489). Oxford: Elsevier.

Singh, B., Jethwa, J. L., Dube, A. K., & Singh, B. (1992). Correlation between observed support pressure and rock mass quality. *Tunnelling and Underground Space Technology*, *7*(1), 59–74.

Singh, M., & Singh, B. (2008). High lateral ratio in jointed rock masses. *Engineering Geology*, *98*, 75–85.

Singh, S. (1997). *Time dependent deformation modulus of rocks in tunnels* (p. 65). M.E. Thesis. Uttarakhand, India: Dept. of Civil Engineering, University of Roorkee.

Terzaghi, K. (1946). R. V. Proctor & T. L. White (Eds.), *Rock defects and loads on tunnel support, introduction to rock tunnelling with steel supports* (p. 271). Youngstown, OH: Commercial Sheering & Stamping Co.

Verman, M. K. (1993). *Rock mass-tunnel support interaction analysis* (p. 258). Ph.D. Thesis. Uttarakhand, India: IIT Roorkee.

World Tunnelling. (1992). Focus on Norway "Norwegian Method of Tunnelling," June issue. In *Proceedings Workshop on Norwegian Method of Tunnelling*. September, New Delhi, India.

Zhang, L., & Einstein, H. H. (2004). Using RQD to estimate the deformation modulus of rock masses. *International Journal of Rock Mechanics and Mining Sciences*, *41*, 337–341.

Chapter 9

Rock Mass Number

My attention is now entirely concentrated on Rock Mechanics, where my experience in applied soil mechanics can render useful services. I am more and more amazed about the blind optimism with which the younger generation invades this field, without paying any attention to the inevitable uncertainties in the data on which their theoretical reasoning is based and without making serious attempts to evaluate the resulting errors.

Annual Summary in Terzaghi's Diary

INTRODUCTION

One of the reasons why rock mass classifications have become popular over the years is that they are easy to use and provide vital information about rock mass characteristics. Classification also leads to making fast decisions during tunneling. Thus, rock mass classification is an amazingly successful approach.

Despite their usefulness, there is some uncertainty about the correctness of the ratings for some of the parameters. How should these uncertainties be managed? With this objective, two rock mass indices—rock mass number (N) and rock condition rating (RCR)—have been adopted. These indices are the modified versions of the two most popular classification systems: N from the Q-system of Barton, Lien, and Lunde (1974) and RCR from the rock mass rating (RMR) system of Bieniawski (1984).

Rock mass number, denoted by N, is the stress-free rock mass quality (Q). Stress-effect was considered indirectly in the form of overburden height (H). Thus, N can be defined by Eq. (9.1), representing basic causative factors in governing the tunneling conditions.

$$N = [RQD/J_n]\,[J_r/J_a]\,[J_w] \tag{9.1}$$

This is needed because of the problems and uncertainties in obtaining the correct rating of Barton's stress reduction factor (SRF) parameter (Kaiser, Mackay, & Gale, 1986; Goel, Jethwa, & Paithankar, 1995a). N is found to be complimentary to the Q-system. Correlations (in Chapter 7) based on N can first be used to identify the ground conditions and then the rating for SRF, because the ground condition and degree of squeezing can be selected to get the Q-value.

RCR is defined as RMR without ratings for the crushing strength of the intact rock material and the adjustment of joint orientation. This is explained in Eq. (9.2).

$$RCR = RMR - (\text{Rating for UCS} + \text{Adjustment of Joint Orientation}) \tag{9.2}$$

RCR, therefore, is free from the uniaxial compressive strength (UCS), which is sometimes difficult to obtain on site. Moreover, parameters N and RCR are equivalent and can be used for a better interrelation.

INTERRELATION BETWEEN Q AND RMR

Interrelations between the two most widely used classification indices, the RMR of Bieniawski (1976) and the Q of Barton et al. (1974), have been proposed by many researchers. Bieniawski (1976) used 111 case histories involving 62 Scandinavian, 28 South African, and 21 other documented case histories from the United States covering the entire range of Q and RMR to propose the following correlation:

$$RMR = 9 \ln Q + 44 \tag{9.3}$$

Based on case histories from New Zealand, Rutledge and Preston (1978) proposed a different correlation as

$$RMR = 5.9 \ln Q + 43 \tag{9.4}$$

Moreno (1980), Cameron-Clarke and Budavari (1981), and Abad et al. (1984) also proposed different correlations between Q and RMR as presented in Eqs. (9.5)–(9.7), respectively.

$$RMR = 5.4 \ln Q + 55.2 \tag{9.5}$$

$$RMR = 5 \ln Q + 60.8 \tag{9.6}$$

$$RMR = 10.5 \ln Q + 41.8 \tag{9.7}$$

Evaluation of the correlations given in Eqs. (9.3) through (9.7) based on 115 case histories, including 77 reported by Bieniawski (1976), 4 from the Kielder experimental tunnel reported by Hoek and Brown (1980), and 34 collected from India, indicated that the correlation coefficients of these approaches are not very reliable. The correlation of Rutledge and Preston (1978) provided the highest correlation coefficient of 0.81, followed by Bieniawski (1976), Abad et al. (1984), Moreno (1980), and Cameron-Clarke and Budavari (1981) in decreasing order as shown in Table 9.1 and Figure 9.1. These correlations, therefore, are not highly reliable for an interrelation between Q and RMR.

The New Approach

Attempts to correlate Q and RMR in Eqs. (9.3) through (9.7) ignore the fact that the two systems are not truly equivalent. It seems, therefore, that a good correlation can be developed if N and RCR are considered.

RCR and rock mass number N from 63 cases were used to obtain a new interrelation. The 63 cases consisted of 36 from India, 4 from the Kielder experimental tunnel (reported

TABLE 9.1 Evaluation of Various Correlations between RMR and Q

Lines in Figure 9.1	Approach	Correlation coefficient
A	Bieniawski (1976)	0.77
B	Rutledge & Preston (1978)	0.81
C	Moreno (1980)	0.55
D	Cameron-Clarke & Budavari (1981)	High scatter
E	Abad et al. (1984)	0.66

Source: Goel et al., 1995b.

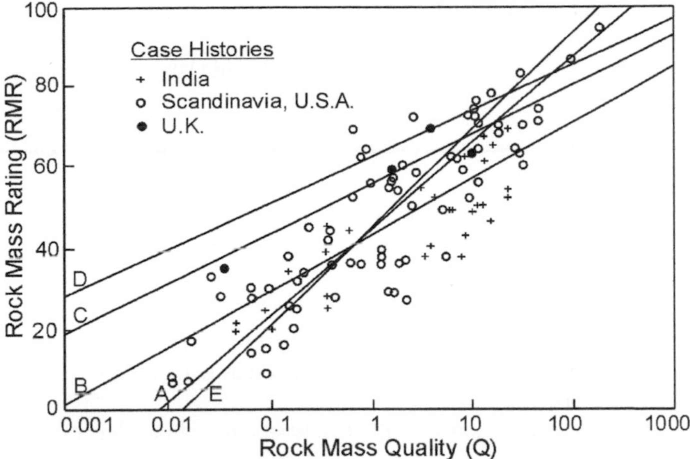

FIGURE 9.1 Correlations between Q and RMR. *(From Goel et al., 1995b)*

by Hoek & Brown, 1980), and 23 from the Norwegian Geotechnical Institute (NGI) (reported by Bieniawski, 1984). Details about the six parameters for Q and information about joint orientation vis-à-vis tunnel axis with respect to these 23 NGI cases were picked up directly from Barton et al. (1974). Estimates of UCS (q_c) of rock material were made from rock descriptions given by Barton et al. (1974) using strength data for comparable rock types from Lama and Vutukuri (1978). Using the obtained ratings for joint orientation and q_c and RMR from Bieniawski (1984), it was possible to estimate values of RCR. Thus, the values of N and RCR for the 63 case histories were plotted in Figure 9.2 and the following correlation was obtained:

$$\text{RCR} = 8 \ln \text{N} + 30, \text{ for } q_c > 5 \text{ MPa} \tag{9.8}$$

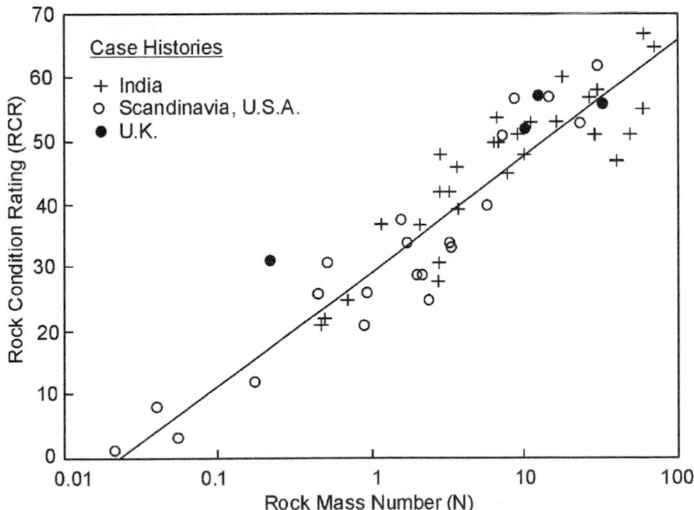

FIGURE 9.2 Correlations between N and RCR. *(From Goel et al., 1995b)*

Equation (9.8) has a correlation coefficient of 0.92, and it is not applicable to the borderline of soil and rock mass according to data from Sari and Pasamehmetoglu (2004). The following example explains how Eq. (9.8) could be used to obtain RMR from Q and vice versa.

Example 9.1

The values of the parameters of RMR and Q collected in the field are given in Table 9.2.

(a) **RMR from Q**

$$N = (RQD\ J_r\ J_w)/(J_n\ J_a) = 26.66 \text{ as shown in Table 9.2}$$

Corresponding to N = 26.66, RCR = 56.26 (Eq. 9.8)
RMR = RCR + (ratings for q_c and joint orientation as per Eq. 9.2)
RMR = 56.26 + [4 + (−)12]
RMR = 48.26 (It is comparable to RMR = 49 obtained from direct estimation as shown in Table 9.2.)

(b) **Q from RMR**
RCR = RMR − (ratings for q_c and joint orientation as per Eq. 9.2)
RCR = 49 − (4 − 12)
RCR = 57
Corresponding to RCR = 57, N = 29.22 (Eq. 9.8)

$$Q = (N/SRF) = 29.22/2.5$$

Q = 11.68 (almost equal to the field estimated value, Table 9.2)
The slight difference in directly estimated values of Q and RMR and those obtained by the proposed interrelation are due to the inherent scatter in Eq. (9.8).

TABLE 9.2 Values of the Parameters of RMR and Q Collected in the Field

RMR system		Q-system	
Parameters for RMR	*Rating*	*Parameters for Q*	*Rating*
RQD (80%)	17	RQD	80
Joint spacing	10	J_n	9
Joint condition	20	J_r	3
		J_a	1
Groundwater	10	J_w	1
RCR =	**57**	**N =**	**26.66**
Crushing strength q_c	+4	SRF	2.5
Joint orientation	(−)12	—	—
RMR =	**49**	**Q =**	**10.6**

PREDICTION OF GROUND CONDITIONS

All the correlations for predicting ground conditions using rock mass number N have been presented in Table 7.6. The main advantage of rock mass number is that it does not assume ground conditions, it predicts them.

In practice, the rock mass is classified into categories I, II, III, and so forth. Accordingly, support systems are prescribed. There are unusual geological conditions at some sections. These possible conditions (flowing ground, running ground, etc.) should also be classified in the contract and a support system should also be suggested. Further, there should be first and second contingency clauses in the same contract for better preparedness.

PREDICTION OF SUPPORT PRESSURE

These correlations are based on measured support pressures and other related parameters from several Indian tunnels that have steel rib support. Detailed field studies have been carried out for eight tunneling projects located in the Himalayas and peninsular India.

Two sets of empirical correlations for estimating support pressure for tunnel sections under non-squeezing and squeezing ground conditions have been developed using N and the measured values of support pressures, the tunnel depth (H), the tunnel radius (a), and the expected tunnel closure (u_a) from 25 tunnel sections (Goel et al., 1995a; Singh et al., 1997). The correlations are described in the following section.

Non-Squeezing Ground Condition

$$p_v(el) = \left[\frac{0.12H^{0.1} \cdot a^{0.1}}{N^{0.33}}\right] - 0.038,\ \text{MPa} \tag{9.9}$$

Kumar (2002) found that Eq. (9.9) is valid for overburden (H) up to 1400 m in the NJPC tunnel in India.

Squeezing Ground Condition

$$p_v(sq) = \left[\frac{f(N)}{30}\right] \cdot 10^{\left[\frac{H^{0.6} \cdot a^{0.1}}{50 \cdot N^{0.33}}\right]},\ \text{MPa} \tag{9.10}$$

where $p_v(el)$ = short-term roof support pressure in non-squeezing ground condition in MPa; $p_v(sq)$ = short-term roof support pressure in squeezing ground condition in MPa; f(N) = correction factor for tunnel closure obtained from Table 9.3, and H and a = tunnel depth and tunnel radius in meters, respectively.

The above correlations were evaluated using measured support pressures, and the correlation coefficients of 0.96 and 0.95 were obtained for Eqs. (9.9) and (9.10), respectively (Goel et al., 1995a). For larger tunnels (diameter up to 9 m) in squeezing ground conditions the estimated support pressures (Eq. 9.10) match the measured values.

EFFECT OF TUNNEL SIZE ON SUPPORT PRESSURE

Prediction of support pressures in tunnels and the effect of tunnel size on support pressure are the two most important problems in tunnel mechanics and have attracted the attention of many researchers. The effect of tunnel size on support pressure presented in this chapter is described in Goel, Jethwa, and Dhar (1996).

TABLE 9.3 Correction Factor for Tunnel Closure in Eq. (9.10)

S. No.	Degree of squeezing	Normalized tunnel closure (%)	f(N)
1	Very mild squeezing ($275\ N^{0.33} \cdot B^{-0.1} < H < 360\ N^{0.33} \cdot B^{-0.1}$)	1–2	1.5
2	Mild squeezing ($360\ N^{0.33} \cdot B^{-0.1} < H < 450\ N^{0.33} \cdot B^{-0.1}$)	2–3	1.2
3	Mild to moderate squeezing ($450\ N^{0.33} \cdot B^{-0.1} < H < 540\ N^{0.33} \cdot B^{-0.1}$)	3–4	1.0
4	Moderate squeezing ($540\ N^{0.33} \cdot B^{-0.1} < H < 630\ N^{0.33} \cdot B^{-0.1}$)	4–5	0.8
5	High squeezing ($630\ N^{0.33} \cdot B^{-0.1} < H < 800\ N^{0.33} \cdot B^{-0.1}$)	5–7	1.1
6	Very high squeezing ($800\ N^{0.33} \cdot B^{-0.1} < H$)	>7	1.7

N = rock mass number; H = tunnel depth in meters; B = tunnel width in meters.
Tunnel closure depends significantly on the method of excavation. In highly squeezing ground condition, heading and benching method of excavation may lead to tunnel closure >8%.
Source: Goel et al., 1995a.

Various empirical approaches for predicting support pressures have recently been developed. Some researchers demonstrated that support pressure is independent of tunnel size (Daemen, 1975; Jethwa, 1981; Barton et al., 1974; Singh et al., 1992), whereas other researchers advocated that support pressure is directly dependent on tunnel size (Terzaghi, 1946; Deere et al., 1969; Wickham, Tiedmann, & Skinner, 1972; Unal, 1983). A review on the effect of tunnel size on support pressure with a concept proposed by Goel (1994) is presented in this chapter.

Review of Existing Approaches

Empirical approaches of estimating support pressure are presented in Table 9.4 to study the effect of tunnel size on support pressure. A discussion is presented in the next section.

Influence of Shape of the Opening

The empirical approaches listed in Table 9.4 were developed for flat roofs and arched roofs. For an underground opening with a flat roof, the support pressure is generally found to vary with the width or size of the opening, whereas in an opening with an arched roof the support pressure is found to be independent of tunnel size (Table 9.4). The RSR system of Wickham et al. (1972) is an exception, probably because the conservative system was not backed by actual field measurements for caverns. The mechanics suggest that the normal forces and therefore the support pressure will be more for rectangular opening with a flat roof by virtue of the detached rock block in the tension zone, which is free to fall.

TABLE 9.4 Important Empirical Approaches and Their Recommendations

Approach	Results based on	Recommendations
Terzaghi (1946)	a. Experiments in sands b. Rectangular openings with flat roof c. Qualitative approach	Support pressure increases with the opening size
Deere et al. (1969)	a. Based on Terzaghi's theory and classification on the basis of RQD	Support pressure increases with the opening size
Wickham et al. (1972) RSR system	a. Arched roof b. Hard rocks c. Quantitative approach	Support pressure increases with the opening size
Barton et al. (1974) Q-system	a. Hard rocks b. Arched roof c. Quantitative approach	Support pressure is independent of the opening size
Unal (1983) using RMR of Bieniawski (1976)	a. Coal mines b. Rectangular openings with flat roof c. Quantitative approach	Support pressure increases with the opening size
Singh et al. (1992)	a. Arched roof (tunnel/cavern) b. Both hard and weak rocks c. Quantitative approach	Support pressure is observed to be independent of the opening size (2–22 m)

Source: Goel et al., 1996.

Influence of Rock Mass Type

Support pressure is directly proportional to the size of the tunnel opening with weak or poor rock masses, whereas in good rock masses the situation is reversed (Table 9.4). Hence, it can be inferred that the applicability of an approach developed for weak or poor rock masses has a doubtful application in good rock masses.

Influence of In Situ Stresses

Rock mass number (N) does not consider in situ stresses, which govern the squeezing or rock burst conditions; instead, the height of overburden is accounted for in Eqs. (9.9) and (9.10) for estimation of support pressures. Thus, in situ stresses are indirectly considered.

Goel et al. (1995a) evaluated the approaches of Barton et al. (1974) and Singh et al. (1992) using the measured tunnel support pressures from 25 tunnel sections. They found that the approach of Barton et al. (1974) is unsafe in squeezing ground conditions and the reliability of the approaches of Singh et al. (1992) and Barton et al. (1974) depend upon the rating of Barton's SRF. Also found is that the approach of Singh et al. (1992) is unsafe for larger tunnels ($B > 9$ m) in squeezing ground conditions (see the section Correlation by Singh et al. (1992) in Chapter 8). Kumar (2002) evaluated many classification systems and found rock mass number to be the best from the case history of the NJPC tunnel in India.

TABLE 9.5 Effect of Tunnel Size on Support Pressure

S. No.	Type of rock mass	Increase in support pressure due to increase in tunnel span or diameter from 3 to 12 m
A. Tunnels with arched roof		
1	Non-squeezing ground conditions	Up to 20% only
2	Poor rock masses/squeezing ground conditions (N = 0.5 to 10)	20–60%
3	Soft-plastic clays, running ground, flowing ground, clay-filled moist fault gouges, slickensided shear zones (N = 0.1 to 0.5)	100–400%
B. Tunnels with flat roof (irrespective of ground conditions)		400%

Source: Goel et al., 1996.

New Concept on Effect of Tunnel Size on Support Pressure

Equations (9.9) and (9.10) have been used to study the effect of tunnel size on support pressure, which is summarized in Table 9.5.

It is cautioned that the support pressure is likely to increase significantly with the tunnel size for tunnel sections excavated in the following situations:

1. Slickensided zone
2. Thick fault gouge
3. Weak clay and shales
4. Soft plastic clays
5. Crushed brecciated and sheared rock masses
6. Clay-filled joints
7. Extremely delayed support in poor rock masses

Further, both Q and N are not applicable to flowing grounds or piping through seams. They also do not consider mineralogy (water sensitive minerals, soluble minerals, etc.).

CORRELATIONS FOR ESTIMATING TUNNEL CLOSURE

The behavior of concrete, gravel, and tunnel-muck backfills, commonly used with steel arch supports, has been studied. Stiffness of these backfills has been estimated using measured support pressures and tunnel closures. These results have been used to obtain effective support stiffness from the combined support system of steel ribs and backfill (Goel, 1994).

Based on measured tunnel closures from 60 tunnel sections, correlations have been developed for predicting tunnel closures in non-squeezing and squeezing ground conditions (Goel, 1994). These correlations are given in Eqs. (9.11) and (9.12).

Non-Squeezing Ground Condition

$$\frac{u_a}{a} = \frac{H^{0.6}}{28 \cdot N^{0.4} \cdot K^{0.35}} \% \tag{9.11}$$

Squeezing Ground Condition

$$\frac{u_a}{a} = \frac{H^{0.8}}{10 \cdot N^{0.3} \cdot K^{0.6}} \% \tag{9.12}$$

where u_a/a = normalized tunnel closure in percentage, K = effective support stiffness ($= p_v \cdot a/u_a$) in MPa, and H and a = tunnel depth and tunnel radius (half of tunnel width) in meters, respectively.

These correlations can also be used to obtain desirable effective support stiffness so that the normalized tunnel closure is contained within 4% (in the squeezing ground).

EFFECT OF TUNNEL DEPTH ON SUPPORT PRESSURE AND CLOSURE IN TUNNELS

In situ stresses are influenced by the depth below the ground surface (see Chapter 28). Support pressure and the closure for tunnels are also influenced by the in situ stresses. Therefore, the depth of the tunnel, or the overburden, is an important parameter while planning and designing tunnels. The effects of tunnel depth or the overburden on support pressure and closure in a tunnel have been studied using Eqs. (9.9) through (9.12) under both squeezing and non-squeezing ground conditions, which are summarized below.

1. Tunnel depth has a significant effect on support pressure and tunnel closure in squeezing ground conditions; however, it has a lesser effect in non-squeezing ground conditions (Eq. 9.9).
2. The effect of tunnel depth is higher on the support pressure than the tunnel closure.
3. The depth effect on support pressure increases with deterioration in rock mass quality, probably because the confinement decreases and the degree of freedom for the movement of rock blocks increases.
4. This study would be helpful to planners and designers when deciding on realigning a tunnel through better tunneling media or a lesser depth or both to reduce the anticipated support pressure and closure in tunnels.

APPROACH FOR OBTAINING GROUND REACTION CURVE

According to Daemen (1975), the ground reaction curve (GRC) is quite useful for designing the supports for tunnels in squeezing ground conditions. An easy-to-use empirical approach for obtaining the GRC has been developed using Eqs. (9.10) and (9.12) for tunnels in squeezing ground conditions. The approach is explained in Example 9.2.

Example 9.2

The tunnel depth (H) and the rock mass number (N) have been assumed as 500 m and 1, respectively, and the tunnel radius (a) as 5 m. The radial displacement of the tunnel is u_a for a given support pressure $p_v(sq)$.

GRC Using Eq. (9.10)

In Eq. (9.10), as described earlier, f(N) is the correction factor for tunnel closure. For different values of permitted normalized tunnel closure (u_a/a), different values of f(N) are proposed in Table 9.3. The first step is to choose any value of tunnel wall displacement (u_a) in column 1 of Table 9.6. Then the correction factor f(N) is found from Table 9.3 as shown in column 2 of Table 9.6. Finally, Eq. (9.10) yields the support pressure in the roof (p_v) as mentioned in column 3 of Table 9.6. Using Table 9.3 and Eq. (9.10), the support pressures p_v(sq) have been estimated for the assumed boundary conditions and for various values of u_a/a (column 1) as shown in Table 9.6. Subsequently, using the value of p_v (column 3) and u_a/a (column 1) from Table 9.6, GRC has been plotted for u_a/a up to 5% (Figure 9.3).

This approach is simple, reliable, and user friendly because the values of the input parameters can be easily obtained in the field.

TABLE 9.6 Showing Calculations for Constructing GRC Using Eq. (9.10)

Assumed u_a/a (%) (1)	Correction factor (f) (2)	p_v(sq) from Eq. (9.10) (MPa) (3)
0.5	2.7	0.86
1	2.2	0.7
2	1.5	0.475
3	1.2	0.38
4	1.0	0.317
5	0.8	0.25

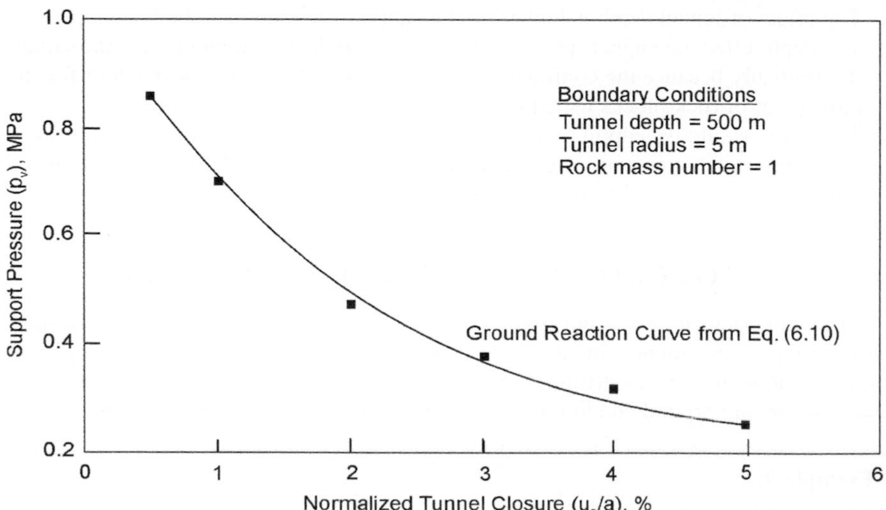

FIGURE 9.3 GRC obtained from Eq. (9.10).

COEFFICIENT OF VOLUMETRIC EXPANSION OF FAILED ROCK MASS

The ground response (reaction) curve depends upon the strength parameters of rock mass and also the coefficient of volumetric expansion of rock mass (k) in the broken zone. Jethwa (1981) estimated values of k as listed in Table 9.7. A higher degree of squeezing was associated with higher k values.

Example 9.3

A rock mass has three joint sets, each spaced at 15 cm. The joints are rough and the joint profile is almost planar. The joint surface of critical joints is altered with a sandy clay coating. The rock mass is moist only. The plan is to design a road tunnel of 9 m diameter at a depth of 350 m. Find out the ground condition likely to be encountered. If it is a squeezing condition, what would be the safe depth to avoid the squeezing condition? Also estimate the support pressure and the supports to be used.

Each joint set of the three joint sets has a joint spacing of 15 cm (joint frequencies = 6 joints per meter). Therefore, volumetric joint count (J_v) = 6 × 3 = 18 (Eq. 4.3). Using volumetric joint, RQD = 115 − 3.3 J_v = 55%. J_n = 9 (three joint sets); J_r = 1.5 (rough planar); J_a= 3.0 (sandy clay coating); J_w = 1.0 (moist only); SRF = 1.0 (competent rock, medium stress); and for a road tunnel, ESR = 1.0. Using Eq. (8.1b), Q = 3.05 (approximately) 3.00. Using Eq. (8.2), vertical support pressure = 0.092 MPa; since SRF = 1, therefore, rock mass number N = Q = 3.0.

For N = 3.0 and tunnel diameter 9 m, the safe tunnel depth to avoid squeezing ground condition is 320 m (Eq. 7.4). Any tunnel having rock cover more than 320 m may face squeezing ground condition in rock masses with N = 3.0. In this example, the tunnel depth is 350 m, hence a squeezing condition is expected. To avoid squeezing, either design the tunnel with a cover of less than 320 m or reduce its diameter. A tunnel diameter of 3.5 m would encounter a non-squeezing ground condition (Eq. 7.4). This is corroborated by an unsupported span of 3.0 m obtained from Barton's approach (Eq. 8.12).

The support pressure (Eq. 9.10) using rock mass number N and considering f(N) value as 1.5 (allowing 1–2% normalized tunnel closure/deformation, Table 9.3) is 0.17 MPa.

TABLE 9.7 Coefficient of Volumetric Expansion of Failed Rock Mass (k) within the Broken Zone

S. No.	Rock type	k
1	Phyllites	0.003
2	Claystones/siltstones	0.01
3	Black clays	0.01
4	Crushed sandstones	0.004
5	Crushed shales	0.005
6	Metabasics (Goel, 1994)	0.006

Source: Jethwa, 1981.

The support pressure obtained using rock mass number, tunnel depth, and tunnel size in squeezing ground condition is almost two times that obtained from Barton's approach.

The difference may occur because of (1) incorrect estimation of the SRF rating, and (2) the effect of tunnel size in squeezing conditions.

In a mild squeezing condition (Table 9.3 and 1–2% normalized tunnel deformation), SRF should be 5.0 (Table 8.6). With this SRF, Q = 3.0/5 = 0.6. Accordingly, the vertical support pressure using Eq. (8.2) is 0.16 MPa, which is almost equal to the support pressure previously obtained from Eq. (9.10). Accordingly, the supports are designed using Figure 8.5 for Q = 0.6 and equivalent dimension = 9/1 = 9. The supports, thus obtained, are 10 cm thick SFRS with 3.5 m long rock bolts at 1.6 m center to center (support category 6).

The above example highlights that the rock mass number (N) approach is found to be complimentary to the Q-system.

REFERENCES

Abad, J., Caleda, B., Chacon, E., Gutierrez, V., & Hidalgo, E. (1984). Application of geomechanical classification to predict the convergence of coal mine galleries and to design their supports. In *5th International Congress on Rock Mechanics* (pp. 15–19). Melbourne, (E), Australia.

Barton, N., Lien, R., & Lunde, J. (1974). Engineering classification of rock masses for the designs of tunnel supports. In *Rock mechanics* (Vol. 6, pp. 189–236). New York: Springer-Verlag.

Bieniawski, Z. T. (1976). Rock mass classifications in rock engineering. In *Proceedings of the Symposium on Exploration for Rock Engineering* (pp. 97–106 in Bieniawski, 1984). Rotterdam: A.A. Balkema.

Bieniawski, Z. T. (1984). *Rock mechanics design in mining and tunneling* (p. 272). Rotterdam: A.A. Balkema.

Cameron-Clarke, I. S., & Budavari, S. (1981). Correlation of rock mass classification parameters obtained from borecore and in situ observations. *Engineering Geology*, *17*, 19–53.

Daemen, J. J. K. (1975). *Tunnel support loading caused by rock failure*. Ph. D. Thesis. Minneapolis, MN: University of Minnesota.

Deere, D. U., Peck, R. B., Monsees, J. E., & Schmidt, B. (1969). *Design of tunnel liners and support system* (Final Report, University of Illinois, Urbana, for Office of High Speed Transportation, Contract No. 3-0152, p. 404). Washington, D.C.: U.S. Department of Transportation.

Goel, R. K. (1994). *Correlations for predicting support pressures and closures in tunnels* (p. 308). Ph.D. Thesis. Maharashtra, India: Nagpur University.

Goel, R. K., Jethwa, J. L., & Dhar, B. B. (1996). Effect of tunnel size on support pressure. Technical Note. *International Journal of Rock Mechanics and Mining Sciences—Geomechanics Abstracts*, *33*(7), 749–755.

Goel, R. K., Jethwa, J. L., & Paithankar, A. G. (1995a). Indian experiences with Q and RMR systems. *Tunnelling and Underground Space Technology*, *10*(1), 97–109.

Goel, R. K., Jethwa, J. L., & Paithankar, A. G. (1995b). Correlation between Barton's Q and Bieniawski's RMR—A new approach. Technical Note. *International Journal of Rock Mechanics and Mining Sciences—Geomechanics Abstracts*, *33*(2), 179–181.

Hoek, E., & Brown, E. T. (1980). *Underground excavations in rocks* (p. 527). Institution of Mining and Metallurgy. London: Maney Publishing.

Jethwa, J. L. (1981). *Evaluation of rock pressure under squeezing rock conditions for tunnels in Himalayas* (p. 272). Ph.D. Thesis. Uttarakhand, India: IIT Roorkee.

Kaiser, P. K., Mackay, C., & Gale, A. D. (1986). Evaluation of rock classifications at B.C. Rail Tumbler Ridge Tunnels. In *Rock Mechanics and Rock Engineering* (Vol. 19, pp. 205–234). New York: Springer-Verlag.

Kumar, N. (2002). *Rock mass characterisation and evaluation of supports for tunnels in Himalaya* (p. 295). Ph.D. Thesis. Uttarakhand, India: WRDM, IIT Roorkee.

Lama, R. D., & Vutukuri, V. S. (1978). *Handbook on mechanical properties of rocks* (Vol. 2, p. 481). Clausthal, Germany: Trans Tech Publications.

Moreno Tallon, E. (1980). *Application de Las Classificaciones Geomechnicas a Los Tuneles de Parjares, II Cursode Sostenimientos Activosen Galeriasy Tunnels*. Madrid: Foundation Gomez-Parto [referred in Kaiser et al. (1986)].

Rutledge, J. C., & Preston, R. L. (1978). Experience with engineering classifications of rock. In *Proceedings of the International Tunnelling Symposium* (pp. A3.1–A3.7). Tokyo.

Sari, D., & Pasamehmetoglu, A. G. (2004). Proposed support design, Kaletepe Tunnel, Turkey. *Engineering Geology*, *72*, 201–216.

Singh, B., Goel, R. K., Jethwa, J. L., & Dube, A. K. (1997). Support pressure assessment in arched underground openings through poor rock masses. *Engineering Geology*, *48*, 59–81.

Singh, B., Jethwa, J. L., Dube, A. K., & Singh, B. (1992). Correlation between observed support pressure and rock mass quality. *Tunnelling and Underground Space Technology*, *7*, 59–75.

Terzaghi, K. (1946). *Rock defeats and load on tunnel supports, introduction to rock tunnelling with steel supports* (R. V. Proctor & T. C. White, Eds.). Youngstown, OH: Commercial Shearing & Stamping Co.

Unal, E. (1983). *Design guidelines and roof control standards for coal mine roofs*. Ph.D. Thesis. Pennsylvania State University [reference Bieniawski (1984)].

Wickham, G. E., Tiedmann, H. R., & Skinner, E. H. (1972). Support determination based on geologic predictions. In *Proceedings of the Rapid Excavation Tunnelling Conference* (pp. 43–64). New York: AIME.

Chapter 10

Rock Mass Index

All things by immortal power near or far, hiddenly to each other are linked.

Francis Thompson

English Victorian Post

INTRODUCTION

There is no single parameter that can fully designate the properties of jointed rock masses. Various parameters have different significance, and only in an integrated form can they describe a rock mass satisfactorily.

Palmstrom (1995) proposed a rock mass index (RMi) to characterize rock mass strength as a construction material. The presence of various defects (discontinuities) in a rock mass that tend to reduce its inherent strength are taken care of in rock mass index (RMi), which is expressed as

$$RMi = q_c \cdot J_P \tag{10.1}$$

where q_c = the uniaxial compressive strength (UCS) of the intact rock material in MPa. J_P = the jointing parameter composed of mainly four jointing characteristics, namely, block volume or density of joints, joint roughness, joint alteration, and joint size. It is a reduction coefficient representing the effect of the joints in a rock mass. The value of J_P varies from almost 0 for crushed rock masses to 1 for intact rocks = s^n Hoek and Brown's criterion (Eq. 13.6). RMi = rock mass index denoting UCS of the rock mass in MPa.

SELECTION OF PARAMETERS USED IN RMi

For jointed rock masses, Hoek, Wood, and Shah (1992) reported that the strength characteristics are controlled by the block shape and size as well as their surface characteristics determined by the intersecting joints. They recommended that these parameters were selected to represent the average condition of the rock mass. Similar ideas have been proposed earlier by Tsoutrelis, Exadatylos, and Kapenis (1990) and Matula and Holzer (1978).

This does not mean that the properties of the intact rock material should be disregarded in rock mass characterization. After all, if joints are widely spaced or if an intact rock is weak, the properties of the intact rock may strongly influence the gross behavior of the rock mass. The rock material is also important if the joints are discontinuous. In addition, the rock description includes the geology and the type of material at the site, although rock properties in many cases are downgraded by joints. Keep in mind that the properties of rocks have a profound influence on the formation and development of joints. Petrological data can make an important contribution toward the prediction of mechanical performance, provided that one looks beyond the rock names at the observations on which they are based (Franklin, Broch, & Walton, 1970). Therefore, it is important to retain the names for the different rock types because they show relative indications of their inherent properties (Piteau, 1970).

These considerations and the study of more than 15 different classification systems have been used by Palmstrom (1995) when selecting the following input parameters for RMi:

1. Size of the blocks delineated by joints—measured as block volume, V_b
2. Strength of the block material—measured as UCS, q_c
3. Shear strength of the block faces—characterized by factors for the joint characteristics, jR and jA (Tables 10.1 and 10.3)
4. Size and termination of the joints—given as their length and continuity factor, jL (Table 10.2)

CALIBRATION OF RMi FROM KNOWN ROCK MASS STRENGTH DATA

It is practically impossible to carry out triaxial or shear tests on rock masses at a scale that is the same size as the underground excavations (Hoek & Brown, 1988). As the RMi is meant to express the compressive strength of a rock mass, a calibration of the same is necessary.

The UCS of intact rock, q_c, is defined and can be determined within a reasonable accuracy. The jointing parameter (J_P), however, is a combined parameter made up

TABLE 10.1 The Joint Roughness Found from Smoothness and Waviness

Small-scale smoothness* of joint surface	Large-scale waviness of joint plane				
*(The ratings in **bold** are similar to Jr in the Q-system)*	*Planar*	*Slightly undulating*	*Undulating*	*Strongly undulating*	*Stepped or interlocking*
Very rough	2	3	4	6	6
Rough	**1.5**	2	**3**	4.5	6
Smooth	**1**	1.5	**2**	3	4
Polished or slickensided*	**0.5**	1	**1.5**	2	3

For filled joints: jR = 1; for irregular joints a rating of jR = 6 is suggested

**For slickensided surfaces the ratings given cover possible movement along the lineation. (For movements across lineation, a rough or very rough rating should be applied for the surface.)*

Source: Palmstrom, 2000.

TABLE 10.2 The Joint Length and Continuity Factor (jL)

Joint length (m)	Term	Type	jL: *Continuous joints*	jL: *Discontinuous joints***
<0.5	Very short	Bedding/foliation parting	3	6
0.1–1.0	Short/small	Joint	2	4
1–10	Medium	Joint	1	2
10–30	Long/large	Joint	0.75	1.5
>30	Very long/ large	Filled joint, seam or shear*	0.5	1

**Often a singularity (special feature), and should in these cases be treated separately.*
***Discontinuous joints end in massive rock mass.*

Source: Palmstrom, 1996, 2000.

of the block volume, V_b, which can be found from field measurements, and the joint condition factor, jC, which is the result of three independent joint parameters (roughness, alteration, and size, Eq. 4.5a).

Results from large scale tests and field measurements of rock mass strength have been used to determine how V_b and jC can be combined to express the jointing parameter, J_P. Calibration has been performed using known test results of the UCS and the inherent parameters of the rock mass. The values for V_b and J_P are plotted in Figure 10.1, and the lines representing jC have been drawn. These lines are expressed as

$$J_P = 0.2\ (jC)^{0.5} \cdot (V_b)^D \tag{10.2}$$

where V_b is given in m^3 and $D = 0.37 \cdot jC^{-0.2}$.

Joint condition factor (jC) is correlated with jR, jA, and jL as follows:

$$jC = jL(jR/jA) \tag{10.3}$$

Various parameters of RMi and their combination in the RMi are shown in Figure 10.2, whereas the ratings of joint roughness (jR), joint size and termination (jL), and joint alteration (jA) are listed in Tables 10.1, 10.2, and 10.3, respectively. Joint roughness (jR) together with joint alteration (jA) define the friction angle as in the Q-system of Barton, Lien, and Linde (1974) in Chapter 8. The classification of RMi is presented in Table 10.4.

For example, jC and J_P are most commonly given as

$$jC = 0.2\ V_b^{0.37} \text{ and } J_P = 0.28\ V_b^{0.32}$$

For jC = 1.75 the jointing parameter can simply be expressed as

$$J_P = 0.25\ (V_b)^{0.33}$$

and for jC = 1 the jointing parameter from Eq. (10.2) is expressed as

$$J_P = 0.2\ V_b^{0.37}$$

TABLE 10.3 Characterization and Rating of the Joint Alteration Factor

Term	Description	jA
	A. Contact between rock wall surfaces	
Clean joints		
Healed or welded joints	Softening, impermeable filling (quartz, epidote, etc.)	0.75
Fresh joint walls	No coating or filling on joint surface, except from staining (rust)	1
Alteration of joint wall		
i. 1 grade more altered	The joint surface exhibits one class higher alteration than the rock	2
ii. 2 grade more altered	The joint surface shows two classes higher alteration than the rock	4
Coating or thin filling		
Sand, silt, calcite, etc.	Coating of friction materials without clay	3
Clay, chlorite, talc, etc.	Coating of softening and cohesive minerals	4

B. Filled joints with partial or no contact between the joint wall surfaces

Type of filling material	Description	Partial wall contact (thin filling <5 mm*)	No wall contact (thick filling or gouge)
Sand, silt, calcite, etc. (non-softening)	Filling of friction material without clay	4	8
Compacted clay materials	"Hard" filling of softening and cohesive materials	6	6–10
Soft clay materials	Medium to low over-consolidation of filling	8	12
Swelling clay materials	Filling material exhibits clear swelling properties	8–12	13–20

**Based on joint thickness division in the RMR system (Bieniawski, 1973).*

Source: Palmstrom, 1996, 2000.

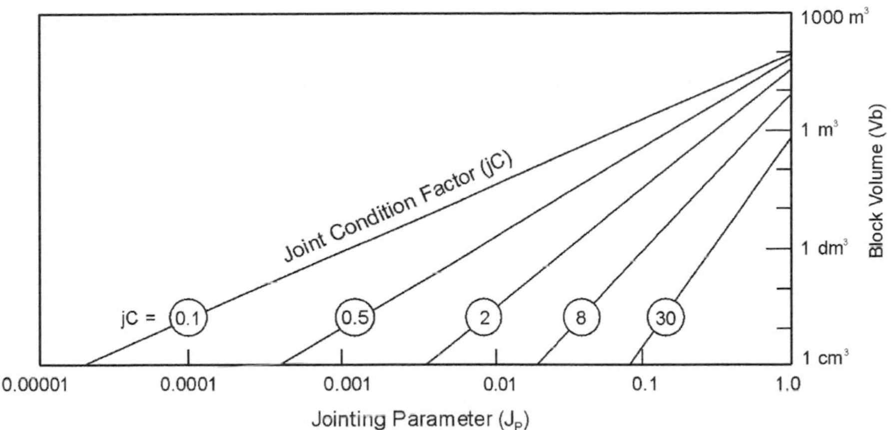

FIGURE 10.1 The graphical combination of block volume (V_b), joint condition factor (jC), and jointing parameter (J_P).

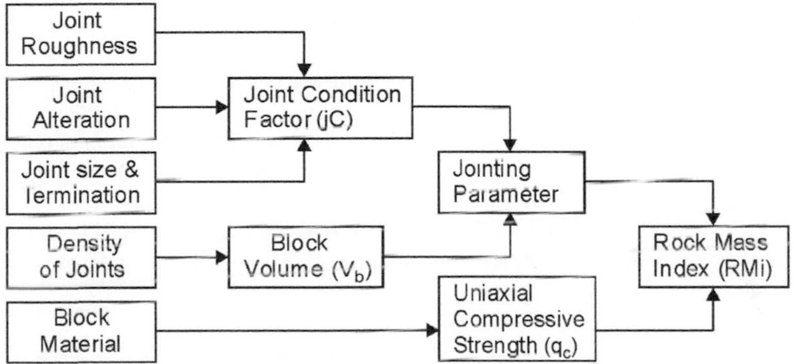

FIGURE 10.2 The combination of the parameters used in RMi. *(From Palmstrom, 1996)*

SCALE EFFECT

Significant scale effects are generally involved when a sample size is enlarged from laboratory size to field size (Figure 3.1). From the calibration described earlier, RMi is related to large samples where the scale effect has been included in J_P. The joint size factor (jL) is also a scale variable. However, for massive rock masses where the jointing parameter $J_P \approx 1$, the scale effect for the UCS (q_c) must be accounted for as q_c is related to the 50 mm sample size. Barton (1990) suggested from data presented by Hoek and Brown (1980) and Wagner (1987) that the actual compressive strength for large field samples with diameter (d, measured in millimeters) may be determined using the following equation (Figure 10.3):

$$q_c = q_{co}(50/d)^{0.2} = q_{co}(0.05/Db)^{0.2} = q_{co} \cdot f \tag{10.4}$$

where q_{co} is the UCS for a 50 mm sample size.

Equation (10.4) is valid for a sample diameter up to several meters, and may, therefore, be applied for massive rock masses. Thus, $f = (0.05/Db)^{0.2}$ is the scale factor for compressive strength. The approximate block diameter in Eq. (10.4) may be found

TABLE 10.4 Classification of RMi

Term		
For RMi	*Related to rock mass strength*	**RMi value**
Extremely low	Extremely weak	<0.001
Very low	Very weak	0.001–0.01
Low	Weak	0.01–0.1
Moderate	Medium	0.1–1.0
High	Strong	1.0–10.0
Very high	Very strong	10–100
Extremely high	Extremely strong	>100

Source: Palmstrom, 1996.

from Db = $(V_b)^{0.33}$, or where a pronounced joint set occurs, simply by applying the spacing of this set.

Figure 10.4 shows the same diagram as Figure 10.1 where measurements other than block volume can also be applied to determine jC. These are shown in the upper left part of the diagram in Figure 10.4. Here, the volumetric joint count (J_v) for various joint sets (and/or block shapes) can be used instead of the block volume. Also, RQD can be used, but its inability to characterize massive rock and highly jointed rock masses leads to a reduced value of J_P.

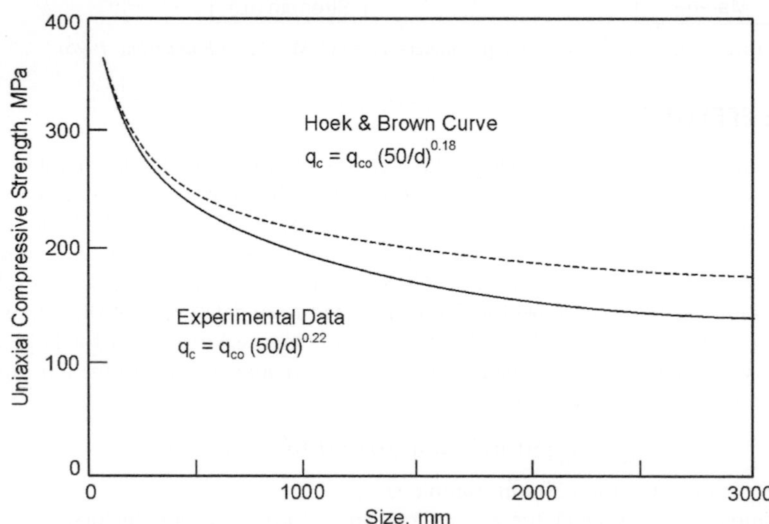

FIGURE 10.3 Empirical equations for scale effect of uniaxial compressive strength. *(From Barton, 1990, based on data from Hoek and Brown, 1980, and Wagner, 1987)*

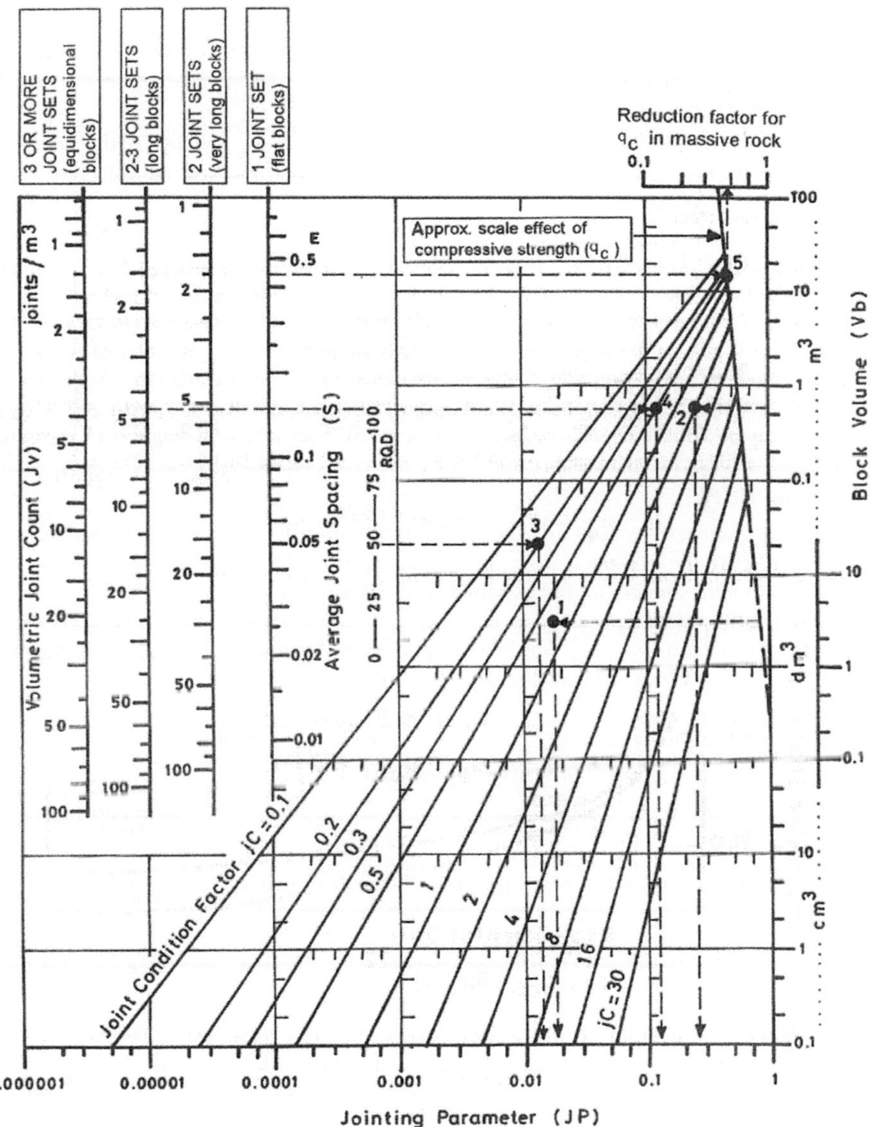

FIGURE 10.4 The jointing parameter J_P found from the joint condition factor jC, various measurements of jointing intensity (V_b, J_v, RQD), and UCS (q_c). *(From Palmstrom, 1996)*

EXAMPLES (PALMSTROM, 1995)

Example 10.1

Block volume has been measured as $V_b = 0.003\ m^3$. From the following condition and using Tables 10.1–10.3, the value of joint condition factor is worked out as $jC = 0.75$ based on:

- Rough joint surfaces and small undulations of the joint wall which give $jR = 3$.
- Clay-coated joints, that is, $jA = 4$; 3–10 m long; continuous joints give $jL = 1$.

On applying the values for V_b and jC in Figure 10.4, a value of $J_P = 0.02$* is found. With a compressive strength of the rock $q_c = 150$ MPa, the value of RMi = 3 (strong rock).

*Using Eq. (10.2), a value of $J_P = 0.018$ is found.

Example 10.2

The block volume $V_b = 0.63\ m^3$. The joint condition factor $jC = 2$ is determined from Tables 10.1–10.3 based on:

- Smooth joint surfaces and planar joint walls which give $jR = 4$.
- Fresh joints, that is, $jA = 1$; and 1–3 m long discontinuous joints, that is, $jL = 3$.

From Figure 10.4 the value $J_P = 0.25$* is found. With a compressive strength $q_c = 50$ MPa, the value of RMi = 12.5 (very strong rock).

*$J_P = 0.24$ is found using Eq. (10.2).

Example 10.3

Values of RQD = 50 and $jC = 0.2$ give $J_P = 0.015$ as shown in Figure 10.4.

Example 10.4

Two joint sets spaced 0.3 m and 1 m and some random joints have been measured. The volumetric joint count $J_v = (1/0.3) + (1/1) + 0.5^* = 4.5$.

With a joint condition factor $jC = 0.5$, the jointing parameter $J_P = 0.12$ (using the columns for 2–3 joint sets in Figure 10.4).

*Assumed influence from the random joints.

Example 10.5

Jointing characteristics: one joint set with spacing S = 0.45 m and $jC = 8$.

For the massive rock; the value of J_P is determined from the reduction factor for compressive strength f = 0.45. For a rock with $q_c = 130$ MPa the value of RMi = 59.6 (very strong rock mass).

APPLICATIONS OF RMi

Figure 10.5 shows the main areas of RMi application together with the influence of its parameters in different fields. RMi values cannot be used directly in classification systems as many of them are composed of their own systems. Some of the input parameters in RMi are similar to those used in the other classifications and may then be applied more or less directly.

The jointing parameter (J_P) in RMi is similar to the constant s ($= J_P^2$) in the Hoek-Brown failure criterion (Eq. 13.6) for rock masses. From V_b and jC, Cai et al. (2004) quantified Geological Strength Index (GSI) as per Eq. (26.5). The rock mass strength characteristics found from RMi can also be applied for numerical characterization in the New Austria Tunneling Method (NATM) as well as for input to prepare ground response (reaction) curves (Table 10. 5).

Palmstrom (1995) claims that the application of RMi in rock support involves a more systematized collection and application of the input data. RMi also uses a clearer definition of the different types of ground. It covers a wider range of ground conditions and includes more variables than the two main classification systems—RMR and the Q-system.

Palmstrom and Singh (2001) suggested correlations between modulus of deformation and RMi ($E_d = 7\ RMi^{0.4}$, GPa for RMi > 1).

BENEFITS OF USING RMi

As claimed by Palmstrom (1996), some of the benefits of the RMi system in rock mechanics and rock engineering are

- Enhances the accuracy of the input data required in rock engineering by its systematic approach of rock mass characterizations.
- Easily used for rough estimates when limited information about the ground condition is available, for example, in early stages of a feasibility design of a project where rough estimates are sufficient.

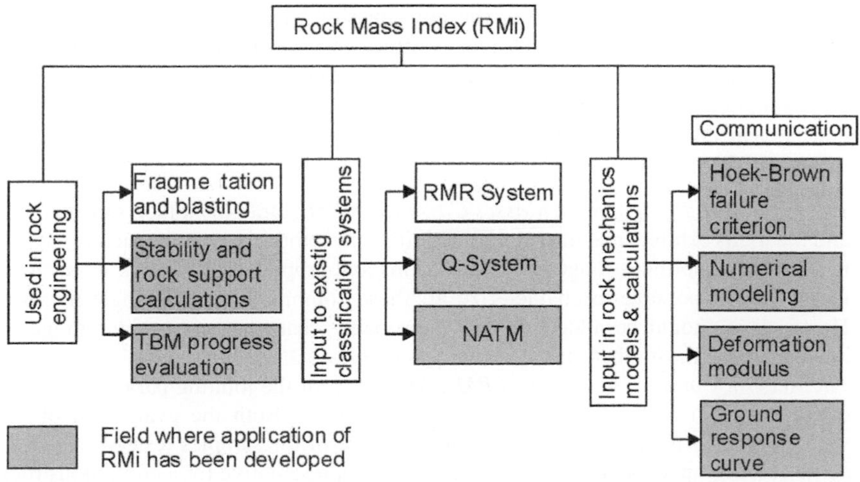

FIGURE 10.5 Main applications of RMi in rock mechanics and rock engineering. *(From Palmstrom, 1996)*

TABLE 10.5 Suggested Numerical Division of Ground According to NATM

S. No.	NATM class	Rock mass/ground properties represented by J_P	Competency factor ($Cg = RMi/\sigma_\theta$)
1	Stable	Massive ground ($J_P > 0.5$)	>2
2	Slightly raveling	$0.2 < J_P < 0.6$	>1
3	Raveling	$0.05 < J_P < 0.2$	>1
4	Strongly raveling	$J_P < 0.05$	0.7–2.0
5	Squeezing	Continuous ground	0.35–0.7
6	Strongly squeezing	Continuous ground	<0.35

σ_θ = maximum tangential stress along tunnel periphery.

- Well suited for comparisons and exchange of knowledge between different locations, as well as in general communication.
- Offers a stepwise system suitable for engineering judgment.
- Easier and more accurate to find the values of s (= $J_P{}^2$ or $J_P{}^{1/n}$) using the RMi system than the methods outlined by Hoek and Brown (1980), which incorporate use of the RMR or the Q-system (see Chapter 26).
- Covers a wide spectrum of rock mass variations and therefore has wider applications than other rock mass classification and characterization systems.
- Using parameters in RMi can improve inputs in other rock mass classification systems and in NATM.

LIMITATIONS OF RMi

As RMi is restricted to express only the compressive strength of rock masses, it is possible to arrive at a simple expression, contrary to the general failure criterion for jointed rock masses developed by Hoek and Brown (1980) and Hoek et al. (1992). Because simplicity is preferred in the structure and in the selection of parameters in RMi, it is clear that such an index may result in inaccuracy and limitations, the most important of which are connected to

The Range and Types of Rock Masses Covered by RMi: Both the intact rock material and the joints exhibit great directional variations in composition and structure, which results in an enormous range in compositions and properties for a rock mass. It is, therefore, not possible to characterize all these combinations in a single number. However, it should be added that RMi probably characterizes a wider range of materials than most other classification systems.

The Accuracy in the Expression of RMi: The value of the jointing parameter (J_P) is calibrated from a few large-scale compression tests. Both the evaluation of the various factors (jR, jA, and V_b) in J_P and the size of the samples tested—which in some cases did not contain enough blocks to be representative for a continuous rock mass—have resulted in certain errors that are connected to the expression developed

for the J_P. In addition, the test results used were partly from dry and partly from wet samples, which may have further reduced the accuracy of the data. The value of RMi can, therefore, be approximate. In some cases the errors in the various parameters may partly neutralize each other. Strength is not a unique property of brittle materials. Bieniawski (1973) realized that widely different values of strengths are mobilized in slopes, foundations, and tunnels. As such, RMR takes the type of structure into account (see Chapter 6), But RMi does not. So mobilization factors are needed in J_P.
The Effect of Combining Parameters That Vary in Range: The input parameters to RMi express a range of variation related to changes in the actual representative volume of a rock mass. Combination of these variables in RMi (and any other classification system) may cause errors.

From the previous discussion, RMi in many cases will be inaccurate in characterizing the strength of such a complex assemblage of different materials and defects that make up a rock mass. For these reasons, RMi is regarded as a relative expression of rock mass strength. Kumar (2002) attempted to compare RMi and the Q-system and found that RMi is very conservative and Eq. (13.9) based on the Q-system gives a better assessment of strength enhancement in tunnels.

REFERENCES

Barton, N. (1990). Scale effects or sampling bias? In *International Workshop on scale effects in rock masses* (pp. 31–55). Rotterdam: Balkema. (Reprinted from A. Pinto da Cunha, Ed.). (1990). Scale effects in rock masses. In *Proceedings of the First International Workshop* (p. 532). Loen, June 7–8. Rotterdam: A. A. Balkema.

Barton, N., Lien, R., & Lunde, J. (1974). Engineering classification of rock masses for the design of rock support. In *Rock mechanics* (Vol. 6, pp. 189–236). New York: Springer-Verlag.

Bieniawski, Z. T. (1973). Engineering classification of jointed rock masses. *Transactions of the South African Institution of Civil Engineers*, *15*(12), 335–344.

Cai, M., Kaiser, P. K., Uno, H., Tasaka, Y., & Minami, M. (2004). Estimation of rock mass deformation modulus and strength of jointed hard rock masses using the GSI system. *International Journal of Rock Mechanics and Mining Sciences*, *41*, 3–19.

Franklin, J. A., Broch, E., & Walton, G. (1970). Logging the mechanical character of rock. *Transactions of the Institute of Mining and Metallurgy*, *A 80*, A1–A9.

Hoek, E., & Brown, E. T. (1980). *Underground excavations in rocks* (p. 527). Institution of Mining and Metallurgy. London: Maney Publishing.

Hoek, E., & Brown, E. T. (1988). The Hoek-Brown failure criterion—A 1988 update. In *15th Canadian Rock Mechanics Symposium* (pp. 31–38).

Hoek, E., Wood, D., & Shah, S. (1992). A modified Hoek-Brown failure criterion for jointed rock masses. In *International Conference EUROCK '92* (pp. 209–214). London.

Kumar, N. (2002). *Rock mass characterisation and evaluation of supports for tunnels in Himalaya* (p. 295). Ph.D. Thesis. Uttarakhand, India: WRDM, ITT, Roorkee.

Matula, M., & Holzer, R. (1978). Engineering topology of rock masses. In *Proceedings of Felsmekanik Kolloquium, Grunlagen ung Andwendung der Felsmekanik* (pp. 107–121). Karlsruhe, Germany.

Palmstrom, A. (1995). Characterising the strength of rock masses for use in design of underground structures. In *Conference of Design and Construction of Underground Structures* (pp. 43–52). New Delhi, India.

Palmstrom, A. (1996). RMi—A system for characterizing rock mass strength for use in rock engineering. *Journal of Rock Mechanics and Tunnelling Technology*, *1*(2), 69–108.

Palmstrom, A. (2000). Recent developments in rock support estimates by the RMi. *Journal of Rock Mechanics and Tunnelling Technology*, *6*(1), 1–24.

Palmstrom, A., & Singh, R. (2001). The deformation modulus of rock masses—Comparison between in situ tests and indirect estimates. *Tunnelling and Underground Space Technology*, *16*, 115–131.

Piteau, D. R. (1970). Geological factors significant to the stability of slopes cut in rock. In *Proceedings of the Symposium on Planning Open Pit Mines* (pp. 33–53). Johannesburg, South Africa.

Tsoutrelis, C. E., Exadatylos, G. E., & Kapenis, A. P. (1990). Study of the rock mass discontinuity system using photoanalysis. In *Proceedings of the Symposium on Mechanics of Jointed and Faulted Rock* (pp. 103–112), Vienna, Austria.

Wagner, H. (1987). Design and support of underground excavations in highly stressed rock. In *Proceedings of the 6th ISRM Congress* (Vol. 3). Montreal, Canada.

Chapter 11

Rate of Tunneling

Most human beings experience a certain amount of fear when confronted with change. The level varies from moderate dislike to intense hatred. One of the few things stronger than fear of change is love of money. Structure the change so that it provides a potential for profit and the change will happen.

At some point in time the urgings of pundits, the theories of scientists and the calculations of engineers have to be translated into something that the miner can use to drive tunnel better, faster and cheaper. We shall call this change.

Excerpts of the report prepared by Robert F. Baker et al.

INTRODUCTION

The excavation of tunnels is affected by many uncertainties. The time of completion of tunneling projects is grossly underestimated in many cases, because proper evaluation of the factors that affect the rate of tunnel excavation is ignored. The factors affecting the blasting and drilling method of tunnel excavation are

1. Variation in ground/job conditions and geological problems encountered
2. Quality of management and managerial problems
3. Various types of breakdowns or holdups

The first of these is very important because the rate of tunnel driving is different for different types of ground conditions; for example, the tunneling rate is lower in poor ground conditions. Depending upon the ground conditions, different methods of excavation are adopted for optimum advance per round so that the excavated rock can be supported within the bridge action period or the stand-up time. Frequent changes in ground conditions seriously affect the tunneling rate because both the support and excavation method need to be changed. This is perhaps the reason why tunnel boring machines (TBMs) are not used very often for tunneling in the lower Himalayas.

The second factor affects the rate of tunneling differently due to different management conditions, even in the same type of ground condition. Poor management condition affects the tunneling rate more adversely than poor rock mass condition.

The third factor pertains to the breakdowns or holdups during various operations in the tunneling cycle. These holdups cause random delays. Based on the data collected from a number of projects, Chauhan (1982) proposed a classification for the realistic assessment of the rate of tunneling, which is presented in the following section.

CLASSIFICATION OF GROUND/JOB CONDITIONS FOR RATE OF TUNNELING

The rate of tunneling is seriously affected by the ground conditions. The factors, under the ground condition, affecting the rate of tunneling include (Terzaghi, 1946; Bieniawski, 1973, 1974; Barton, Lien, & Lunde, 1974):

1. Geology, such as type of rock, rock quality designation (RQD), joint system, dip and strike of strata, the presence of major fault or thrust zones and their frequencies and type, and rock mass properties
2. Method of excavation including blast pattern and drilling arrangement
3. Type of support system and its capacity
4. Inflow of water
5. Presence of inflammable gases
6. Size and shape of tunnel
7. Construction adits whether horizontal or inclined, their grade size, and length
8. High temperature in very deep tunnels ($H > 1000$ m) or thermic regions

Based on the previous factors affecting the rate of tunneling, ground conditions are classified into three categories: good, fair, and poor (Table 11.1). This means that for good ground conditions the rate of tunneling is higher and for poor ground conditions the rate of tunneling is lower. The job/ground conditions in Table 11.1 are presented in the order of their weightage to the rate of tunneling.

CLASSIFICATION OF MANAGEMENT CONDITIONS FOR RATE OF TUNNELING

The rate of tunneling may vary in the same ground condition depending upon management quality. The factors affecting management conditions include:

1. Overall job planning, including selection of equipment and the decision-making process
2. Training of personnel
3. Equipment availability including parts and preventive maintenance
4. Operating supervision
5. Incentives to workers
6. Coordination
7. Punctuality of staff
8. Environmental conditions
9. Rapport and communication at all levels

These factors affect the rate of tunneling both individually and collectively. Each factor is assigned a weighted rating. The maximum rating possible in each subgroup has also been assigned out of a possible 100 (Table 11.2), which represents ideal conditions. At a particular site the ratings of all the factors are added to obtain a collective classification rating for the management condition. Using this rating, the management condition has been classified into good, fair, and poor as shown in Table 11.3. The proposed classification system for management is valid for tunnels longer than 500 m, which are excavated by the conventional drilling and blasting method.

TABLE 11.1 Classification of Ground/Job Conditions

		Ground/job conditions		
S. No.	**Parameter**	***Good***	***Fair***	***Poor***
1	Geologic structure	Hard, intact, massive stratified or schistose, moderately jointed, blocky and seamy	Very blocky and seamy squeezing at moderate depth	Completely crushed, swelling and squeezing at great depth
2(a)	Point load strength index	>2 MPa	1–2 MPa	Index cannot be determined but is usually less than 1 MPa
2(b)	Uniaxial compressive strength	>44 MPa	22–44 MPa	<22 MPa
3	Contact zones	Fair to good or poor to good rocks	Good to fair or poor to fair rocks	Good to poor or fair to poor rocks
4	Rock quality designation (RQD)	60–100%	25–60%	<25%
5(a)	Joint formation	Moderately jointed to massive	Closely jointed	Very closely jointed
5(b)	Joint spacing	>0.2 m	0.05–0.2 m	<0.05 m
6(a)	Joint orientation	Very favorable, favorable, and fair	Unfavorable	Very unfavorable
6(b)	Strike of tunnel axis and dip with respect to tunnel driving	(i) Perpendicular; 20 to 90° along dip; 45 to 90° against dip	(i) Perpendicular; 20 to 45° against dip	(i) Parallel; 45 to 90°
		(ii) Parallel; 20 to 45°	(ii) Irrespective of strike; 0 to 20°	—
7	Inflammable gases	Not present	Not present	May be present

Continued

TABLE 11.1 Classification of Ground/Job Conditions—Cont'd

S. No.	Parameter	Ground/job conditions		
		Good	*Fair*	*Poor*
8	Water inflow	None to slight	Moderate	Heavy
9	Normal drilling depth/round	>2.5 m	1.2 m–2.5 m	<1.2 m
10	Bridge action period	>36 hrs	8–36 hrs	<8 hrs

The geologist's predictions based on investigation data and laboratory and site tests include information on parameters at S. Nos. 1 to 6. This information is considered adequate for classifying the job conditions approximately.

Source: Chauhan, 1982.

TABLE 11.2 Ratings for Management Factors for Long Tunnels

S. No.	Subgroup	Item	Maximum rating for		Remarks for improvement in management condition
			Item	*Subgroup*	
1	Overall job planning	(i) Selection of construction plant and equipment including estimation of optima size and number of machines required for achieving ideal progress	7		
		(ii) Adoption of correct drilling pattern and use of proper electric delays	6		
		(iii) Estimation and deployment of requisite number of workers and supervisors for ideal progress	5		
		(iv) Judicious selection of construction method, adits, location of portals, etc.	4		Horizontal adits sloping at the rate of 7% toward portal to be preferred to inclined adits or vertical shafts.
		(v) Use of twin rail track	2		
		(vi) Timely shifting of California switch at the heading	2	26	
2	Training of personnel	(i) Skill of drilling crew in the correct holding, alignment, and thrust application on drilling machines	4		Proper control of drilling and blasting will ensure a high percentage of advance from the given drilling depth and also good fragmentation of rock, which facilitates mucking operation.
		(ii) Skill of muck loader operator	4		
		(iii) Skill of crew in support erection	3		A skilled crew should not take more than a half hour for erection of one set of steel rib supports.

Continued

TABLE 11.2 Ratings for Management Factors for Long Tunnels—Cont'd

S. No.	Subgroup	Item	Maximum rating for *Item*	Maximum rating for *Subgroup*	Remarks for improvement in management condition
		(iv) Skill of blastman	2		
		(v) Skill of other crews	2	15	
3	Equipment availability and preventive maintenance	Time lost in tunneling cycle due to breakdowns of equipment including derailments, etc.			
		(i) up to 1 hour	12–15		
		(ii) 1–2 hours	9–11		
		(iii) 2–3 hours	6–8		
		(iv) >3 hours	0–5	15	
4	Operation supervision	(i) Supervision of drilling and blasting (effectiveness depends on location, depth, and inclination of drill holes; proper tamping; and use of blasting delays)	7		Improper drilling may result in producing: (i) Unequal depth of holes, which results in lesser advance per meter of drilling depth
					(ii) Wrong alignment of hole, which may lead to:
					(a) Overbreak due to wrong inclination of periphery holes
					(b) Secondary blasting due to wrong inclination of other than periphery holes
					Improper tamping of blast hole charge and wrong use of blasting delays result in improper blasting effects

		(ii) Supervision of muck loading/hauling system	3		Especially in a rail haulage system in which rapid feeding of mine cars to a loading machine at the heading is essential for increasing productivity of loader.
		(iii) Supervision of rib erection, blocking, and packing	3		
		(iv) Other items of supervision such as scaling, layout, etc.	2	15	
5	Incentive to workers	(i) Progress bonus	5		Define the datum monthly progress as that value which delineates good and fair management conditions for particular job conditions. Introduce bonus slabs for every additional 5 m progress and distribute the total monthly bonus thus earned among the workers on the basis of their importance, skill, and number of days worked during the month. The amount for each slab should be fixed so that these are progressive and each worker should get about 50% of his monthly salary as a progress bonus, if ideal monthly progress is achieved.
		(ii) Incentive bonus	2		This should be given for certain difficult and hazardous manual operations like rib erection/shear zone treatment, etc.
		(iii) Performance bonus	1		This should be given to the entire tunnel crew equally if the quarterly progress target is achieved.
		(iv) Achievement bonus	1	9	It is to be given for completion of whole project on schedule. It should be given to the whole construction crew and may be equal to one year's interest on capital cost.

Continued

TABLE 11.2 Ratings for Management Factors for Long Tunnels—Cont'd

S. No.	Subgroup	Item	Maximum rating for		Remarks for improvement in management condition
			Item	*Subgroup*	
6	Coordination	(i) Coordination of activities of various crews inside the tunnel	5		Coordination between designers and construction engineers should be given top priority. Designers should be boldly innovative.
		(ii) Use of CPM for overall perspective and control of the whole job	4	9	Safety saves money. Contingency and emergency plans should be ready before tunneling.
7	Environmental conditions and housekeeping	Proper lighting, dewatering, ventilation, provision of safety wear to workers, and general job cleanliness	4	4	
8	Punctuality of staff	(i) Prompt shift change-over at the heading	4		
		(ii) Loss of up to 1/3 hour in shift change-over	3		
		(iii) Loss of more than 1/3 hour in shift change-over	0–2	4	
9	Rapport and communication	Commitment, good rapport, and communication at all levels of working including top management and government level including human relations	3	3	Team spirit is the key to success in underground construction. The contractors have to be encouraged to succeed.

Source: Chauhan, 1982.

TABLE 11.3 Rating for Different Management Conditions

S. No.	Management condition	Rating
1	Good	80–100
2	Fair	51–79
3	Poor	≤50

Source: Chauhan, 1982.

It may be noted that the rate of tunneling can be easily improved by improving the management condition, which is manageable, unlike the ground conditions, which cannot be changed easily. So, it is necessary to pay at least equal, if not more, attention to the management condition than to the ground condition. Hence, there is an urgent need for management consultancy to improve the tunneling rate.

The key to success of tunnel engineers is the evolution of a flexible method of construction of the support system. The on-the-spot strengthening of a support system is done by spraying additional layers of shotcrete/SFRS or using long rock bolts in the unexpectedly poor geological conditions. This is a sound strategy of management in tunneling within the complex geological situations. Affection is the key to success in the management. Young engineers love challenging works. There should be no hesitation in throwing challenges to young engineers. Otherwise these young engineers may lose interest in routine management.

COMBINED EFFECT OF GROUND AND MANAGEMENT CONDITIONS ON RATE OF TUNNELING

A combined classification system for ground conditions and management conditions has been developed by Chauhan (1982). Each of the three ground conditions has been divided into three management conditions, thus nine categories have been obtained considering both ground and management conditions. The field data of six tunneling projects in the Indian Himalayas have been divided into these nine categories to study the combined effect. Each category has three performance parameters, including:

1. Actual working time (AWT)
2. Breakdown time (BDT)
3. Advance per round (APR)

A matrix of job and management factors has been developed from the data to evaluate tunnel advance rate (Table 11.4).

Ground and management factors in the matrix are defined as a ratio of actual monthly progress to achievable monthly progress under a corresponding set of ground and management conditions. Knowing the achievable production for a tunneling project, these factors could hopefully yield values of expected production under different management and geological conditions on each project.

In squeezing ground conditions, the rate of tunneling would be only 13% of the theoretical rate for poor management condition. Past experience suggests that

TABLE 11.4 Ground and Management Factors

	Management conditions		
Ground conditions	*Good*	*Fair*	*Poor*
Good	0.78	0.60	0.44
Fair	0.53	0.32	0.18
Poor	0.30	0.21	0.13

Source: Chauhan, 1982.

management tends to relax in good tunneling conditions and becomes alert and active in poor rock conditions.

Further studies are needed to update Tables 11.2 through 11.4 for modern tunneling technology, but trends are expected to be similar.

The management of projects funded by World Bank is an ideal example. They appoint international rock mechanics experts for their hydroelectric projects. In major state-funded projects, international experts on rock mechanics should be appointed on the Board of Consultants, because they help achieve self-reliance. Modern tunneling contracts contain clauses for contracting companies to arrange for the classification of rock masses, the decision of supports, and instrumentation by competent rock engineers or engineering geologists. Further, contractually there should be first and second contingency plans for better preparedness during tunneling hazards.

TUNNEL MANAGEMENT (SINGH, 1993)

Management is an art, demanding strength of character, intelligence, and experience. Deficiencies in management are, therefore, difficult to remove. Experience is not what happens to you, it is what you do with what happens to you. Everyone is potentially a high performer. Motivation comes from the top. What glorifies self-respect automatically improves one's efficiency. Often interference by the manager mars the initiative of young engineers. Feedback is essential to improve performance, just like feedback is very important for stability of the governing system in electronics. Efficient, clear communication of orders to concerned workers and their feedback is essential for management success. Computer networks and cell phones are used today for better informal rapport at a project site. The modern management is committed visible management. The defeatist attitude should be defeated. The leader should have the willpower to complete the vast project. There should be respect for individuals in the organization. The happier the individual, the more successful he will be. *If you want to be happy your whole life, love your work*. The right persons at the right place according to their interests contribute to success of projects (according to Dr. V.M. Sharma, AIMIL, India).

Tunnel construction is a complex, challenging, and hazardous profession. It demands a high skill of leadership, technology, and communication. On-the-spot decisions are needed in tunneling crises. Mutual respect between government, engineers, and contractors is necessary especially during privatization. Usually bad news does not travel upward to the executive management. The basic ingredient in any tunneling project management is trust. Quality consciousness should be the culture of a construction agency. Is quality work possible in government when there is lack of creative freedom? Work of good quality is possible by framing proper specifications in a contract document. The contractor's point of view is that payments should be made early for quick reinvestment. Unfortunately construction industries are unorganized in many countries. With the increasing trend for global organization, efficiency will go upward in the future.

Because no two construction jobs are alike, it is very difficult to evolve a system (of stockpiles of materials, fleet of tunneling machines, etc.) for a new project site. Construction problems vary so much from job to job that they defy management, machines, and known methods. Then a contractor uses ingenuity to design tools and techniques that will lead to success in tunneling. Machines may be used for a variety of other purposes with slight modifications. Excellent companies are really close to their customers (engineers) and regard them highly. Their survival depends upon the engineer's satisfaction.

Critical path analysis, if properly applied and used, can be a great help to any construction agency, especially in a tunneling job. Use of software for critical path analysis for cost control is most effective and economical, and then coordination among workers becomes easy. Naturally a management organization becomes more efficient during a crisis. Cost-effective consciousness must permeate all ranks of engineers and workers. Organization set-up is the backbone of a long tunneling project.

The completion of a hydro project is delayed when long lengths of tunnels have to be created in weak and complex geological conditions, so the idea of a substantial bonus for early completion is becoming more widespread.

POOR TENDER SPECIFICATIONS

Tendering for tunneling projects remains speculative since actual ground conditions encountered during construction often do not match the conditions shown in the tender specifications, particularly in the Himalayas, young mountains, and complex geological environments. The practice of adopting payment rates according to actual ground condition does not exist. Insufficient geological, hydrogeological, and geotechnical investigations and poor estimates invariably lead to owner–contractor conflicts, delays in projects, arbitration, and escalation of project cost to three times the original estimate. The following are some of the main reasons attributed to this poor tunneling scenario in developing nations:

1. Inadequate geological investigations and absence of rock mechanics appreciation before inviting a tender bid, which results in major geological surprises during execution.
2. Lack of proper planning and sketchy and incompetent preparation of designs at the pre-tender stage.
3. Unrealistic projection of cost estimates and cost benefit ratio and completion schedules at initial stages.
4. Inadequate infrastructure facilities at the site.
5. Unrealistic and unfair contract conditions and poor profit margins that lead to major disputes and delays in dispute resolution.
6. Lack of motivation and commitment on the part of owners, especially of government departments and public sector agencies.
7. Lack of specific provisions in the tender document regarding modern technology.
8. Lack of teamwork between the owner, the contractor, the geologist, and the rock mechanics expert.

9. Risk sharing between contractor and owner is generally unfair.
10. Lack of appropriate indigenous construction technology is seen in developing nations.

It is important to emphasize that although sufficient expertise is available in tunneling technology, the administration seldom takes advantage of the intellectual resources in the right perspective at the right time.

CONTRACTING PRACTICE

On some occasions, it is the inexperience or incompetence of the contractor that delays a project. Lack of strategy, weak project team, and inadequate attention from the top management sometimes also result in delays and slippage. In some cases, contractors are found ill equipped, cash poor, and lacking in professionalism. Just to grab the project deal, they compromise on rates. Finding very low profits when the work starts, they raise unreasonable claims and disputes to improve profit margin, which results in disputes followed with arbitration, delays, and time and cost overruns in some developing countries.

The following measures are suggested to avoid delays in project schedules and cost escalation due to contractors:

1. In the pre-bid meeting, an objective evaluation of potential contractors should be made and inefficient contractors should be eliminated at this stage.
2. The contract should be awarded to a group of contractors, each an expert in specific activities such as design, tunneling machines, construction, rock mechanics, geology, and so forth. Using this process, the project authorities will have the services of a team of competent contractors.
3. Contractors should hire trained and experienced staff and should upgrade technology on a continuous basis. They should take assistance during project commissioning from technical experts of R&D organizations. This will equip the contractors to handle major geological surprises, substantiate their claims, and economize their routine operations.

QUALITY MANAGEMENT BY INTERNATIONAL TUNNELING ASSOCIATION

Oggeri and Ova (2004) suggested the following principles of quality management for tunneling:

1. Quality in tunneling means knowledge. Knowledge is necessary to correctly fulfill the requirement of the design. Knowledge is necessary to better learn and "copy" what previous designers have done.
2. Experience, good contracts, professionalism, self-responsibility, and simple rules are required to reach the objectives of design and perform properly.
3. Successful planning is the key to a successful project.
4. Transfer of information both upward and downward in an organization, in a format understood by all, is the key issue.
5. *There is direct, linear relation between project quality and project cost.*
6. Design a strategy of tunneling in all possible ground conditions.
7. Tunneling projects are well suited for "on-the-job training," since large projects use state-of-the-art technology. Engineers should participate in international tunneling conferences and meet specialists and report their difficulties.

8. If a process is innovative, a testing program prior to the production should be conducted.
9. All along the project, coordinating technical features, economical results, contractual agreements, environmental effects, and safety standards is necessary to achieve significant results.
10. Correct choice is essential for the type of contract, conditions of the contract, financing, and procurement procedures for equipment.
11. Knowledge is transferred not only between parties during project phases, but to parties after completion of a project as well, including universities and other technical organizations.

REFERENCES

Barton, N., Lien, R., & Lunde, J. (1974). Engineering classification of rock masses for the design of tunnel supports. In *Rock mechanics* (Vol. 6, No. 4, pp. 189–236). New York: Springer-Verlag.

Bieniawski, Z. T. (1973). Engineering classification of jointed rock masses. *Transactions of the South African Institution of Civil Engineers, 15*, 335–342.

Bieniawski, Z. T. (1974). Geomechanics classification of rock masses and its application in tunnelling. In *Proceedings of the 3rd International Congress on Rock Mechanics* (Vol. VIIA, pp. 27–32). Denver, Colorado: ISRM.

Chauhan, R. L. (1982). *A simulation study of tunnel excavation*. Ph.D. Thesis. Uttarakhand, India: IIT Roorkee.

Oggeri, C., & Ova, G. (2004). Quality in tunnelling. *Tunnelling and Underground Space Technology, 19*, 239–272.

Singh, J. (1993). *Heavy construction planning, equipment and methods* (p. 1084). New Delhi: Oxford and IBH Publishing Co. Pvt. Ltd.

Terzaghi, K. (1946). Rock defects and load on tunnel supports. In R. V. Proctor & T. L. White (Eds.), *Introduction to rock tunnelling with steel supports* (p. 271). Youngstown, OH: Commercial Shearing & Stamping Co.

Chapter 12

Support System in Caverns

I believe that the engineer needs primarily the fundamentals of mathematical analysis and sound methods of approximation.

Th. Von Karman

SUPPORT PRESSURE

Large underground openings are called "caverns." Caverns are generally sited in good rock masses where the rocks are massive and dry, and the ground condition would be either self-supporting or non-squeezing (and generally $Q > 1$, $E_d > 2$ GPa except in the shear zones, but $H < 350 \cdot Q^{1/3}$ m).

To assess roof and wall support pressures the approaches discussed in Chapter 8 are reliable and can be adopted. The approach of Goel, Jethwa, and Paithankar (1995) in Chapter 9 has been developed for tunnels with diameters up to 12 m; therefore, its applicability for caverns with a diameter of more than 12 m is yet to be evaluated. The modified Terzaghi's theory of Singh, Jethwa, and Dube (1995a), as discussed in Chapter 5, may also be used to estimate the roof support pressures.

The 3D finite element analysis of the powerhouse cavern of the Sardar Sarovar hydroelectric project in India illustrates that the wall support pressures are smaller than the roof support pressures, the stiffness of the wall shotcrete is lower than the roof shotcrete. The value of p_{wall} away from the shear zone is approximately 0.07 to 0.11 p_{roof}, whereas in the area of the 2 m wide shear zone p_{wall} is about 0.20 to 0.50 p_{roof}. The predicted support pressures in the roof both away from and near the shear zone are approximately equal to the empirical ultimate support pressures for surrounding rock mass quality and mean value of rock mass quality, respectively (Samadhiya, 1998), as discussed in Chapter 2 in the section Treatment for Tunnels.

Roof support requirements (including bolt length and their spacing) can be estimated from the empirical approaches of Cording, Hendron, and Deere (1971); Hoek and Brown (1980); Barton et al. (1980); and Barton (1998). These approaches are based on the rule of thumb and do not include the rock mass type and the support pressure for designing the bolt length. It is pertinent to note that none of these approaches, except Barton's method and the modified Terzaghi's theory of Singh et al. (1995a), provide a criterion for estimating the support pressure for caverns.

The philosophy of rock reinforcement is to stitch rock wedges together and prevent them from sliding down from the roof and the walls. Empirical approaches based on rock mass classifications provide realistic bolt lengths in weak zones when compared with the results of the numerical analysis. In view of this, Singh et al. (1995b) presented the following approach to designing anchors/rock bolts for cavern walls in non-squeezing ground conditions. Park, Kim, and Lee (1997) used this design concept for four food storage caverns in Korea. The Himachal Pradesh (TM) software package based on this approach may be used for designing support systems for walls and roofs. It has been used successfully at the Ganwi mini hydel project in H.P. and several other projects in India. The software can also be used for tunnels in both non-squeezing and squeezing ground conditions.

WALL SUPPORT IN CAVERNS

The reinforced rock wall column (L > 15 m) has a tendency to buckle under tangential stress (Bazant, Lin, & Lippmann, 1993) due to the possibility of vertical crack propagation behind the reinforced rock wall (Figure 12.1). The length of anchors/rock bolts should be adequate to prevent the buckling of the rock wall column and hence the vertical crack propagation.

Thus, equating the buckling strength of the reinforced rock column (assuming both ends are fixed) and the average vertical (tangential) stress on the haunches along the bolt length, we obtain

$$\frac{l'_w}{L} > \left[\frac{F_{wall} \times 12\sigma_\theta}{4 \times \pi^2 E_d}\right]^{1/2} \tag{12.1}$$

$$l_w = l'_w + \frac{FAL}{2} + \frac{s_{bolt}}{4} - s_{rock} + d \tag{12.2}$$

where σ_θ = effective average tangential stress on haunches and is ≈ 1.5× overburden pressure; l_w = length of bolts/anchors in wall; l'_w = effective thickness of reinforced rock column ($l_w \geq l'_w$); and d = depth of damage of rock mass due to blasting (1–3 m). E_d = modulus of deformation of reinforced rock mass, which may be approximately equal to modulus of deformation of natural rock mass and

$$0.3\ H^{\alpha}\ 10^{(RMR-20)/38}\ \text{GPa (Verman, 1993)} \tag{12.3}$$

and

$$H^{0.2} \cdot Q^{0.36}\ \text{GPa for } Q < 10 \text{ (Singh et al., 1998)} \tag{12.4}$$

α = 0.16–0.30 (more for weak rocks) and F_{wall} = mobilization factor for buckling.

$$F_{wall} = 3.25\ p_{wall}^{0.10}\ \text{(for pretensioned bolts)}, \tag{12.5}$$

and

$$= 9.5\ p_{wall}^{-0.35}\ \text{(for anchors)}, \tag{12.6}$$

FAL = fixed anchor length to give pull-out capacity p_{bolt} (higher for poor rocks); s_{bolt} = spacing of bolts/anchors (= spacing of rows of bolts) and the square root of area of rock mass supported by one bolt; s_{rock} = average spacing of joints in rock mass; and L = height of the wall of cavern.

Singh, Fairhurst, and Christiano (1973), with the help of a computer model, showed that the ratio of the moment of inertia of bolted layers to that of unbolted layers increases with both a decrease in thickness and the modulus of deformation of rock layers. The experiments of Fairhurst and Singh (1974) also confirmed this prediction for ductile

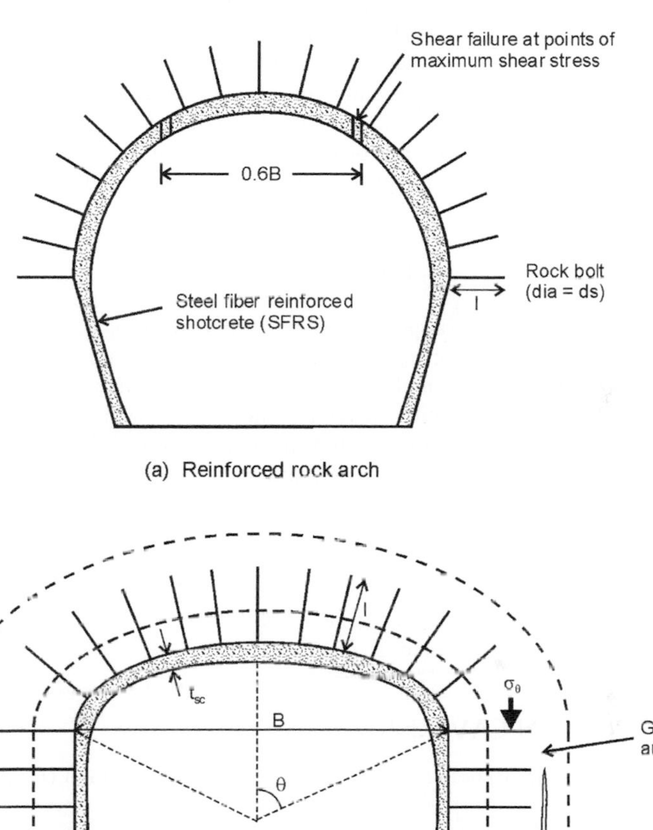

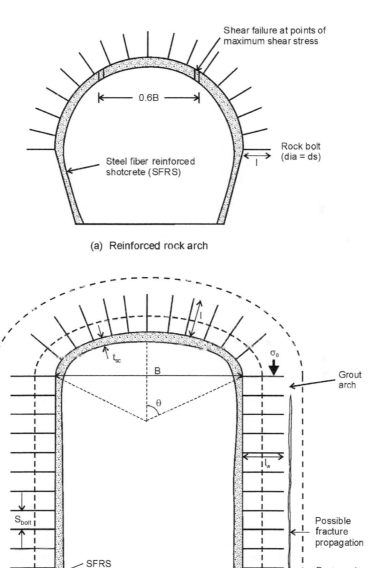

FIGURE 12.1 Design of support system for underground openings: (a) reinforced rock arch and (b) reinforced rock frame.

layers. The mobilizing factor for anchors (Eq. 12.6) simulates this tendency empirically as F_{wall} decreases with a decrease in rock mass quality and p_{wall}. In other words, rock anchors are more effective than pretensioned bolts in poor rock masses, as strains in both the rock mass and the anchors are higher in poor rocks.

The same length of bolts should be used in the roof as used in the walls, since the tangential force from the roof arch will also be transmitted to the rock wall column.

Stability of reinforced haunches is automatic because of the presence of a critically oriented joint. If steel ribs are used to support the roof, additional reinforcement of haunches is required. (Failure of haunches due to heavy thrust of the large steel ribs has been observed in caverns and larger tunnels in poor rock conditions.) The thickness of shotcrete should be checked for shearing failure as follows:

$$u_w + p_{wall} \leq \frac{2q_{sc} \,.\, t_{wsc}}{L \cdot F_{wsc}} \tag{12.7}$$

where p_{wall} = ultimate wall support pressure (t/m^2), 0.28 p_{roof} near major shear zones, and 0.09 p_{roof} in caverns; u_w = average seepage pressure in wall (t/m^2) and 0 in grouted rock columns; t_{wsc} = thickness of shotcrete or steel fiber reinforced shotcrete (SFRS) in wall; F_{wsc} = mobilization factor for shotcrete in wall and $\cong 0.60 \pm 0.05$; $L \cdot F_{WSC}$ = span between points of maximum shear stress in wall shotcrete; q_{sc} = shear strength of shotcrete = 300 t/m^2 (3.0 MPa), and shear strength of SFRS = 550 t/m^2 (5.5 MPa), and is 0.2 $\times$ observed uniaxial compressive strength (UCS) of shotcrete or SFRS.

In Eq. (12.7), the support capacity of wall rock bolts is not accounted for because they prevent the buckling of the wall columns of the rock mass. If longer bolts are provided in the walls, shotcrete of a lesser thickness may be recommended. Further research is needed to improve Eq. (12.7), which is conservative.

ROOF SUPPORT IN CAVERNS

The recommended angle (θ) between the vertical and the spring point (Figure 12.1b) is given by

$$\sin\theta = \frac{1.3}{B^{0.16}} \leq 1 \tag{12.8}$$

where B is the width of the roof arch in meters.

The ultimate roof support capacity is given by a semi-empirical theory (Singh et al., 1995) for both tunnels and caverns:

$$p_{ult} + u = p_{sc} + p_{bolt} \tag{12.9}$$

where p_{ult} = ultimate support pressure estimated from Eq. (8.9) ($f' = 1$) in t/m^2; u = seepage pressure in the roof rock after commissioning of the hydroelectric project in t/m^2 and is 0 in nearly dry rock mass; and p_{sc} = support capacity of shotcrete/SFRS in t/m^2 and

$$\frac{2\, t_{sc} \cdot q_{sc}}{F_{sc} \cdot B} \tag{12.10}$$

where $F_{sc} = 0.6 \pm 0.05$ (higher for caverns) and $F_{sc} \cdot B$ = horizontal distance between vertical planes of maximum shear stress in the shotcrete in the roof (Figure 12.1a).

$$p_{bolt} = \frac{2.1' \, q_{cmrb} \cdot \sin\theta}{F_s \cdot B} = \text{support capacity of reinforced arch} \quad (12.11)$$

p_{bolt} = capacity of each rock anchor/bolt tension in t/m^2 and q_{cmrb} = UCS of reinforced rock mass in t/m^2 and

$$\left[\frac{P_{bolt}}{s_{bolt}^2} - u\right] \cdot \frac{1+\sin\phi_j}{1-\sin\phi_j} > 0 \quad (12.12)$$

s_{bolt} = spacing of rock bolts/anchors in meters.

$$1' = 1 - \frac{FAL}{2} - \frac{s_{bolt}}{4} + s_{rock} \quad (12.13)$$

$$1 = \text{length of rock bolt in roof} \quad (12.14)$$

$\tan\phi_j = \frac{J_r}{J_a} < 1.5$ and J_{rm}/J_{am} near shear zones; F_s = mobilization factor for rock bolts, 3.25 $p_{ult}^{0.10}$ for pretension bolts, and 9.5 $p_{ult}^{-0.35}$ for rock anchors and full-column grouted rock bolts; J_{rm} = mean joint roughness number near shear zone (see the section Treatment for Tunnels in Chapter 2); and J_{am} = mean joint alteration number near shear zone (see the section Treatment for Tunnels in Chapter 2).

These mobilization factors have been back analyzed from tables of support systems of Barton et al. (1974) and the chart for SFRS (Figure 8.5). Later, Thakur (1995) confirmed the previous design criteria from 120 case histories. Alternatively, Figure 8.2 may be used for selection of an SFRS support system in the feasibility design. A study for 10 years in a hydroelectric project (see the section correlation by Singh et al., 1992, in Chapter 8) showed that the ultimate support pressure for water-charged rock mass with erodible joint filling may increase up to 6 times the short-term support pressure due to the seepage erosion. This is unlikely to happen in hydroelectric caverns in strong rocks with very low permeability (<0.1 lugeon).

TM software can be used to design a support system for tunnels and caverns with and without shear zones (Singh & Goel, 2002). At the detailed design stage, UDEC/3DEC software packages are recommended for a rational design of support systems and to discover the best sequence of excavation to restrain progressive failure of rock mass. Appendix II gives the bond strength of grouted bolts needed for these programs. Maximum tensile stress occurs at junctions of openings, and tensile stresses also exist in the roof and the walls. Hence, there is the need for proper study to ensure that the rock mass is adequately reinforced to absorb critical tensile stresses.

The strong bond between shotcrete and rock mass is the key to success in stabilizing a cavern, because it drastically reduces bending stresses in the shotcrete lining.

STRESS DISTRIBUTION IN CAVERNS

Stress distribution should be studied carefully. The 2D stress analysis of deep caverns of the Tehri Dam project in India shows that the stress concentration factor ($\sigma_\theta/\gamma \cdot H$) at haunch is about 2.5 initially and decreases to about 1.5 when the cavern is excavated down below the haunches to the bottom of the cavern. The 3D stress analysis of the shallow cavern of the Sardar Sarovar project in India shows that a final stress concentration factor at haunch is only about 1.1 (Samadhiya, 1998). In both the cases the extent of the distressed zone goes beyond 2L as the low shear stiffness of joints does not allow high shear stresses in the rock mass. The 3D distribution of shear stresses in the shotcrete at the Sardar Sarovar project suggests that the horizontal distance between vertical planes of maximum shear stresses is $B \cdot F_{sc}$, where F_{sc} is approximately 0.60 ± 0.05 (Samadhiya, 1998).

OPENING OF DISCONTINUITIES IN ROOF DUE TO TENSILE STRESS

In the Himalayan region, thin bands of weak rocks are found within good rock masses. Sometimes these thin bands are just above the roof. Separation between a stronger rock mass above and the weak bands below it takes place where the overall tensile stress is more than the tensile strength (q_{tj}) of the weak band. As such, longer rock bolts are needed soon after excavation to stop this separation and stabilize the roof. Thus, tensile strength (q_{tj}) needs to be estimated for the minimum value of Q in the band and the adjoining rock mass (Chapter 13 and Eq. 13.21).

ROCK REINFORCEMENT NEAR INTERSECTIONS

In mine roadways, Tincelin (1970) recommended a 25% increase in the length of rock bolts near intersections. In caverns, the length of rock bolts for both the wall of the cavern and an intersecting tunnel can be increased by about 35% in the vicinity of intersections with the tunnels. This ensures that the rock mass in tension is effectively reinforced. Example 8.2 describes a design example for the intersection of two canal tunnels.

RADIAL DISPLACEMENTS

Based on a large number of case histories, Barton (1998) found the following approximate correlations for absolute radial displacement (δ) in the crown of the roof and center of the wall away from shear zone/weak zones (for B/Q = 0.5 to 250):

$$\delta_v = \frac{B}{100\,Q}\sqrt{\frac{\sigma_v}{q_c}} \tag{12.15}$$

$$\delta_h = \frac{H_t}{100\,Q}\sqrt{\frac{\sigma_h}{q_c}} \tag{12.16}$$

where δ_v, δ_h = radial displacement in roof and wall, respectively; σ_v, σ_h = in situ vertical stress and horizontal stress normal to the wall of the cavern, respectively; B = span of the cavern; H_t = total height of the cavern; Q = average rock mass quality; and q_c = UCS of the rock material.

PRECAUTIONS

1. For D-shaped tunnels, $\theta = 90°$.
2. The directional rock bolts should be designed for tackling loads due to the wheels of the crane on the haunches.
3. Support must be installed within the stand-up time (Figure 6.1).

While adopting the empirical approaches, it must be ensured that the ratings for the joint sets, joint spacing, rock quality designation (RQD), and so forth are scaled down for the caverns if initial ratings are obtained from the drifts. This is done because a few joint sets and weak intrusions in a drift could be missed. The rock mass quality should be downgraded in the area of a shear zone and a weak zone (see the section Treatment for Tunnels in Chapter 2). A mean value of deformation modulus E_m should be substituted for E_d in Eq. (12.1) for estimating the length of wall anchors. Similarly, a mean value of rock mass quality (Q_m) and joint roughness number (J_{rm}) should be used in Eq. (8.9) for assessment of the ultimate support pressure.

Stresses in the shotcrete lining and rock anchors may be reduced significantly by delaying subsequent layers of shotcrete (except initial layers), but no later than the stand-up time. Instrumentation for the measurements of stress and deformation in the roof and the walls of a cavern or in tunnels is a must to ensure a safe support system. Instrumentation would also provide feedback for improvements in the designs of such future projects. Location of instrumentation should be judiciously selected depending upon the weak zones, rock mass quality, and intersection of openings.

Example 12.1

Two parallel road tunnels are constructed for six lanes in basalt. The tunnels are D-shaped with diameter (B) of approximately 16 m and with 2 m high side walls with clear spacing of 20 m. The maximum overburden (H) is 165 m. The rock mass parameters are RMR = 73, Q = 10, $J_a = 1.0$, $J_r = 3.0$, and $J_w = 1.0$ (minor seepage from side walls). The construction engineers want a rapid rate of tunneling and life of the support system should be 100 years. The UCS of SFRS is 30 MPa and its flexural strength is 3.7 MPa.

The short-term support pressure in the roof may be assessed by following correlation (Eq. 6.6) for the arch opening, given by Goel and Jethwa (1991):

$$p_{roof} = \frac{7.5B^{0.1}\ H^{0.5} - RMR}{20\ RMR} = \frac{7.5 \times 16^{0.1} \times 165^{0.5} - 73}{20 \times 73} = 0.037\ \text{MPa}$$

The ultimate support pressure is read by the chart (Figure 8.2) of Barton et al. (1974) as follows (the dotted line is observed to be more reliable than correlation).

$$p_{roof} = 0.9 \times 1 \times 1\text{kg/cm}^2 \text{ or } 0.09\ \text{MPa}$$

(The rock mass is in non-squeezing ground condition ($H < 350\ Q^{1/3}$) and so $f' = 1.0$. The overburden is less than 320 m, so $f = 1.0$.)

It is proposed to provide the SFRS (and no rock bolts for faster rate of tunneling). The SFRS thickness (t_{fsc}) is given by the following correlation (using Eq. 12.10):

$$t_{fsc} \quad \frac{0.6\ B\ p_{roof}}{2\ q_{fsc}} = \frac{0.6 \times 1600 \times 0.09}{2 \times 5.5} = 8\ \text{cm}$$

$$= 16\ \text{cm (near portals)}$$

The tensile strength of SFRS is considered to be about one-tenth of its UCS, so its shear strength (q_{sc}) will be approximately $2 \times 30/10 = 6.0$ MPa, but we will say 5.5 MPa (uniaxial tensile strength is generally less than its flexural strength). Past experience reflects the same information.

The life of SFRS is the same as concrete in a polluted environment of approximately 50 years. Life may be increased to 60 years by providing an extra cover of 5 cm of

SFRS. If SFRS is damaged later, the corroded part should be scratched and a new layer of shotcrete should be sprayed that will last for 100 years. For this the recommended thickness of SFRS is $t_{fsc} = 13$ cm $= 21$ cm (near portals).

Example 12.2

The width of the pillar is more than the sum of the half-widths of adjoining openings in the non-squeezing grounds. The width of the pillar is also more than the total height of the larger of two caverns (18 m); hence the proposed separation of 20 m is safe (Hoek, 2007).

The following precautions need to be taken:

1. Loose pieces of rocks should be scraped thoroughly before shotcreting for better bonding between the two surfaces.
2. Unlined drains should be created on both sides of each tunnel to drain out the groundwater and then should be covered by reinforced cement concrete (RCC) slabs for road safety.
3. Tunnel exits should be decorated with art and arrangements should be made for bright lighting to illuminate the tunnels.

REFERENCES

Barton, N. (1998). Quantitative description of rock masses for the design of NMT reinforcement. In *International Conference on Hydropower Development in Himalayas* (pp. 379–400). Shimla, India.

Barton, N., Lien, R., & Lunde, J. (1974). Engineering Classification of Rock Masses for the Designs of Tunnel Supports. *Rock Mechanics, 6*, 189–236. Springer-Verlag.

Barton, N., Loset, F., Lien, R., & Lune, J. (1980). Application of Q-system in design decisions concerning dimensions and appropriate support for underground installation. In *Subsurface Space* (pp. 553–561). New York: Pergamon.

Bazant, Z. P., Lin, F. B., & Lippmann, H. (1993). Fracture energy release and size effect on borehole breakout. *International Journal of Numerical and Analytical Methods in Geomechanics, 17*, 1–14.

Cording, E. J., Hendron, A. J., & Deere, D. U. (1971). Rock engineering for underground caverns. In *Symposium on Underground Chambers* (pp. 567–600). Phoenix, Arizona: ASCE.

Fairhurst, C., & Singh, B. (1974). Roof bolting in horizontally laminated mine roof. *Engineering and Mining Journal*, 80–90.

Goel, R. K., Jethwa, J. L., & Paithankar, A. G. (1995). Indian experiences with Q and RMR systems. *Tunnelling and Underground Space Technology, 10*(1), 97–109.

Hoek, E. (2007). *Practical rock engineering*. www.rocscience.com.

Hoek, E., & Brown, E. T. (1980). *Underground excavations in rocks* (p. 527). Institution of Mining and Metallurgy. London: Maney Publishing.

Park, E. S., Kim, H. Y., & Lee, H. K. (1997). A study on the design of the shallow large rock cavern in the Gonjiam underground storage terminal. In *Proceedings of the 1st Asian Rock Mechanics Symposium on Environmental & Strategy Concerns in Underground Construction* (pp. 345–351). Seoul, Korea.

Samadhiya, N. K. (1998). *Influence of shear zone on stability of cavern* (p. 334). Ph.D. Thesis. Uttarakhand, India: Dept. of Civil Engineering, IIT Roorkee.

Singh, B., Fairhurst, C., & Christiano, P. P. (1973). Computer simulation of laminated roof reinforced with grouted bolt. In *Proceedings of the IGS Symposium on Rock Mechanics and Tunnelling Problems* (pp. 41–47). Kurukshetra, India.

Singh, B., & Goel, R. K. (2002). *Software for engineering control of landslide and tunnelling hazards* (p. 344). Rotterdam. A. A. Balkema (Swets & Zeitlinger).

Singh, B., Goel, R. K., Mehrotra, V. K., Garg, S. K., & Allu, M. R. (1998). Effect of intermediate principal stress on strength of anisotropic rock mass. *Tunnelling and Underground Space Technology, 13*(1), 71–79.

Singh, Bhawani, Jethwa, J. L., Dube, A. K., & Singh, B. (1992). Correlation between Observed Support Pressure and Rock Mass Quality. *Tunnelling & Underground Space Technology*, *7*(1), 59–74. Elsevier.

Singh, B., Jethwa, J. L., & Dube, A. K. (1995a). A classification system for support pressure in tunnels and caverns. *Journal of Rock Mechanics and Tunnelling Technology, 1*(1), 13–24.

Singh, B., Viladkar, M. N., & Samadhiya, N. K. (1995b). A semi-empirical method of the design of support systems in underground openings. *Tunnelling and Underground Space Technology, 3*, 375–383.

Thakur, B. (1995). *Semi-empirical method for design of supports in underground excavations* (p. 126). M.E. Thesis. Uttarakhand, India: IIT Roorkee.

Tincelin, E. (1970). Roof bolting recommendations. In *Parley of Cooperation and Industrial Promotion for Exploration and Exploitation of Mineral Deposits and Mineral Processing*. Sydney, Australia, May.

Verman, M. (1993). *Rock mass-tunnel support interaction analysis* (p. 258). Ph.D. Thesis. Uttarakhand, India: IIT Roorkee.

Chapter 13

Strength Enhancement of Rock Mass in Tunnels

The behaviour of macroscopic systems is generally described by non-linear laws. (The non-linear laws may explain irreversible phenomena like instabilities, dualism, unevolving societies, cycles of growth and decay of societies. The linear laws are only linear approximation of the non-linear laws at a point in time and space.)

Ilya Prigogine
Nobel Laureate

CAUSES OF STRENGTH ENHANCEMENT

Instrumentation and monitoring of underground openings in a complex geological environment is the key to success. Careful back analysis of the data observed in the initial stages of excavation provides valuable knowledge of the constants of the selected constitutive model, which may then be used in the forward analysis to predict performance of the support system. Back analysis of data from many project sites has shown a significant enhancement of rock mass strength around tunnels. Rock masses surrounding a tunnel perform much better than theoretical expectations, except near thick and plastic shear zones, faults, thrusts, and intra-thrust zones, and in water-charged rock masses.

Rock masses have shown constrained dilatancy in tunnels, so failure does not occur along rough joints due to interlocking and tightly packed rock blocks are not free to rotate, unlike soil grains. The strength of a rock mass in tunnels thus tends to be equal to the strength of a rock material (Pande, 1997).

Empirical criteria of rock mass failure are trusted more than theoretical criteria. In 1997, Sheorey evaluated them critically. However, designers like the linear approximation for practical applications.

EFFECT OF INTERMEDIATE PRINCIPAL STRESS ON TANGENTIAL STRESS AT FAILURE IN TUNNELS

The intermediate principal stress (σ_2) along the tunnel axis may be of the order of half the tangential stress (σ_1) in deep tunnels (Figure 13.1). Lade and Kim (1988) suggested the following polyaxial failure criterion in terms of the first and third stress invariants for soils, concrete, and rocks.

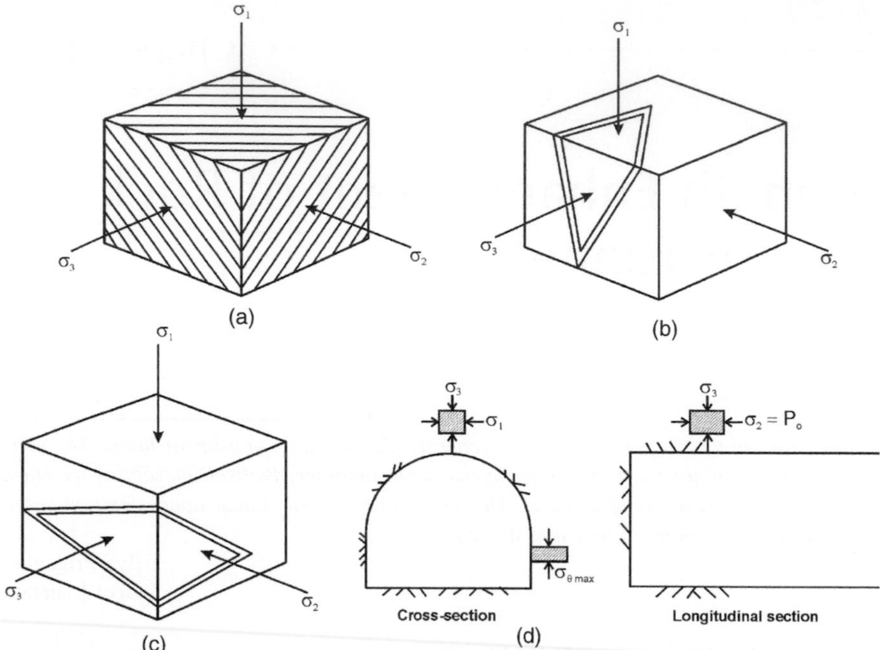

FIGURE 13.1 (a) Anisotropic rock material with one joint set (slate, schist, etc.), (b) mode of failure in rock mass with two joint sets, (c) $p_{horizontal} >> p_{vertical}$, and (d) direction of s_1, σ_2, and σ_3 in the tunnel.

$$\left[\frac{I_1^3}{I_3} - 27\right]\left[\frac{I_1}{p_a}\right]^m = \eta \tag{13.1}$$

where $I_1 = \sigma_1 + \sigma_2 + \sigma_3$, $I_3 = \sigma_1 \cdot \sigma_2 \cdot \sigma_3$, and p_a = atmospheric pressure.

The parameters m and η are dimensionless constants of a material. Analysis of many sets of data of concrete and rocks generally indicates that $m < 1.5$.

According to Wang and Kemeny (1995), σ_2 has a strong effect on σ_1 at failure even if σ_3 is equal to zero. Their polyaxial laboratory tests on hollow cylinders led to the following strength criterion:

$$\frac{\sigma_1}{q_c} = 1 + A[e^{\sigma_3/\sigma_2}] \cdot \left[\frac{\sigma_2}{q_c}\right]^{1-f \cdot e^{(\sigma_3/\sigma_2)}} \tag{13.2}$$

$$\therefore \ \sigma_1 \approx q_c + (A + f) \cdot (\sigma_3 + \sigma_2) \text{ for } \sigma_3 << \sigma_2$$

where f = material constant (0.10–0.20); A = material constant (0.75–2.00); and q_c = average uniaxial compressive strength (UCS) of rock material ($\sigma_2 = \sigma_3 = 0$) for various orientations of planes of weakness.

In the case of unsupported tunnels, $\sigma_3 = 0$ on its periphery. So, Eq. (13.2) simplifies to

$$\frac{\sigma_1}{q_c} = 1 + A\left[\frac{\sigma_2}{q_c}\right]^{(1-f)}$$

It may be inferred that σ_2 will enhance σ_1 at failure by 75–200% when $\sigma_2 \approx$ qc. Strength enhancement may be much more as propagation of fracture will be behind the excavated face (Bazant, Lin, & Lippmann, 1993). Murrell (1963) suggested 100% increase in σ_1 at failure when $\sigma_2 = 0.5\ \sigma_1$ and $\sigma_3 = 0$. Thus, the effective confining pressure appears to be an average of σ_2 and σ_3 and not just equal to σ_3 in the anisotropic rocks and weak rock masses.

Hoek (1994) suggested the following modified criterion for estimating the strength of jointed rock masses at high confining stresses (e.g., around $\sigma_3 > 0.10\ q_c$)

$$\sigma_1 = \sigma_3 + q_c\left[m\left(\frac{\sigma_3}{q_c}\right) + s\right]^n \tag{13.3}$$

where σ_1 and σ_3 = maximum and minimum effective principal stresses, respectively; m = Hoek-Brown rock mass constant (same as m_b in Chapter 26); s and n = rock mass constants; s = 1 for rock material, n = 0.5 and 0.65 − (GSI/200) ≤ 0.60 for GSI < 25 (or use Eq. 26.9 for any GSI); q_c = UCS of the intact rock core of standard NX size; and GSI = geological strength index ≈ RMR'$_{89}$ − 5 for RMR > 23 (see Chapter 26),

$$(m/m_r) = s^{1/3} \text{ for GSI} > 25 \text{ (see Chapter 26)} \tag{13.4}$$

where m_r = Hoek-Brown rock material constant.

The Hoek and Brown (1980) criterion in Eq. (13.3) is applicable to rock slopes and opencast mines with weathered and saturated rock mass. They have suggested values of m and s for Eq. (13.3). The Hoek and Brown (1980) criterion may be improved as a poly-axial criterion after replacing σ_3 (within bracket in Eq. 13.3) by effective confining pressure $(\sigma_2 + \sigma_3)/2$ as mentioned previously for weak and jointed rock masses. It may be noted that parameters m_r and q_c should be calculated from the upper bound Mohr's envelope of triaxial test data on rock cores in the case of anisotropic rock materials (Hoek, 1998).

According to Hoek (2007), rock mass strength is as follows (for disturbance factor D = 0):

$$q_{cmass} = (0.0034\ m_r^{0.8})q_c\ \{1.029 + 0.025\exp(-0.1m_r)\}^{GSI} \tag{13.5}$$

The Hoek and Brown (1980) criterion assumes isotropic rock and rock mass condition and should only be applied to those rock masses in which there are many sets of closely spaced joints with similar joint surface characteristics. Therefore, the rock mass may be considered to be an isotropic mass; however, the joint spacing should be much smaller than the size of the structure of the opening.

When one of the joint sets is significantly weaker than the others, the Hoek and Brown criterion Eq. (13.3) should not be used, as the rock mass behaves as an anisotropic mass. In these cases, the stability of the structure should be analyzed by considering a failure mechanism involving the sliding or rotation of blocks and wedges defined by intersecting discontinuities (Hoek, 2007). Singh and Goel (2002) presented software for wedge analysis for rock slopes (SASW) and WEDGE and UWDGE for tunnels and caverns.

Keep in mind that most of the strength criteria are not valid at low confining stresses and tensile stresses, as modes of failure are different. Hoek's criterion is applicable for high confining stresses only where a single mode of failure by faulting takes place; hence the quest for a better model to represent jointed rock masses.

UNIAXIAL COMPRESSIVE STRENGTH OF ROCK MASS

Equation 13.3 defines that uniaxial compressive strength of a rock mass is given by

$$q_{cmass} = q_c\, s^n \tag{13.6}$$

Equation (13.6) underestimates mobilized rock mass strength in tunnels. To use Eq. (13.3) in tunnels, a value of constant s must first be obtained from Eqs. (13.6) and (13.9) as follows:

$$s = \left[(7\,\gamma\, Q^{1/3})/q_c\right]^{1/n} \tag{13.7}$$

Ramamurthy (1993) and co-workers (Roy, 1993; Singh & Rao, 2005) conducted extensive triaxial tests on dry models of jointed rock mass using plaster of Paris ($q_c = 9.46$ MPa). They varied joint frequency, inclination of joints, thickness of joint fillings, and so forth and simulated a wide variety of rock mass conditions. Their extensive test data suggest the following approximate correlation for all rock masses:

$$q_{cmass}/q_c = [E_{mass}/E_r]^{0.7} \tag{13.8}$$

where, q_{cmass} = UCS of model of jointed rock mass in σ_1 direction; q_c = UCS of model material (plaster of Paris) and UCS of in situ block of rock material after size correction; E_{mass} = average modulus of deformation of jointed rock mass model ($\sigma_3 = 0$) in σ_1 direction; and E_r = average modulus of deformation of model material in the laboratory ($\sigma_3 = 0$).

The power in Eq. (13.8) varies from 0.5 to 1.0. Griffith's theory of failure suggests that the power is 0.5, whereas Sakurai (1994) felt the power in Eq. (13.8) is about 1.0 for jointed rock masses. Further research at the Indian Institute of Technology (IIT) in Delhi suggests that power in Eq. (13.8) is in the range of 0.56 and 0.72 (Singh & Rao, 2005). It appears that the power of 0.7 in Eq. (13.8) is realistic. Equation (13.8) can be used reliably to estimate UCS of a rock mass (q_{cmass}) from the values of E_{mass} or E_d obtained from uniaxial jacking tests within openings and slopes.

Considerable strength enhancement of the rock mass in tunnels has been observed by Singh et al. (1997). Based on the analysis of data collected from 60 tunnels, they recommended that the mobilized rock mass strength of the actual or disturbed rock mass is

$$q_{cmass} = 7\,\gamma\, Q^{1/3},\ \text{MPa (for } Q < 10,\ 100 > q_c > 2\text{MPa},\ SRF = 2.5,\ J_w = 1) \tag{13.9}$$

$$q_{cmass} = \left[(5.5\,\gamma\, N^{1/3})/B^{0.1}\right],\ \text{MPa (as per Eq. 7.5)} \tag{13.10}$$

where γ = unit weight of rock mass (gm/cc); N = actual rock mass number, that is, stress-free Barton's Q soon after the underground excavation; Q = actual (disturbed) rock mass quality soon after the underground excavation and corrected for SRF = 2.5; B = tunnel span or diameter in meters and SRF = 2.5 at the time of peak failure of in situ rock mass. See the section Correlation by Singh et al. (1992); in Chapter 8.

Equation (13.8) also shows that there is significantly high enhancement in the strength of rock mass. Kalamaras and Bieniawski (1995) suggested the following relationship between q_{cmass} and RMR:

$$q_{cmass} = q_c \cdot \exp\left[\frac{RMR - 100}{24}\right] \quad (13.11)$$

Barton (2002) modified Eq. (13.9) on the conservative side for calculating Q_{TBM} for tunnel boring machines (TBMs; according to Eq. 14.2):

$$\sigma_{cm} = 5\,\gamma\,(Q \cdot q_c/100)^{1/3}\ \text{MPa} \quad (13.12a)$$

where $q_c = I_s/25$ for anisotropic rocks in MPa (schists, slates, etc.) and I_s = standard point load strength index of rock cores (corrected for size effect for NX size cores). Barton (2005) clarified that Eq. (13.12) should be used only for Q_{TBM} (Chapter 14).

The correlations of Barton (2002) for the underground openings are

$$c_p \cong \frac{RQD}{J_n} \cdot \frac{1}{SRF} \cdot \frac{q_c}{100},\ \text{MPa (SRF} = 2.5) \quad (13.12b)$$

$$\tan\phi_p \cong \frac{J_a \cdot J_w}{J_r} + 0.1 \quad \text{Eq}(13.12c)$$

$$\therefore\ c_p \cdot \tan\phi_p \cong \frac{1}{k}$$

$$\cong Q_c - \frac{Q \cdot q_c}{100}$$

where c_p = peak cohesion of rock mass in MPa; ϕ_p = peak angle of internal friction of rock mass; k = permeability of rock mass in lugeon (10^{-7} m/sec); and Q_c = normalized rock mass quality.

The last term in Eq. (13.12c) is added by Choudhary (2007) because ϕ_p for the rock mass is more than ϕ_j for its joints due to the interlocking of rock blocks. He analyzed 11 cases of squeezing in tunnels in the Himalayas in India, and found Eqs. (13.12b), (13.12c), and (13.14) to be realistic (with SRF = 2.5 in the elastic zones).

Based on block shear tests, Singh et al. (1997) proposed the following correlation for estimating the UCS of the saturated rock mass for use in rock slopes in hilly areas:

$$q_{cmass} = 0.38\,\gamma \cdot Q^{1/3},\ \text{MPa} \quad (13.13)$$

Equation (13.13) suggests that the UCS of rock mass would be low on slopes. This is probably because joint orientation becomes a very important factor for slopes due to unconstrained dilatancy and low intermediate principal stress, unlike tunnels. Further, failure takes place along joints near slopes. In slopes of deep opencast mines, joints may be tight and of smaller length. The UCS of such a rock mass may be much higher and may be found from Hoek's criterion (Eq. 13.5) for analysis of the deep-seated rotational slides.

Equations (13.8) and (13.9) are intended only for 2D stress analysis of underground openings. The strength criterion for 3D analysis is presented in the next section.

REASON FOR STRENGTH ENHANCEMENT IN TUNNELS AND A NEW FAILURE THEORY

Consider a cube of rock mass with two or more joint sets as shown in Figure 13.1. If high intermediate principal stress is applied on the two opposite faces of the cube, then the chances of wedge failure are more than the chances of planar failure as found in the triaxial tests. The shear stress along the line of intersection of joint planes will be proportional to $\sigma_1 - \sigma_3$ because σ_3 will try to reduce shear stress. The normal stress on both the joint planes will be proportional to $(\sigma_2 + \sigma_3)/2$. Hence, the criterion for peak failure at low confining stresses may be as follows ($\sigma_3 < q_c/2$, $\sigma_2 < q_c/2$, and SRF > 0.05):

$$\sigma_1 - \sigma_3 = q_{cmass} + A[(\sigma_2 + \sigma_3)/2], \tag{13.14}$$

$$q_{cmass} = q_c \left[\frac{E_d}{E_r}\right]^{0.70} \cdot \left[\frac{d}{S_{rock}}\right]^{0.20}, \tag{13.15}$$

$$\Delta = \frac{\phi_p - \phi_r}{2}, \text{ beyond failure} \tag{13.16}$$

where q_{cmass} = average UCS of undisturbed rock mass for various orientation of principal stresses; σ_1, σ_2, σ_3 = final compressive and effective principal stresses equal to in situ stress plus induced stress minus seepage pressure; A = average constants for various orientation of principal stress (value of A varies from 0.6 to 6.0), $2 \cdot \sin\phi_p/(1 - \sin\phi_p)$, and $A_i + 2(1 - SRF)$ for rock mass with fresh joints; A_i = a value for intact rock material; SRF = q_{cmass}/q_c (strength reduction factor); ϕ_p = peak angle of internal friction of rock mass, is $\cong \tan^{-1}[(J_r J_w/J_a) + 0.1]$ at a low confining stress, is < peak angle of internal friction of rock material, and = 14° – 57°; S_{rock} = average spacing of joints; q_c = average UCS of rock material for core of diameter d (for schistose rock also); Δ = peak angle of dilatation of rock mass at failure; ϕ_r = residual angle of internal friction of rock mass = $\phi_p - 10° \geq 14°$; E_d = modulus of deformation of undisturbed rock mass ($\sigma_3 = 0$); and E_r = modulus of elasticity of the rock material ($\sigma_3 = 0$).

The peak angle of dilatation is approximately equal to $(\phi_p - \phi_r)/2$ for rock joints (Barton & Brandis, 1990) at low σ_3. This correlation (Eq. 15.8) may be assumed for jointed rock masses also. It is assumed that no dilatancy takes place before the peak failure so that strain energy is always positive during the deformation. The proposed strength criterion reduces to Mohr-Coulomb's criterion for triaxial conditions.

The significant rock strength enhancement in underground openings is due to σ_2 or in situ stress along tunnels and caverns, which pre-stresses rock wedges and prevents their failure both in the roof and the walls. However, σ_3 is released due to stress-free excavation boundaries (Figure 13.1d). In the rock slopes σ_2 and σ_3 are nearly equal and negligible. Therefore, there is insignificant or no enhancement of the strength. As such, block shear tests on a rock mass give realistic results for rock slopes and dam abutments only, because $\sigma_2 = 0$ in this test. Equation (13.14) may give a general criterion of undisturbed jointed rock masses for underground openings, rock slopes, and foundations.

Another cause of strength enhancement is higher UCS of rock mass (q_{cmass}) due to higher Ed because of constrained dilatancy and restrained fracture propagation near the excavation face only in underground structures. In rock slopes, E_d is found to be much less due to complete stress release and low confining pressure because of σ_2 and σ_3 and the long length of weathered filled up joints. So, q_{cmass} will also be low near rock slopes for the same Q-value (Eq. 13.13). Mohr-Coulomb criterion (Eq. 26.12) is valid for poor rock mass where $q_{cmass} < 0.05q_c$ or 1 MPa.

Through careful back analysis, both the model and its constants should be deduced. Thus, A, E_d, and q_{cmass} should be estimated from the feedback of instrumentation data at the beginning of the construction stage. With these values, forward analysis should be attempted carefully as mentioned earlier. At present, a non-linear back analysis may be difficult, and it does not give unique (or most probable) parameters.

The proposed strength criterion is different from Mohr-Coulomb's strength theory (Eq. 26.12), which works well for soils and isotropic materials. There is a basic difference in the structure of soil and rock masses. Soils generally have no pre-existing planes of weaknesses so planar failure can occur on a typical plane with dip direction toward σ_3. However, rocks have pre-existing planes of weaknesses like joints and bedding planes, and as such, failure occurs mostly along these planes of weaknesses. In the triaxial tests on rock masses, planar failure takes place along the weakest joint plane. In a polyaxial stress field, a wedge type of failure may be the dominant mode of failure if $\sigma_2 >> \sigma_3$. Therefore, Mohr-Coulomb's theory needs to be modified for anisotropic and jointed rock masses.

The new strength criterion is proved by extensive polyaxial tests on anisotropic tuff (Wang & Kemeny, 1995) and six other rocks. It is interesting to note that the constant A is the same for biaxial, triaxial, and polyaxial tests (Singh et al., 1998). Further, the effective in situ stresses (upper bound) on ground level in mountainous areas appear to follow Eq. (13.14) ($q_{cmass} = 3$ MPa, A = 2.5), which indicates a state of failure of earth crust near the water-charged ground due to the tectonic stresses.

The output of the computer program SQUEEZE shows that the predicted support pressures are of the order of those observed in 10 tunnels in the squeezing ground condition in the Himalayas in India. There is a rather good cross-check between the theory of squeezing and the observations except in a few cases. Thus, Eqs. (13.14) and (13.15) assumed in the theory of squeezing are again justified partially (Singh & Goel, 2002).

In the NJPC project tunnel excavated under high rock cover of 1400 m through massive to competent gneiss and schist gneiss, the theory predicted rock burst condition ($J_r/J_a = 0.75$, i.e., > 0.5). According to site geologists Pundhir, Acharya, and Chadha (2000), initially a cracking noise was heard followed by the spalling of 5–25 cm thick rock columns/slabs and rock falls. This is mild rock burst condition. Another cause of rock burst is the Class II behavior of gneiss according to the tests at IIT. According to Mohr-Coulomb's theory most severe rock bursts or squeezing conditions were predicted under rock cover more than 300 m ($q_c = 27$ MPa and $q_{cmass} = 15.7$ MPa). Mild rock burst conditions were actually met where overburden was more than 1000 m. However, polyaxial theory (Eq. 13.14) suggested mild rock burst conditions above overburden of 800 m. Thus, polyaxial theory of strength is validated further by the SQUEEZE program (Singh & Goel, 2002). Recently, Rao, Tiwari, and Singh (2003) developed the polyaxial testing system. Their results were replotted and parameter A was found to increase slightly from 3.8 to 4.2 for dips of joints from 0 to 60°, although q_{cmass} changed drastically. Thus, the suggested hypothesis appears to be applicable approximately for the rock masses with three or more joint sets.

Poor Rock Masses

Squeezing is found to occur in tunnels in the nearly dry weak rocks where overburden H is more than 350 $Q^{1/3}$m. The tangential stress at failure may be about $2\gamma H$ assuming hydrostatic in situ stresses. Thus, mobilized compressive strength is $2\ \gamma\ 350Q^{1/3} = 700\ \gamma Q^{1/3}$ T/m^2 (Eq. 13.9).

Singh originally proposed Eq. (13.9) in a lecture at the Workshop of Norwegian Method of Tunneling in New Delhi, India, in 1993 and reported it later after confirmation (Singh et al., 1997). Since the criterion for squeezing is found to be surprisingly independent of UCS ($q_c < 50$ MPa), in their opinion no correction for UCS (q_c) is needed for weak rocks.

Many investigators agreed with Eq. (13.9) (Grimstad & Bhasin, 1996; Barla, 1995; Barton, 1995; Choubey, 1998; Aydan, Dalgic, & Kawamoto, 2000; and others). It may be argued that q_{cmass} should be the same for given RQD, J_n, J_r, and J_a values irrespective of overburden depth and water pressure in joints. High overburden and water pressure can cause long-term damage to the rock mass due to induced fractures, opening of fractures, softening, seepage erosion, and so forth. Hence, Eq. (13.9) is justified logically if Q is obtained soon after excavation in the nearly dry, weak rock masses.

Eleven cases of tunnels in the squeezing ground have also been analyzed by Singh and Goel (2002). In poor rocks, the peak angle of internal friction (ϕ_p) is back analyzed and related as follows:

$$\tan\phi_p = \frac{J_r}{J_a} + 0.1 \leq 1.5 \text{ (for } J_w = 1) \tag{13.17}$$

The addition of 0.1 accounts for interlocking of rock blocks. It may be visualized that interlocking occurs more often in jointed rock mass than in soils due to low void ratio. Further, Kumar (2000) showed theoretically that the angle of internal friction of laminated rock mass is slightly higher than the sliding angle of friction of its joints.

Failure of Inhomogeneous Geological Materials

With inhomogeneous geological material, the process of failure is initiated by its weakest link (zone of loose soil and weak rock, crack, bedding plane, soft seam, etc.). Thus, natural failure surfaces are generally three-dimensional (perhaps four-dimensional), which start from this weakest link and propagate toward a free surface (or face of excavation). As such the intermediate principal stress (σ_2) plays an important role and governs the failure and the constitutive relations of the naturally inhomogeneous geological materials (both in rock masses and soils) in the field. Since micro-inhomogeneity is unknown, assumption of homogeneity is popular among engineers. Therefore, intuition states that the effective confining stress is about $[(\sigma_2 + \sigma_3)/2]$ in naturally inhomogeneous soils as well as fault-gouges.

Failure in an inhomogeneous geological material is progressive, whereas a homogeneous rock fails suddenly. Hence, the advantage of inhomogeneous materials offered by nature is that they give advance warning of the failure process by starting slowly from the weakest zone.

Failure of Laminated Rock Mass

Laminated rock mass is generally found in the roof of underground coal mines and in the bottom of opencast coal mines. The thin rock layers may buckle under high horizontal in situ stresses first and then rupture progressively by violent brittle failure. Therefore, the assumption of shear failure along joints is not valid. As such, the proposed hypothesis of effective confining stress $[(\sigma_2 + \sigma_3)/2]$ may not be applicable in the unreinforced and laminated rock masses. The suggested hypothesis appears applicable for the rock masses with three or more joint sets.

CRITICAL STRAIN OF ROCK MASS

The basic concept of structure design cannot be applied in tunnels, because stresses and strains are not reliably known. Critical strain is a better measure of failure.

Critical strain (ε_{cr}) is defined as the ratio between UCS (q_{cmass}) and the modulus of deformation (E_d) of rock mass (Sakurai, 1997). He found that the critical strain is nearly independent of joints, water content, and temperature. Singh, Singh, and Choudhari (2007) reported the following correlation for the critical strain and verified the same using 30 case histories:

$$\varepsilon_{cr} = \frac{5.84\, q_c^{0.88}}{Q^{0.12}\, E_r^{0.63}}\ (\text{percent}) \geq \frac{100\, q_c}{E_r} \tag{13.18}$$

$$\varepsilon_{cr} \geq \varepsilon_\theta = 100\, u_a/a$$

where ε_{cr} = critical strain of the rock mass in percentage; ε_θ = tangential strain around opening in percentage, = (observed deflection of crown in downward direction/radius of tunnel), and = 100 u_a/a (Figure 13.2); E_r = tangent modulus of the rock material (in MPa); q_c = UCS of rock material (in MPa); and Q = rock mass quality.

In Japan there were few construction problems in tunnels where $\varepsilon_\theta < \varepsilon_{mass}$ or ε_r. Critical strain appears to be somewhat size dependent.

Predictions and actual observations differ greatly in tunnels, and joints need more attention from engineers. It is easier to observe strains than stresses in the rock mass. Sakurai (1997) classified the hazard warning level into three stages in relation to degree of stability as shown in Figure 13.2. He observed that where strains in the roof ($\varepsilon_\theta = u_a/a$) were less than the warning level I, there were no problems in the tunnels, but tunneling problems were encountered where strains approached warning level III. Swarup, Goel, and Prasad (2000) confirmed these observations in 19 tunnels in weak rocks in the Himalayas.

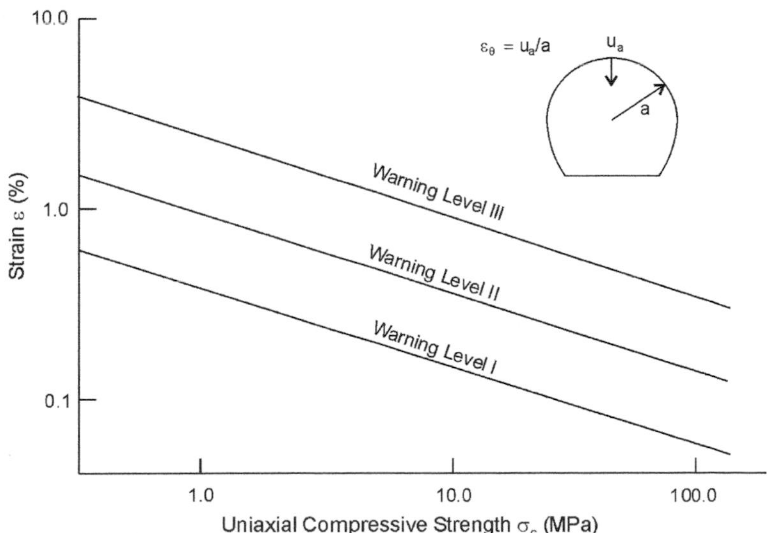

FIGURE 13.2 Hazard warning levels for assessing the stability of tunnels. *(From Sakurai, 1997)*

CRITERION FOR SQUEEZING GROUND CONDITION

Equation (13.14) suggests the following criterion for squeezing/rock burst ($\sigma_1 = \sigma_\theta$, $\sigma_3 = 0$, and $\sigma_2 = P_o$ along tunnel axis in Figure 13.1d):

$$\sigma_\theta > q_{c\ mass} + \frac{A \cdot P_o}{2} = q'_{cmass} \tag{13.19}$$

Palmstrom (1995) observed that σ_θ/q_{cmass} or σ_θ/RMi may be much higher than 1, that is, 1.5 to 3 for squeezing. Thus, his experience confirmed the proposed criterion (13.19), which shows that squeezing may occur when the constant A is small (<1.5). There is a need for in situ triaxial test data for further proof.

Eleven tunnels in the Himalayas showed that squeezing ground conditions are generally encountered where the peak angle of internal friction (ϕ_p) is less than 30°, J_r/J_a is less than 0.5, and overburden is higher than 350 $Q^{1/3}$ m in which Q is Barton's (disturbed) rock mass quality with SRF = 2.5. The predicted support pressures using Eq. (13.14) agree better with observed support pressure in the roof and walls than those by Mohr-Coulomb's theory (Chaturvedi, 1998).

ROCK BURST IN BRITTLE ROCKS

Kumar (2002) observed the behavior of the 27 km long NJPC tunnel and found that the mild rock burst occurred where A is more than 2.0 and $J_r/J_a > 0.5$. In 15 sections with rock cover (H) of more than 1000 m, his studies validated Eq. (13.19) for approximately predicting mild rock burst/slabbing conditions and estimating rock mass strength q_{cmass}. If $\sigma_\theta/q'_{cmass} > 0.6$, then spalling was observed in the blocky rock mass. He also inferred from 50 tunnel sections that the ratio between tangential stress and mobilized biaxial strength (σ_θ/q'_{cmass}) is a better criterion for predicting the degree of squeezing condition than Mohr-Coulomb's theory (σ_θ/q_{cmass}). Figure 7.3 also showed that the rock burst may occur where the normalized rock cover $HB^{0.1} > 1000$ m, N > 1.5, and $J_r/J_a > 0.5$.

For safe tunneling, understanding the "post-peak" behavior of a rock mass is often critical (Figure 3.2; the section Homogeneity and Inhomogeneity in Chapter 3). Unfortunately, costly mistakes are often made because of the lack of understanding of the actual complex and brittle behavior under high in situ stress, overreliance on analyses, or lack of experience in low stress conditions. In rock burst conditions, it is necessary to adopt a robust engineering approach that focuses on a flexible construction process and ensures that all construction machines work well. Additional uncertainties can be managed by adopting an observational design-as-you-go approach.

Failure of underground openings in hard and brittle rocks is a function of the in situ stress magnitudes and the characteristics of the rock mass; that is, the intact rock strength and the fracture network (Figure 13.3). At low in situ stress magnitudes, the failure process is controlled by the continuity and distribution of the natural fractures in the rock mass. However, as in situ stress magnitudes increase, the failure process is dominated by new stress-induced fractures growing parallel to the excavation boundary. This fracturing is generally referred to as "brittle failure." Initially, at intermediate depths, these failure regions are localized near the tunnel perimeter, but at great depth the fracturing envelopes the whole boundary of the excavation (Figure 13.3). Unlike ductile materials in which shear slip surfaces can form while continuity of material is maintained, brittle failure deals with materials for which continuity must first be disrupted before kinematically feasible failure mechanisms can form (Martin, Kaiser, & McCreath, 1999).

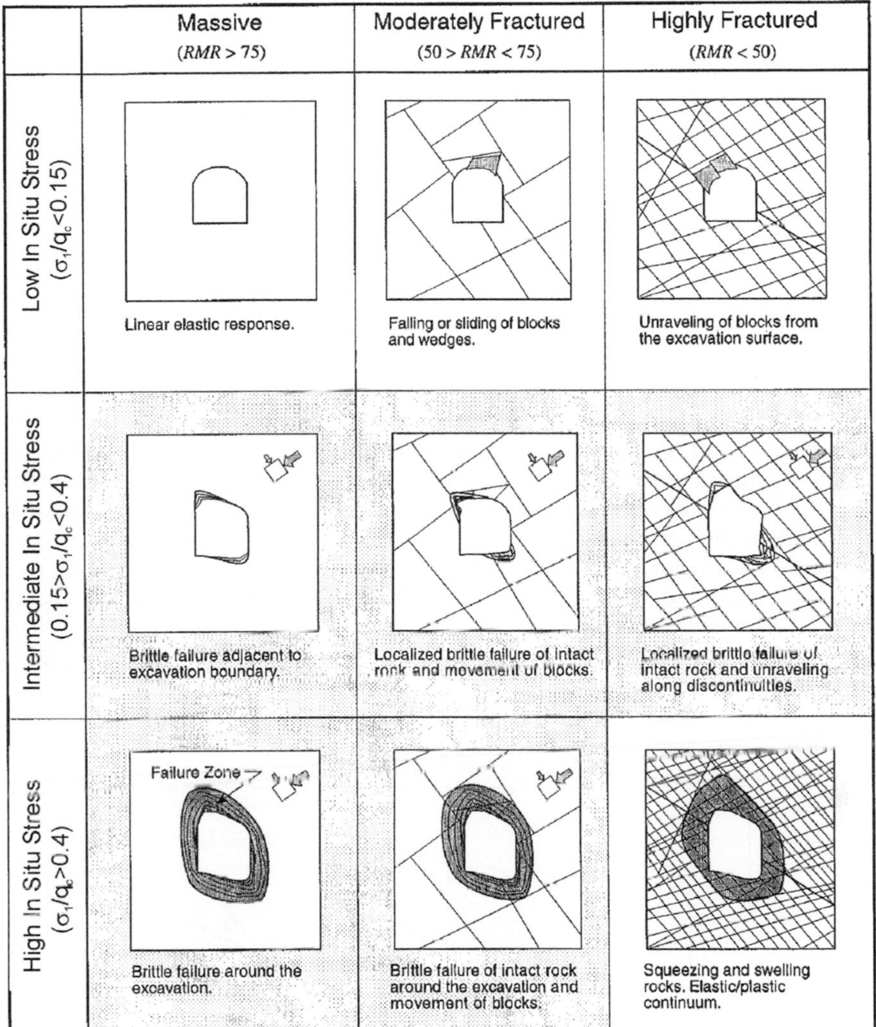

FIGURE 13.3 Examples of tunnel instability and brittle failure (highlighted gray squares) as a function of rock mass rating (RMR) and the ratio of the maximum far-field stress (σ_1) to the UCS (q_c). *(Modified from Hoek et al., 1995, and Martin et al., 1999)*

In brittle failing rock mass where stress-induced failure leads to the creation of a zone of fractured rock or cavity near excavation (Figure 13.3), tunneling basically involves three aspects: (1) retention of broken rock near excavation, (2) control of deformations due to bulking of fractured rocks, and (3) dissipation of strain energy if failure occurs violently. The fracturing may degrade rock mass quality drastically, but the process of fracturing or spalling is fortunately mostly self-stabilizing similar to squeezing grounds. The depth (d_f) of local spalling is correlated with maximum tangential stress (σ_θ) as follows (Kaiser, 2006):

$$\frac{d_f}{a} = C_1 \frac{\sigma_\theta}{q_c} - C_2 \qquad (13.20)$$

where $C_1 = 1.37\ (1 + 0.4v)$; $C_2 = 0.57$; v = peak particle velocity due to remote seismic event in m/sec; and a = radius of tunnel.

Thus the length of (resin) bolt is equal to d_f + FAL (fixed anchor length). It is advised to use yielding bolts that can deform up to 80 mm. It is better to excavate in many steps to reduce the strain energy released as it causes rock burst. Moreover, highly stressed tunnel faces should be shaped convexly to remove potentially unstable rock before it can cause serious safety hazards to workers near the tunnel face (Kaiser, 2006).

TENSILE STRENGTH ACROSS DISCONTINUOUS JOINTS

The length of joints is generally less than 5 m in tunnels in young rock masses except for bedding planes. Discontinuous joints thus have tensile strength. Mehrotra (1996) conducted 44 shear block tests on both nearly dry and saturated rock masses. He also obtained non-linear strength envelopes for various rock conditions. These strength envelopes were extrapolated carefully in tensile stress regions so that they were tangential to Mohr's circle for uniaxial tensile strength as shown in Figure 13.4. It was noted that the non-linear strength envelopes for both nearly dry and saturated rock masses converged to nearly the same uniaxial tensile strength across discontinuous joints (q_{tj}) within the blocks of rock masses. It is related to Barton's rock mass quality (Figure 13.5) as follows:

$$q_{tj} = 0.029\ \gamma\ Q^{0.31},\ \text{MPa} \qquad (13.21)$$

where γ is the unit weight of the rock mass in g/cc (T/m^3). In case of tensile stresses, the criterion of failure is as shown in Eq. (13.22).

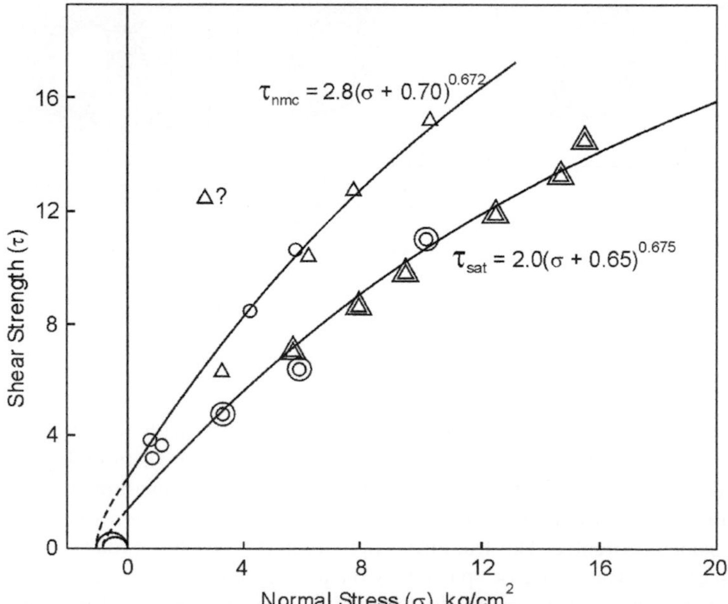

FIGURE 13.4 Estimation of tensile strength of rock mass from Mohr's envelope. *(From Mehrotra, 1992)*

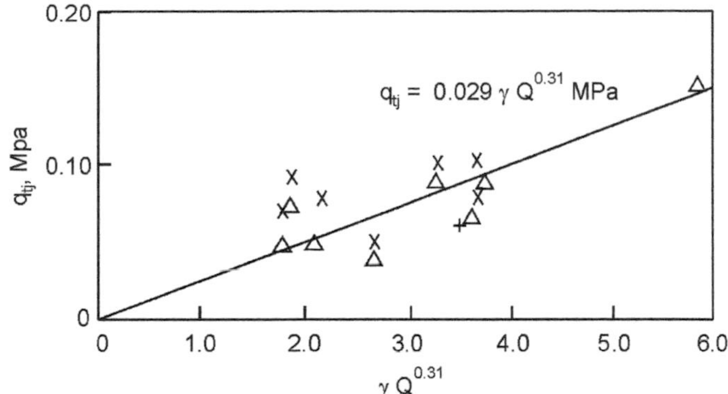

FIGURE 13.5 Plot between q_{tj} and $\gamma \cdot Q^{0.31}$ (γ in g/cc or T/m^3).

$$-\sigma_3 = q_{tj} \tag{13.22}$$

The tensile strength across discontinuous joints is not zero as generally assumed; it is found to be significant in hard rocks.

The tensile stress in the tunnel roof of span B will be of the order of γB in the vertical direction. Equating this with q_{tj}, the span of self-supporting tunnels obtained from Eq. (13.21) would be 2.9 $Q^{0.31}$ m. Barton, Lien, and Lunde (1974) found the self-supporting span to be 2 $Q^{0.4}$ m. This comparison is very encouraging. Thus, it is understood that the wedge analysis considering q_{tj} and in situ stress along the tunnel axis may result in a more accurate value of the self-supporting tunnel span. Equation (13.21) may also be used in distinct element software.

DYNAMIC STRENGTH OF ROCK MASS

It appears logical to assume that dynamic strain at failure should be of the same order as the static strain at failure for a given confining stress. Dynamic strain at failure should be proportional to modulus of elasticity of rock mass (E_e) and static strain at failure should be proportional to E_d. Therefore, the following hypothesis for dynamic strength enhancement is proposed.

$$q_{cmdyn}/q_{cmass} = (E_e/E_d)^{0.7} \tag{13.23}$$

where q_{cmdyn} = dynamic strength of rock mass.

In seismic analysis of concrete dams, dynamic strength enhancement may be quite high, particularly for a weathered rock mass, because the instantaneous modulus of elasticity (E_e from Eq. 8.19) will be much higher than the long-term modulus of deformation E_d (Eq. 8.18). Extensive research is urgently needed to obtain more realistic correlations for dynamic strength enhancement.

RESIDUAL STRENGTH PARAMETERS

Mohr-Coulomb's theory will be applicable to residual failure as a rock mass would be reduced to non-dilatant soil-like condition. The mobilized residual cohesion (c_r) is approximately equal to 0.1 MPa and is not negligible unless tunnel closure is more than 5.5% of its diameter. The mobilized residual angle of internal friction (ϕ_r) is approximately 10 degrees less than the peak angle of internal friction (ϕ_p), but more than 14 degrees. Rock mechanics helps to judge the support system (Singh & Goel, 2002).

REFERENCES

Aydan, O., Dalgic, S., & Kawamoto, T. (2000). Prediction of squeezing potential of rocks in tunnelling through a combination of an analytical method and rock mass classification. *Italian Geotechnical Journal, XXXIV*(1), 41–45.

Barla, G. (1995). Squeezing rocks in tunnels. *ISRM News Journal, 2*(3&4), 44–49.

Barton, N. (1995). The influence of joint properties in modelling jointed rock masses. In *Eighth International Rock Mechanics Congress* (Vol. 3, pp. 1023–10320). Tokyo, Japan.

Barton, N. (2002). Some new Q-value correlations to assist in site characterisation and tunnel design. *International Journal of Rock Mechanics and Mining Sciences, 39*, 185–216.

Barton, N. (2005). Personal communication to R.K. Goel.

Barton, N., & Brandis, S. (1990). Review of predictive capabilities of JRC-JCS model in engineering practice. Reprinted from N. R. Barton & O. Stephansson (Eds.), *Rock Joints Proceedings of a Regional Conference of the International Society for Rock Mechanics* (p. 820). Leon.

Barton, N., Lien, R., & Lunde, J. (1974). Engineering classification of rock masses for the design of tunnel support. In *Rock mechanics* (Vol. 6, pp. 189–236). New York: Springer-Verlag.

Bazant, Z. P., Lin, F. B., & Lippmann, H. (1993). Fracture energy release and size effect in borehole breakout. *International Journal for Numerical and Analytical Methods in Geomechanics, 17*, 1–14.

Chaturvedi, A. (1998). *Strength of anisotropic rock masses* (p. 82). M.E. Thesis. Uttarakhand, India: Department of Civil Engineering, IIT Roorkee.

Choubey, V. D. (1998). Potential of rock mass classification for design of tunnel supports—Hydroelectric Projects in the Himalayas. In *International Conference on Hydro Power Development in Himalayas* (pp. 305–336). Shimla, India.

Choudhary, J. S. (2007). *Closure of underground openings in jointed rocks* (p. 324). Ph.D. Thesis. Uttarakhand, India: Department of Civil Engineering, IIT Roorkee.

Grimstad, E., & Bhasin, R. (1996). Stress strength relationships and stability in hard rock. *Proceedings of the Conference on Recent Advances in Tunnelling Technology* (Vol. I, pp. 3–8). New Delhi, India.

Hoek, E. (1994). Strength of rock and rock masses. *ISRM News Journal, 2*, 416.

Hoek, E. (1998). Personal Discussions with Prof. Bhawani Singh on April 4 at Tehri Hydro Development Corporation Ltd., Rishikesh, India.

Hoek, E. (2007). *Practical rock engineering* (Chap. 12). www.rocscience.com.

Hoek, E., & Brown, E. T. (1980). *Underground excavations in rocks* (p. 527). Institution of Mining and Metallurgy. London: Maney Publishing.

Hoek, E., Kaiser, P. K., & Bawden, W. F. (1995). *Support of underground excavations in hard rock* (p. 215). Rotterdam: A.A. Balkema.

Kaiser, P. K. (2006). Rock mechanics considerations for construction of deep tunnels in brittle rock. In C. F. Leung & Y. X. Zhou (Eds.), *Proceedings of the ISRM International Symposium 2006 and 4th Asian Rock Mechanics Symposium on Rock Mechanics in Underground Construction* (pp. 47–58). World Scientific Publishing Co. Singapore.

Kalamaras, G. S., & Bieniawski, Z. T. (1995). A rock strength concept for coal seams incorporating the effect of time. In *Proceedings of the 8th International Congress on Rock Mechanics* (Vol. 1, pp. 295–302).

Kumar, N. (2002). *Rock mass characterisation and evaluation of supports for tunnels in Himalaya* (p. 289). Ph.D. Thesis. Uttarakhand, India: WRDM, IIT Roorkee.

Kumar, P. (2000). Mechanics of excavation in jointed underground medium. In *Symposium On Modern Techniques in Underground Construction* (pp. 49–75). New Delhi: CRRI, ISRMTT.

Lade, P. V., & Kim, M. K. (1988). Single hardening constitutive model for frictional materials—Part III: Comparisons with experimental data. *Computers and Geotechnics, 6*, 31–47.

Martin, C. D., Kaiser, P. K., & McCreath, D. R. (1999). Hoek–Brown parameters for predicting the depth of brittle failure around tunnels. *Canadian Geotechnical Journal, 36*, 136–151.

Mehrotra, V. K. (1992). *Estimation of engineering parameters of rock mass* (p. 267). Ph.D. Thesis. Uttarakhand, India: IIT Roorkee.

Mehrotra, V. K. (1996). Failure envelopes for jointed rocks in Lesser Himalaya. *Journal of Rock Mechanics and Tunnelling Technology*, *2*(1), 59–74.

Murrell, S. A. K. (1963). A criterion for brittle fracture of rocks and concrete under triaxial stress and the effect of pore pressure on the criteria. In C. Fairhurst (Ed.), *Fifth Symposium on Rock Mechanics* (pp. 563–577). Oxford, UK: Pergamon.

Palmstrom, A. (1995). Characterising the strength of rock masses for use in design of underground structures. In *Conference of Design and Construction of Underground Structures* (pp. 43–52). New Delhi.

Pande, G. N. (1997). *SQCC Lecture on Application of the Homogenisation Techniques in Soil Mechanics and Structure*. Uttarakhand, India: IIT Roorkee, September 26.

Pundhir, G. S., Acharya, A. K., & Chadha, A. K. (2000). Tunnelling through rock cover of more than 1000 m—A case study. In *International Conference on Tunnelling Asia 2000* (Vol. 1, pp. 3–8). New Delhi, India.

Ramamurthy, T. (1993). Strength and modulus responses of anisotropic rocks. In *Comprehensive rock engineering* (Chap. 13, pp. 313–329). New York: Pergamon.

Rao, K. S., Tiwari, R. P., & Singh, J. (2003). Development of a true triaxial system (TTS) for rock mass testing. In *Conference on Geotechnical Engineering for Infrastructural Development, IGC-2003* (Vol. I, pp. 51–58). Uttarakhand, India: IIT Roorkee.

Roy, N. (1993). *Engineering behaviour of rock masses through study of jointed models* (p. 365). Ph.D. Thesis. New Delhi: Civil Engineering Department, IIT.

Sakurai, S. (1994). Back analysis in rock engineering. *ISRM News Journal*, *2*(2), 17–22.

Sakurai, S. (1997). Lessons learned from field measurements in tunnelling. *Tunnelling and Underground Space Technology*, *12*(4), 453–460.

Sheorey, P. R. (1997). *Empirical rock failure criterion* (p. 176). London: Oxford & IBH Publishing Co. and Rotterdam: A. A. Balkema.

Singh, B., & Goel, R. K. (2002). *Software for engineering control of landslide and tunnelling hazard* (p. 344). Rotterdam: A. A. Balkema (Swets & Zeitlinger).

Singh, B., Goel, R. K., Mehrotra, V. K., Garg, S. K., & Allu, M. R. (1998). Effect of intermediate principal stress on strength of anisotropic rock mass. *Tunnelling and Underground Space Technology*, *13*(1), 71–79.

Singh, Bhawani, Jethwa, J. L., Dube, A. K., & Singh, B. (1992). Correlation between observed support pressure and rock mass quality. *Tunnelling & Underground Space Technology*, *7*(1), 59–74 Elsevier.

Singh, Bhawani, Viladkar, M. N., Samadhiya, N. K., & Mehrotra, V. K. (1997). Rock mass strength parameters mobilized in tunnels. *Tunnelling and Underground Space Technology*, *12*(1), 47–54.

Singh, M., & Rao, K. S. (2005). Empirical methods to estimate the strength of jointed rock masses. *Engineering Geology*, *77*, 127–137.

Singh, M., Singh, B., & Choudhari, J. (2007). Critical strain and squeezing of rock mass in tunnels. *Tunnelling and Underground Space Technology*, *22*, 343–350.

Swarup, A. K., Goel, R. K., & Prasad, V. V. R. (2000). Observational approach for stability of tunnels. In *Tunnelling Asia 2000* (pp. 38–44). New Delhi.

Wang, R., & Kemeny, J. M. (1995). A new empirical failure criterion under polyaxial compressive stresses. In J. J. K. Daemen & R. A. Schultz (Eds.), *Rock Mechanics: Proceedings of the 35th U. S. Symposium* (p. 950). Lake Tahoe, June 4–7.

Chapter 14

Rock Mass Quality for Open Tunnel Boring Machines

Any manager of a project must understand that his success depends on the success of the contractor. The contractors have to be made to succeed. They have many problems. We cannot always talk within the rigid boundaries of a contract document. No, without hesitation. I go beyond the contract agreement document.

E. Sreedharan
Managing Director, Delhi Metro Rail Corporation

INTRODUCTION

Tunnel boring machines (TBMs) have extreme rates of tunneling of 15 km/year and 15 m/year and sometimes even less. The expectation of fast tunneling places great responsibility on those evaluating geology and hydrogeology along a planned tunnel route. When rock conditions are reasonably good, a TBM may be two to four times faster than the drill and blast method. The problem lies in the extremes of rock mass quality, which can be both too bad and too good (no joints), where alternatives to TBM may be faster (Barton, 1999). The basic advantages of TBMs are high safety with low overbreaks, little disturbance to surrounding rock mass, and low manpower. However, set-up and dismantling time are significant and the range of available tunnel cross-section shapes is limited (Okubo, Fukui, & Chen, 2003). Engineers should not use TBMs where engineering geological investigations have not been done in detail and rock masses are very heterogeneous. Contractors can design TBMs according to the given rock mass conditions, which are nearly homogeneous.

There have been continuous efforts to develop a relationship between rock mass characterization and essential machine characteristics such as cutter load and cutter wear, so that surprising rates of advance become the expected rates. Even from a 1967 open TBM, Robbins (1982) reported 7.5 km of advance in shale during four months. Earlier in the same project, 270 m of unexpected glacial debris took nearly seven months. The advance rate (AR) of 2.5 m/h has declined to 0.05 m/h in the same project. This can be explained by engineering rock mass classification. The TBM should not be used in squeezing ground conditions, rock burst conditions, and flowing grounds, because it is likely to get stuck or damaged.

Barton (2000a) incorporated a few parameters in the Q-system that influence the performance of a TBM to obtain Q_{TBM} (i.e., rock mass quality for an open TBM). Using Q_{TBM}, Barton (2000a) believed that the performance of TBMs in a particular type of rock mass may be estimated. His approach is presented in this chapter.

Q AND Q_{TBM}

The Q-system was developed by Barton et al. in 1974 from drill and blast tunnel case records, which now total 1250 cases (Grimstad & Barton, 1993). Q-values stretch over six orders of magnitude of rock mass quality. Continuous zones of squeezing rock and clay can have a Q = 0.001, while virtually unjointed hard massive rock can have a Q = 1000. Both conditions are extremely unfavorable for TBM advance: one stops the machine for extended periods and requires heavy pre-treatment and supports, and the other slows average progress to 0.2 m/h over many months due to multiple daily cutter shifts (Barton, 1999).

The general trends for penetration rate (PR) with uninterrupted boring and actual AR measured over longer periods is shown in Figure 14.1. Highlighted here is the penetration rate of a TBM, which may be high, but the real AR depends on tunnel support needs and conveyor capacity. The Q-value goes a long way to explain the different magnitudes of PR and AR, but it is not sufficient without modification and the addition of some machine–rock interaction parameters.

A new method has been developed by Barton (1999) for estimating PR and AR using Q-value and Q_{TBM}, which is strongly based on the familiar Q parameters with additional machine–rock mass interaction parameters. Together, these give a potential 12 orders of magnitude range of Q_{TBM}. The real value depends on the cutter force.

There are four basic classes of rock tunneling conditions that need to be described in a quantitative way:

1. Jointed, porous rock, easy to bore, frequent support
2. Hard, massive rock, tough to bore, frequent cutter change, no support
3. Overstressed rock, squeezing, stuck machine, needs over-boring, heavy support
4. Faulted rock, overbreak, erosion of fines, long delays for drainage, grouting, temporary steel support, and backfilling.

The conventional Q-value, together with the cutter life index (CLI; Johannessen & Askilsrud, 1993) and quartz content help to explain some of the delays involved. The Q-value can also be used to select support once differences between drill and blast and TBM logging are correctly quantified in the central threshold area of the Q diagram (Figure 8.5).

In relation to the line separating supported and unsupported excavations in the Q-system support chart, a TBM tunnel gives an apparent (and partially real) increase

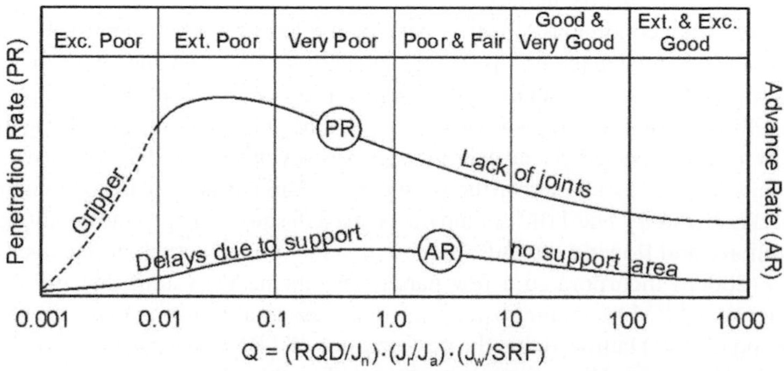

FIGURE 14.1 A conceptual relation between Q, PR, and AR for open TBM. *(From Barton, 2000a)*

in the Q-value of about 2 to 5 times in this region. This is where the TBM tunnel supports are reduced. When the Q-value is lower than in the *central threshold area* (support categories 8 and 9 in Figure 8.5), the TBM tunnel shows similar levels of overbreak or instability as the drill and blast tunnel, and final support derived from the Q-system applies. However, the levels may be preceded by (non-reinforcing temporary) steel sets and lagging (and void formation), each of which require due consideration while designing a support.

The Q_{TBM} is defined in Figure 14.2, and some adjectives at the top of the figure suggest the ease or difficulty of boring. (Note the differences in the Q-value adjectives used in Figure 14.1, which describe rock mass stability and need of tunnel support.) The components of Q_{TBM} are as follows:

$$Q_{TBM} = \frac{RQD_0}{J_n} \times \frac{J_r}{J_a} \times \frac{J_w}{SRF} \times \frac{\sigma_{cm} \text{ or } \sigma_{tm}}{F^{10}/20^9} \times \frac{20}{CLI} \times \frac{q}{20} \times \frac{\sigma_\theta}{5} \tag{14.1}$$

where RQD_0 = RQD (%) interpreted in the tunneling direction. RQD_0 is also used when evaluating the Q-value for rock mass strength estimation; J_n, J_r, J_a, J_w, and SRF = ratings of Barton et al. (1974) and are unchanged (Chapter 8); F = average cutter load (tnf) through the same zone, normalized by 20 tnf (the reason for the high power terms will be discussed later); σ_{cm} or σ_{tm} = compressive and tensile rock mass strength estimates (MPa) in the same zone; CLI = cutter life index (e.g., 4 for quartzite, 90 for limestone); Q = quartz content in percentage terms; and σ_θ = induced biaxial stress on tunnel face (approximately MPa) in the same zone, normalized to an approximate depth of 100 m (= 5 MPa).

The best estimates of each parameter should be assembled on a geological/structural longitudinal section of the planned (or progressing) tunnel. It may be noted that the Q-value should not be calculated using correlations with the rock mass rating (RMR).

The rock mass strength estimate incorporates the Q-value (but with oriented RQD_0), together with the rock density (from an approach by Singh, 1993). The Q-value is

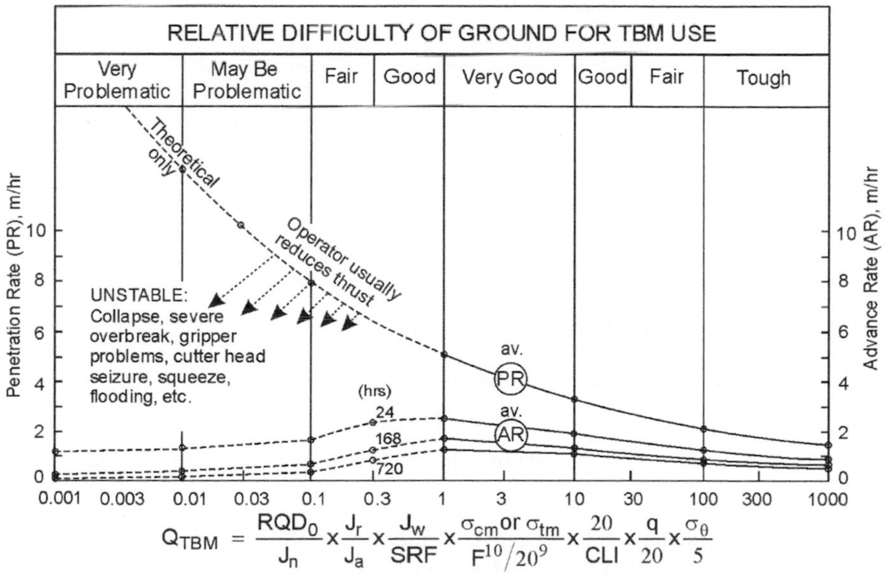

FIGURE 14.2 Suggested relation between PR, AR, and Q_{TBM}. *(From Barton, 2000b)*

normalized by uniaxial strength (q_c) different from 100 MPa (typical hard rock) as defined in Eq. (14.3a) and is normalized by point load strength (I_{50}) different from 4 MPa. A simplified (q_c/I_{50}) conversion of 25 is assumed. Relevant I_{50} anisotropy in relation to the direction of tunneling should be quantified by point load tests in strongly foliated or schistose rocks. The choice between σ_{cm} and σ_{tm} depends on the angle between the tunnel axis and the major discontinuities or foliations of the rock mass to be bored (Barton, 2000b). Use σ_{cm} when the angle is more than 45 degrees and σ_{tm} when the angle is less than 45 degrees. The penetration rate is more when the angle is zero degree.

$$\sigma_{cm} = 5 \cdot \gamma Q_c^{1/3} \tag{14.2}$$

$$\sigma_{tm} = 5 \cdot \gamma Q_t^{1/3} \tag{14.3a}$$

where

$$Q_c = Q \cdot q_c/100 \tag{14.3b}$$

$$Q_t = Q \cdot q_t/100 \tag{14.3c}$$

and $= Q \cdot (I_{50}/4)$ and γ = density in gm/cm^3.

Equations (14.2) and (14.3a) for the estimation of σ_{cm} and σ_{tm} are proposed only for Q_{TBM} where they are useful as a relative measure for comparing with the cutter force (Barton, 2005).

Example 14.1

Slate $Q \approx 2$ (poor stability); $q_c \approx 50$ MPa; $I_{50} \approx 0.5$ MPa; $\gamma = 2.8$ gm/cm^3; $Q_c = 1$; and $Q_t = 0.25$. Therefore, $\sigma_{cm} \approx 14$ MPa and $\sigma_{tm} \approx 8.8$ MPa.

The slate is bored in a favorable direction, hence consider σ_{tm} and $RQD_0 = 15$ (i.e., <RQD). Assume that the average cutter force = 15 tnf; CLI = 20; q = 20%; and $\sigma_\theta = 15$ MPa (approximately 200 m depth). The cleavage joints have $J_r/J_a = 1/1$ (smooth, planar, unaltered). The estimate of Q_{TBM} is as follows:

$$Q_{TBM} = \frac{15}{6} \times \frac{1}{1} \times \frac{0.66}{1} \times \frac{8.8}{15^{10}/20^9} \times \frac{20}{20} \times \frac{20}{20} \times \frac{15}{5} = 39$$

According to Figure 14.2, $Q_{TBM} \approx 39$ should give fair penetration rates (about 2.4 m/h). If the average cutter force was doubled to 30 tnf, Q_{TBM} would reduce to a much more favorable value of 0.04 and PR would increase (by a factor $2^2 = 4$) to a potential 9.6 m/h. However, the real advance rate would depend on tunnel support needs and on conveyor capacity (Barton, 1999).

PENETRATION AND ADVANCE RATES

The ratio between AR and PR is the utilization factor U,

$$AR = PR \cdot U \tag{14.4}$$

The decelerating trend of all the data may be expressed in an alternative format:

$$AR = PR \cdot T^m \tag{14.5}$$

where T is time in hours and the negative gradient (m) values are cited in Table 14.1.

The values of m given in Table 14.1 may be refined in the future as more and more cases of TBM tunnels become available (Barton, 1999).

TABLE 14.1 Deceleration Gradient (−)m and Its Approximate Relation to Q-Value

Q	0.001	0.01	0.1	1	10	100	1000
m_1	−0.9	−0.7	−0.5	−0.22	−0.17	−0.19	−0.21
Unexpected events or expected bad ground. Many stability and support-related delays and gripper problems. Operator reduces PR. This increases Q_{TBM}.				Most variation of (−)m may be due to rock abrasiveness, i.e., cutter life index (CLI), quartz content, and porosity are important. PR depends on Q_{TBM}.			

Subscript 1 is added to m for evaluation by Eq. (14.6).

CUTTER WEAR

The final gradient (−)m can be modified by the abrasiveness of the rock, which is based on a normalized value of CLI (see *www.drillability.com*). Values less than 20 rapidly reduce cutter life, and values over 20 tend to increase cutter life. A typical value of CLI for quartzite might be 4 and for shale 80. Because quartz content (q%) and porosity (n%) may accentuate cutter wear, they are also included in Eq. (14.6) to fine-tune the gradient.

It is also necessary to consider the tunnel size and support needs when measuring cutter wear. Although large tunnels can be driven almost as fast (or even faster) as small tunnels in similar good rock conditions (Dalton, DeVita, & Macaitis, 1993), more support-related delays occur if the rock is consistently poor in the larger tunnel. Therefore, a normalized tunnel diameter (D) of 5 m is used to slightly modify the gradient (m). (Q_{TBM} is already adjusted for tunnel size by the use of the AR cutter force.)

The fine-tuned gradient (−)m is estimated as follows (Barton, 1999):

$$m \approx m_1 \left(\frac{D}{5}\right)^{0.20} \left(\frac{20}{CLI}\right)^{0.15} \left(\frac{q}{20}\right)^{0.10} \left(\frac{n}{2}\right)^{0.05} \tag{14.6}$$

Sometimes PR comes too fast due to logistics and muck handling. There may be a local increase in gradient from 1 hour to 1 day because a more rapid fall occurs in AR.

PENETRATION AND ADVANCE RATES VERSUS Q_{TBM}

The development of a workable relationship between PR and Q_{TBM} was based on trial and error using case records (Barton, 2000a). Striving for a simple relationship, and rounding decimal places, the following correlation was obtained for open TBM:

$$PR \approx 5\,(Q_{TBM})^{-0.2} \tag{14.7}$$

From Eq. (14.5) we can, therefore, also estimate AR as follows:

$$AR \approx 5\,(Q_{TBM})^{-0.2} \cdot T^{m} \tag{14.8}$$

We can also check the operative Q_{TBM} value by back calculation from penetration rate:

$$Q_{TBM} \approx (5/PR)^{5} \tag{14.9}$$

ESTIMATING TIME FOR COMPLETION

The time (T) taken to penetrate a length of tunnel (L) with an average AR is L/AR. From Eq. (14.5) we can derive the following:

$$T = \left(\frac{L}{PR}\right)^{\frac{1}{1+m}} \tag{14.10}$$

Equation (14.10) also demonstrates instability in fault zones, until (−)m is reduced pre- or post-treatment.

Example 14.2

Slate: $Q_{TBM} \approx 39$ (from previous calculations with 15 tnf cutter force). From Eq. (14.7), $PR \approx 2.4$ m/h. Since $Q = 2$, $m_1 = -0.21$ from Table 14.1. If the TBM diameter is 8 m and if $CLI = 45$, $q = 5\%$, and $n = 1\%$, then $m \approx -0.21 \times 1.1 \times 0.89 \times 0.87 \times 0.97 = -0.17$ from Eq. (14.6). If 1 km of slate with similar orientation and rock quality is encountered, it will take the following time to bore it, according to Eq. (14.10):

$$T = (1000/2.4)^{(1/0.83)} = 1433 \text{ hours} \approx 2 \text{ months}$$

i.e., $AR \approx 0.7$ m/h, as also found by using Eq. (14.8) and $T = 1433$ hours.

A working model for estimating open TBM PRs and ARs for different rock conditions, lengths of tunnel, and time of boring was presented. It may be used for prediction and back analysis. Since the model is new, Barton (1999) emphasized that improvements and corrections may be possible as case records become available. Q_{TBM} has been applied successfully in 37 tunnels. Shielded TBM is very useful in metro tunnels. The expensive double-shielded TBMs have been successful in boring through complex geological conditions at shallow depths. Their PR is faster than open TBMs in weak rock masses ($q_c < 45$ MPa).

RISK MANAGEMENT

Okubo et al. (2003) developed a comprehensive expert system, based on a unified knowledge base, for predicting the PRs of TBMs in Japan. The primary reasons for lower PRs are complex ground conditions, inexperience of operators, and shortage of haulage

TABLE 14.2 Difficult Conditions for TBM

Tunnel length	Below 500 m
Excavation diameter	Below 2 m and above 10 m
Minimum radius of curvature	Below 50 m
Gradient	Above 30°
Uniaxial compressive strength	Below 5 MPa or above 250 MPa

Source: Okubo et al., 2003

capacity. The ground conditions in Japan are difficult to forecast due to rapidly changing groundwater levels and the prevalence of fracture zones (shear zones). Table 14.2 describes other difficult conditions for TBMs.

Further, Barton (2004) suggested that probe hole, an efficient drainage, and pre-grouting ahead of the tunnel face are three of the most effective ways to reduce risk, but this may be difficult in TBM tunneling.

Recently dual-mode shield TBMs, developed by M/s Herrenknecht in Germany, bore through in all soil, boulders, and weak rocks (in non-squeezing ground) under a high groundwater table. The advantage of fully shielded TBMs with a pre-cast segment erector is that there is no unsupported ground behind the shield. This is why TBMs have failed in poor ground yet dual-shield TBMs have succeeded. These same TBMs have been used successfully in underground Delhi metros. The details are described by Singh and Goel (2006).

REFERENCES

Barton, N. (1999). TBM performance estimation in rock using Q_{TBM}. *Tunnels and Tunnelling International*, *31*(9), 30–34.

Barton, N. (2000a). *TBM tunnelling in jointed and faulted rock* (p. 173). The Netherlands: A.A. Balkema.

Barton, N. (2000b). Employing the Q_{TBM} prognosis model. *Tunnels and Tunnelling International*, 20–23.

Barton, N. (2004). Risk and risk reduction in TBM rock tunnelling. In Y. Ohnishi & K. Aoki (Eds.), *Proceedings of the ISRM International. Symposium on Contribution of Rock Mechanics to the New Century* (Vol. 1, pp. 29–38). Millpress Japan.

Barton, N. (2005). Personal communication with R.K. Goel.

Barton, N., Lien, R., & Lunde, J. (1974). Engineering classification of rock masses for the design of tunnel supports. In *Rock mechanics* (Vol. 6, No. 4, pp. 189–236). New York: Springer-Verlag.

Bhasin, R. (2004). Personal communication with Bhawani Singh, IIT Roorkee.

Dalton, F. E., DeVita, L. R., & Macaitis, W. A. (1993). TARP tunnel boring machine performance. In L. D. Bowerman & J. E. Monsees (Eds.), *Proceedings of the RETC Conference* (pp. 445–451). Boston, MA: SME.

Grimstad, E., & Barton, N. (1993). Updating of the Q-system for NMT. In: *Proceedings of the International Symposium on Sprayed Concrete — Modern Use of Wet Mix Sprayed Concrete for Underground Support*, Oslo: Fagernes, Norwegian Concrete Association.

Johannessen, S., & Askilsrud, O. G. (1993). Meraaker hydro tunnelling the "Norwegian Way." In L. D. Bowerman & J. E. Monsees (Eds.), *Proceedings of the RETC Conference* (pp. 415–427). Boston, MA: Society of Mining.

Okubo, S., Fukui, K., & Chen, W. (2003). Expert system for applicability of tunnel boring machines in Japan. *Rock Mechanics and Rock Engineering*, *36*(4), 305–322.

Robbins, R. J. (1982). The application of tunnel boring machines to bad rock conditions. In W. Wittke (Ed.), *Proceedings of the ISRM Symposium*, Aachen, Germany (Vol. 2, pp. 827–836). Rotterdam: A. A. Balkema.

Singh, B. (1993). *Norwegian Method of Tunnelling Workshop*. Lecture at CSMRS, New Delhi, India.

Singh, B., & Goel, R. K. (2006). J. A. Hudson (Ed.), *Tunnelling in weak rocks* (p. 489). Amsterdam: Elsevier.

Chapter 15

Strength of Discontinuities

Failure is success if we learn from it.

Malcolm S. Forbes

INTRODUCTION

Rock mass is a heterogeneous, anisotropic, discontinuous mass. When civil engineering structures like dams are founded on rock, they transmit normal and shear stresses on discontinuities in rock mass. Failure may be initiated by sliding along a joint plane near or along the foundation or along the abutments of a dam. For a realistic assessment of the stability of structure with wedge, estimation of the shear resistance of a rock mass along any desired plane of potential shear or along the weakest discontinuity becomes essential. The shear strength of discontinuities depends upon the alteration of joints or the discontinuities, the roughness, the thickness of infillings or the gouge material, the moisture content, and so forth.

The mechanical difference between contacting and non-contacting joint walls usually results in widely different shear strengths and deformation characteristics. For unfilled joints, the roughness and compressive strength of the joint walls are important, whereas with filled joints the physical and mineralogical properties of the gouge material separating the joint walls are of primary concern (Chapter 24).

To quantify the effect of these parameters on the strength of discontinuities, various researchers have proposed different parameters and correlations for obtaining strength parameters. Barton, Lien, and Lunde (1974), probably for the first time, considered joint roughness (J_r) and joint alteration (J_a) in their Q-system to account for the strength of clay-coated discontinuities in the rock mass classification. Later, Barton and Choubey (1977) defined two parameters — joint wall roughness coefficient (JRC) and joint wall compressive strength (JCS) — and proposed an empirical correlation for friction of rock joints without fillings, which can be used for accurately predicting shear strength.

JOINT WALL ROUGHNESS COEFFICIENT

The wall roughness of a joint or discontinuity is potentially a very important component of its shear strength, especially with undisplaced and interlocked features (e.g., unfilled joints). The importance of wall roughness declines as thickness of aperture filling or the degree of any previous shear displacement increases.

JRC_0 (JRC at laboratory scale) may be obtained by visually matching actual roughness profiles with the set of standard profiles proposed by Barton and Choubey (1977). As such, the joint roughness coefficients are suggested for ten types of roughness

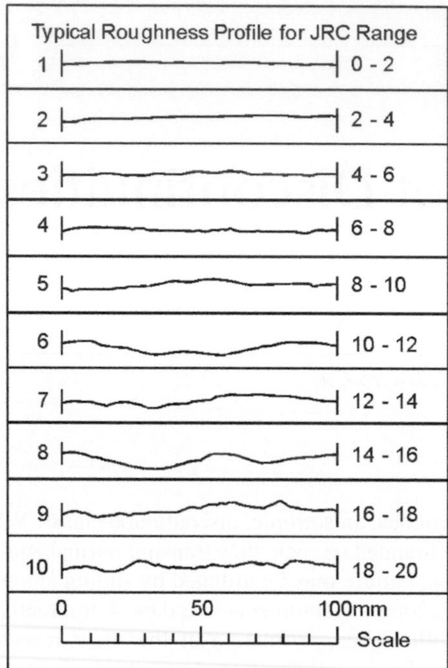

FIGURE 15.1 Standard profiles for visual estimation of JRC. *(From Barton and Choubey, 1977)*

profiles of joints (Figure 15.1). The core sample is intersected by joints at angles varying from 0 to 90° to the axis. Joint samples can vary from a meter or more in length (depending upon the core length) to 100 mm (core diameter). Most samples are expected to range from 100 to 300 mm in length.

The recommended approximate sampling frequency for the above profile-matching procedure is 100 samples per joint set per 1000 m of core. The two most adverse prominent sets should be selected, which must include the adverse joint set selected for J_r and J_a characterization.

Roughness amplitude along a joint length (i.e., a and L measurements), will be made in the field for estimating JRC_n (JRC at a natural large scale). The maximum amplitude of roughness (in millimeters) is usually estimated or measured on profiles of at least two lengths along the joint plane, for example, 100 mm and 1 m length.

It has been observed that the JRC_n can also be obtained from JRC_o using the following equation:

$$JRC_n = JRC_o\,(L_n/L_o)^{-0.02\ JRC_o} \tag{15.1}$$

where L_o is the laboratory scale length (100 mm) and L_n represents the natural larger scale length. The chart of Barton (1982) presented in Figure 15.2 is easier to use for evaluating JRC_n according to the amplitude of asperities and the length of joint profile, which are studied in the field.

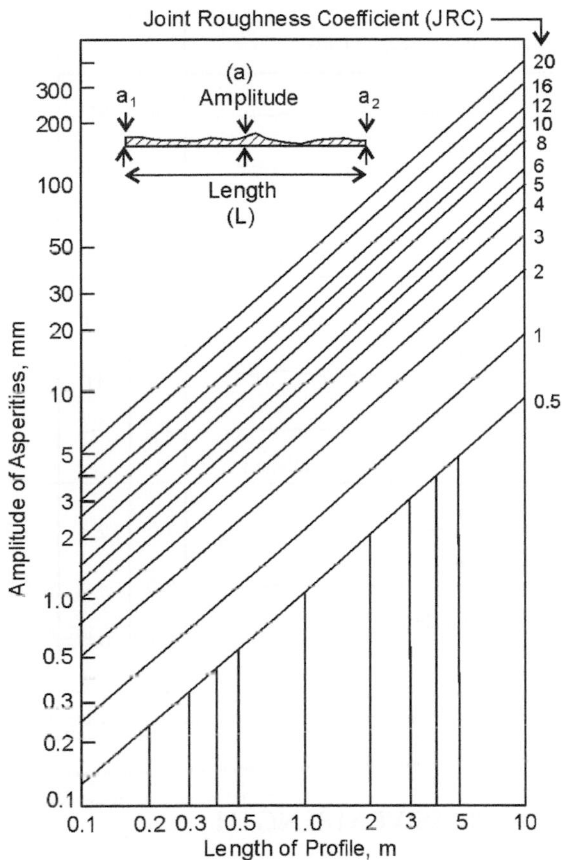

FIGURE 15.2 Assessment of JRC from amplitude of asperitics and length of joint profile. *(From Barton, 1982)*

Relationship between J_r and JRC Roughness Descriptions

The description of roughness in the Q-system given by the parameter J_r and the JRC are related. Figure 15.3 has been prepared by Barton (1993) for engineers who use these rock mass descriptions. The ISRM (1978) suggested methods for visual description of joint roughness profiles which have been combined with profiles given by Barton et al. (1980) and with Eq. (15.1), to produce some examples of the quantitative description of joint roughness provided by these parameters. Increasing experience leads to better visual assessment of JRC based on Figure 15.3.

The roughness profiles shown in Figure 15.3 are assumed to be at least 1 m in length. The column of J_r values would be used in the Q-system, while the JRC values for a 20- and 100-cm block size could be used to generate appropriate shear stress displacement and dilation-displacement curves.

Relation between J_r and JRC_n		J_r	JRC_{20}	JRC_{100}
I	Rough	4	20	11
II	Smooth	3	14	9
III	Slickensided	2	11	8
	Stepped			
IV	Rough	3	14	9
V	Smooth	2	11	8
VI	Slickensided	1.5	7	6
	Undulating			
VII	Rough	1.5	2.5	2.3
VIII	Smooth	1.0	1.5	0.9
IX	Slickensided	0.5	0.5	0.6
	Planar			

FIGURE 15.3 Suggested methods for the quantitative description of different classes of joints using J_r and JRC_n. (Subscripts refer to block size in centimeters.)

JOINT WALL COMPRESSIVE STRENGTH

The JCS of a joint or discontinuity is also an important component of its shear strength, especially with undisplaced and interlocked discontinuities such as unfilled joints (Barton & Choubey, 1977). Similar to JRC, the wall strength JCS decreases as aperture, filling thickness, or the degree of any previous shear displacement increases. JCS, therefore, does not need to be evaluated for thickly (>10 mm) filled joints.

In the field, JCS is measured by performing Schmidt hammer (L-type) tests on the two most prominent joint surfaces where it is smooth and averaging the highest 10 rebound values. JCS_o, the small-scale value of wall strength relative to a nominal joint length (L_o) of 100 mm, may be obtained from the Schmidt hammer rebound value (r) or by using Figure 15.4.

$$JCS_O = 10^{(0.00088\ r\ \gamma\ +\ 1.01)},\ \text{MPa} \tag{15.2}$$

where r = rebound number on smooth weathered joint and γ = dry unit weight of rocks (kN/m^3). If the Schmidt hammer is not used vertically downward, the rebound values need to be corrected to match the values given in Table 15.1.

The joint wall compressive strength may be equal to the uniaxial compressive strength (UCS) of rock material for unweathered joints; otherwise it should be estimated indirectly from the Schmidt hammer index test. The Schmidt hammer gives wrong results on rough joints; therefore, it is advisable not to use Schmidt hammer rebound for JCS with rough joints. Lump tests on saturated small lumps of asperities give a better

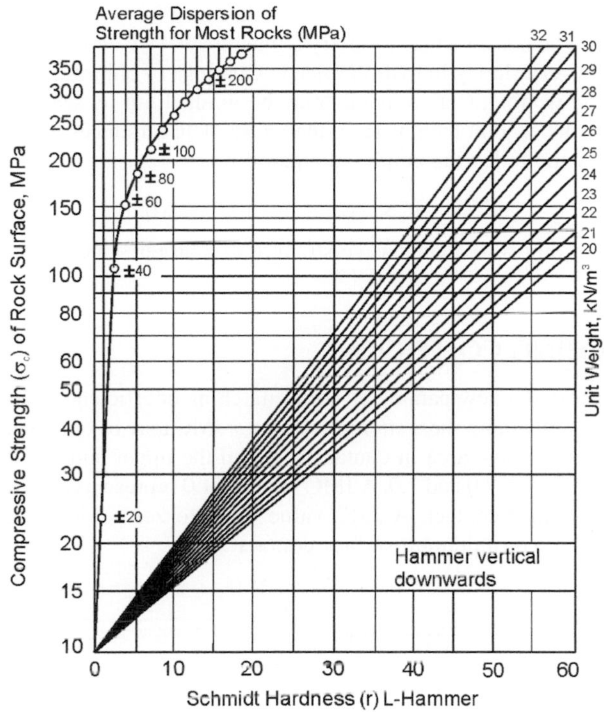

FIGURE 15.4 Correlation chart for compressive strength with rock density and Schmidt hammer rebound number on smooth surfaces. *(From Miller, 1965)*

TABLE 15.1 Corrections for the Orientation of Schmidt Hammer

Rebound	Downward		Upward		Horizontal
r	$\alpha = -90°$	$\alpha = -45°$	$\alpha = +90°$	$\alpha = +45°$	$\alpha = 0°$
10	0	−0.8	—	—	−3.2
20	0	−0.9	−8.8	−6.9	−3.4
30	0	−0.8	−7.8	−6.2	−3.1
40	0	−0.7	−6.6	−5.3	−2.7
50	0	−0.6	−5.3	−4.3	−2.2
60	0	−0.4	−4.0	−3.3	−1.7

Source: Barton and Choubey, 1977.

UCS or JCS_o. Quartz-coated joints in weak rock can give a high Schmidt hammer rebound number, which is a surface property (Bhasin, 2004). Calcite and gypsum infillings may dissolve very slowly in hydro projects. Coatings of chlorite, talc, and graphite reduce strength on wetting. Clay minerals may be washed out by seepage.

For larger blocks or joint lengths (L_n), the value of JCS reduces to JCS_n, where the two are related by the following empirical equation:

$$JCS_n = JCS_o \, (L_n/L_o)^{-0.03 \, JRC_o}, \text{ MPa} \tag{15.3}$$

where JCS_n is the joint wall compressive strength at a larger scale.

JOINT MATCHING COEFFICIENT

Zhao (1997) suggested a new parameter, joint matching coefficient (JMC), in addition to JRC and JCS, for obtaining shear strength of joints. JMC may be obtained by observing the approximate percentage area in contact between the upper and the lower walls of a joint with a value between 0 and 1.0. A JMC value of 1.0 represents a perfectly matched joint with 100% surface contact. A JMC value close to zero indicates a totally mismatched joint with no or minimum surface contact.

RESIDUAL ANGLE OF FRICTION

The effective basic or residual friction angle (ϕ_r) of a joint is an important component of its total shear strength, whether the joint is rock-to-rock interlocked or clay filled. The importance of ϕ_r increases as the clay coating or filling thickness increases up to a critical limit.

An experienced field observer can make a preliminary estimate of ϕ_r. The quartz-rich rocks and many igneous rocks have ϕ_r between 28 and 32°, whereas, mica-rich rock masses and rocks with considerable weathering have somewhat lower values of ϕ_r.

In the Barton-Bandis (1990) joint model, an angle of primary roughness is added to obtain the field value of effective peak friction angle for a natural joint (ϕ_j) without fillings

$$\phi_j = \phi_r + i + JRC \log_{10} (JCS/\sigma) < 70°; \text{ for } \sigma / JCS < 0.3 \tag{15.4}$$

where JRC accounts for secondary roughness in laboratory tests, i represents the angle of primary roughness (undulations) of a natural joint surface and is generally $\leq 6°$, and σ is the effective normal stress across joints.

The value of ϕ_r is important as roughness (JRC) and wall strength (JCS) are reduced through weathering. Residual frictional angle ϕ_r may also be estimated by the equation:

$$\phi_r = (\phi_b - 20°) + 20 \, (r/R) \tag{15.5}$$

where ϕ_b is the basic frictional angle obtained by sliding or tilt tests on dry, planar (but not polished), or cored surface of the rock ($\phi_p = \phi_r = \phi_b$ as JR C= 0: Table 15.2) (Barton & Choubey, 1977). R is the Schmidt rebound on fresh, dry, unweathered smooth surfaces of the rock and r is the rebound number on the smooth natural, perhaps weathered and water-saturated joints ($J_w = 1.0$).

According to Jaeger and Cook (1969), enhancement in the dynamic angle of sliding friction ϕ_r of smooth rock joints may be only about 2 degrees.

TABLE 15.2 Basic Friction Angles of Various Unweathered Rocks Obtained from Flat and Residual Surfaces

Rock type	Moisture condition	Basic friction angle, ϕ_b (degrees)
	A. Sedimentary rocks	
Sandstone	Dry	26–35
Sandstone	Wet	25–33
Sandstone	Wet	29
Sandstone	Dry	31–33
Sandstone	Dry	32–34
Sandstone	Wet	31–34
Sandstone	Wet	33
Shale	Wet	27
Siltstone	Wet	31
Siltstone	Dry	31–33
Siltstone	Wet	27–31
Conglomerate	Dry	35
Chalk	Wet	30
Limestone	Dry	31–37
Limestone	Wet	27–35
	B. Igneous rocks	
Basalt	Dry	35–38
Basalt	Wet	31–36
Fine-grained granite	Dry	31–35
Fine-grained granite	Wet	29–31
Coarse-grained granite	Dry	31–35
Coarse-grained granite	Wet	31–33
Porphyry	Dry	31
Porphyry	Wet	31
Dolerite	Dry	36
Dolerite	Wet	32
	C. Metamorphic rocks	
Amphibolite	Dry	32

Continued

TABLE 15.2 Basic Friction Angles of Various Unweathered Rocks Obtained from Flat and Residual Surfaces—Cont'd

Rock type	Moisture condition	Basic friction angle, ϕ_b (degrees)
	C. Metamorphic rocks	
Gneiss	Dry	26–29
Gneiss	Wet	23–26
Slate	Dry	25–30
Slate	Dry	30
Slate	Wet	21

Source: Barton and Choubey, 1977.

SHEAR STRENGTH OF JOINTS

Barton and Choubey (1977) proposed the following accurate, non-linear correlation for shear strength of natural joints.

$$\tau = \sigma \cdot \tan\left[\phi_r + JRC_n \log_{10}(JCS_n/\sigma)\right] \quad (15.6)$$

where τ is the shear strength of joints, JRC_n may be obtained easily from Figure 15.3, JCS_n from Eq. (15.3), and the rest of the parameters were defined earlier. Under very high normal stress levels ($\sigma >> q_c$ or JCS_n) the JCS_n value increases to the triaxial compressive strength ($\sigma_1 - \sigma_3$) of the rock material in Eq. (15.6) (Barton, 1976). It may be noted that at high normal pressure ($\sigma = JCS_n$), no dilatation takes place as all the asperities are sheared.

The effect of mismatching joint surface on its shear strength has been proposed by Zhao (1997) in his JRC–JCS shear strength model as

$$\tau = c_j + \sigma \cdot \tan\left[\phi_r + JMC \cdot JRC_n \log_{10}(JCS_n/\sigma)\right] \quad (15.7)$$

and dilatation (Δ) across joints is as follows

$$\Delta \approx \frac{1}{2} \cdot JMC \cdot JRC_n \cdot \log_{10}\left(\frac{JCS_n}{\sigma}\right)$$
$$\therefore \Delta \approx \left(\frac{\phi_j - \phi_r}{2}\right), \text{ beyond failure} \quad (15.8)$$

The minimum value of JMC in Eq. (15.8) is 0.3. The cohesion along discontinuity is c_j. Field experience shows that natural joints are not continuous as assumed in theory and laboratory tests; there are rock bridges in between them. The shear strength of these rock bridges adds to the cohesion of the overall rock joint (0–0.1 MPa). The real discontinuous joint should be simulated in the theory or computer program. Further, it may be assumed that dilatancy (Δ) is negligible before peak failure so the net work done by shear stress and (−) normal stress is always positive. Analysis must ensure that no strain energy is generated during dilatant behavior.

For highly jointed rock masses, failure takes place along the shear band (kink band) and not along the critical discontinuity, due to rotation of rock blocks at a low confining stress in rock slopes with continuous joint sets. The apparent angle of friction may be significantly lower in slender blocks. Laboratory tests on models with three continuous joint sets show some cohesion c_j (Singh, 1997). More attention should be given to strength of discontinuity in the jointed rock masses.

For joints filled with gouge or clay-coated joints, the following correlation of shear strength is used for low effective normal stresses (Barton & Bandis, 1990)

$$\tau = \sigma \cdot (J_r/J_a) \tag{15.9}$$

Indaratna and Haque (2000) presented new models of rock joints. They showed a minor effect of stress path on ϕ_j, as peak slip is more evident in constant normal stiffness than in the conventional constant normal loading at low normal stresses.

Sinha and Singh (2000) proposed an empirical criterion for shear strength of filled joints. The angle of internal friction is correlated to the plasticity index (PI) of normally consolidated clays (Lamb & Whitman, 1979). The same may be adopted for thick and normally consolidated clayey gouge in the rock joints as follows (see Chapter 24):

$$\sin \phi_j = 0.81 - 0.23 \log_{10} PI \tag{15.10}$$

Choubey (1998) suggested that the peak strength parameters should be used when designing a rock bolt system and retaining walls, where control measures do not permit large deformations along joints. For long-term stability of unsupported rock and soil slopes, residual strength parameters of rock joints and soil should be chosen in the analyses, respectively, as large displacement may eventually reduce the shear strength of the rock joint to its residual strength.

There is a wide statistical variation in the shear strength parameters found from direct shear tests. For design purposes, average parameters are generally evaluated from median values rejecting values that are too high and too low.

Barton, Bandis, & Bakhtar (1985) related the hydraulic aperture (e) to the measured (geometric) aperture (t) of rock joints when shear displacement is less than $0.75 \times$ peak slip:

$$e = \frac{JRC^{2.5}}{(t/e)^2} \tag{15.11}$$

where t and e are measured in μm. The permeability of rock mass may then be estimated, assuming laminar flow of water through two parallel plates with spacing (e) for each joint.

DYNAMIC SHEAR STRENGTH OF ROUGH ROCK JOINTS

Jain (2000) performed a large number of dynamic shear tests on dry rock joints at Nanyang Technological University (NTU) in Singapore. He observed that significant dynamic normal stress (σ_{dyn}) is developed across the rough rock joints; hence there is high rise in the dynamic shear strength. Thus, the effective normal stress (σ') in Eq. (15.7) may be

$$\begin{aligned} \sigma'_{dyn} &= \sigma_{static} - u_{static} + \sigma_{dyn} - u_{dyn} \\ &\geq \sigma'_{static} \end{aligned} \tag{15.12}$$

It is also imagined that negative dynamic pore water pressure (u_{dyn}) will develop in the water-charged joints due to dilatancy. This phenomenon is likely to be similar to undrained shearing of dilatant and dense sand or over-consolidated clay. Further research is needed to develop correlations for σ_{dyn} and u_{dyn} from dynamic shear tests on rock joints. There is likely to be significant increase in the dynamic shear strength of rock joints due to shearing of more asperities.

THEORY OF SHEAR STRENGTH AT VERY HIGH CONFINING STRESS

Barton (1976) suggested a theory of the critical state of rock materials at very high confining stresses. It appears that the Mohr's envelopes representing the peak shear strength of rock materials (intact) eventually reach a point of saturation (zero gradient on crossing a certain critical state line).

Figure 15.5 integrates all the three ideas on shear strength of discontinuities. The effective sliding angle of friction is about $\phi_r + i$ at low effective normal stresses, where i = angle of asperities of a rough joint. The shear strength (τ) cannot exceed shear strength of the asperities ($= c + \sigma \tan\phi_r$), where ϕ_r = effective angle of internal friction of the ruptured asperities of rock material. The non-linear Eq. (15.7) (with JCS = triaxial strength of rock) is closer to the experimental data than the bilinear theoretical relationship.

There is a critical limit to the shear strength of the rock joint that cannot be higher than the shear strength of weaker rock material at very high confining stress. Figure 15.5 illustrates this idea with the τ = constant saturation (critical state) line. It follows that the (sliding) angle of friction is nearly zero at very high confining stresses, which exist at great depth in the earth plates along inter-plate boundaries. It is interesting to note that the sliding angle of friction at great depth (>40 km) is back analyzed to be as low as 5 degrees in the Tibet Himalayan plate (Shankar, Kapur, & Singh, 2002). Re-crystallization of soft minerals is likely to occur creating smooth surface. The sliding angle of friction between earth plate and underlying molten rock is assumed to be zero, as the coefficient of friction between a fluid and any solid surface is governed by the minimum shear strength of the material. It is

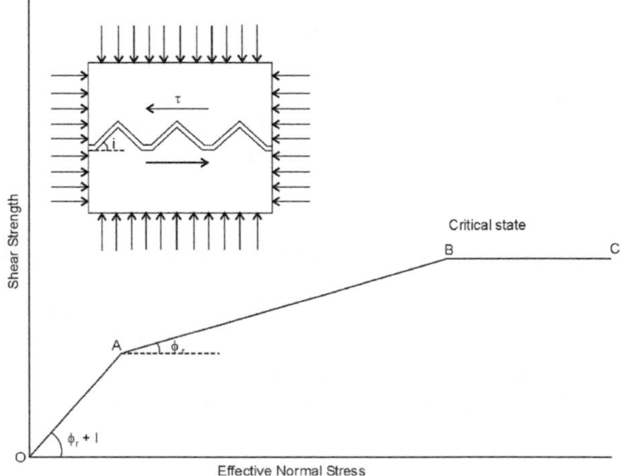

FIGURE 15.5 Shear strength of discontinuities at very high confining pressure (OA is sliding above asperities, AB is shearing of rock asperities, and BC is critical state of rock material at very high confining stress).

now necessary to perform shear tests at both very high confining stresses and high temperatures to find a generalized correlation between τ and σ along mega-discontinuities.

The less frictional resistance along the inter-continental and colliding plate boundaries, the less chance of locked up elastic strain energy in the large earth plates; hence there is less chance of great earthquakes in that area. The highest earthquake occurred in the Tibetan plateau and was only about 7.0 M on the Richter scale.

NORMAL AND SHEAR STIFFNESSES OF ROCK JOINTS

The values of static normal and shear stiffness are used in the finite element method and the distinct element method of analysis of rock structures. Singh and Goel (2002) listed their suggested values based on back analysis of uniaxial jacking tests in the United States and India. Appendix I lists these values.

Barton and Bandis (1990) also found correlation for shear stiffness. The shear stiffness of a joint is defined as the ratio between shear strength τ in Eq. (15.7) and the peak slip. The peak slip may be taken equal to (S/500) $(JRC/S)^{0.33}$, where S is equal to the length of a joint or simply the spacing of joints. Laboratory tests also indicate that the peak slip is nearly a constant for any given joint, irrespective of the normal stress. The normal stiffness of a joint may be 10 to 30 times its shear stiffness. This is the reason why the shear modulus of jointed rock masses is considered to be very low when compared to an isotropic elastic medium (Singh, 1973). The dynamic stiffness is likely to be significantly more than static values. The P-wave velocity and the dynamic normal stiffness may increase after saturation.

REFERENCES

Barton, N. (1976). The shear strength of rock and rock joints. *International Journal of Rock Mechanics and Mining Sciences—Geomechanics Abstracts*, *13*, 255–279.

Barton, N. (1982). Shear strength investigations for surface mining. In *3rd International Conference on Surface Mining* (Chap. 7, pp. 171–196). Vancouver: SME.

Barton, N. (1993). Predicting the behaviour of underground openings in rock. In *Proceedings of the Workshop on Norwegian Method of Tunnelling, CSMRS-NGI Institutional Cooperation Programme* (pp. 85–105). New Delhi, India, September.

Barton, N., Bandis, S., & Bakhtar, K. (1985). Strength deformation and conductivity coupling of rock joints. *International Journal of Rock Mechanics and Mining Sciences—Geomechanics Abstracts*, *22*, 121–140.

Barton, N., & Brandis, S. (1990). Review of predictive capabilities of JRC-JCS model in engineering practice. In N. R. Barton & O. Stephansson (Eds.), *Rock Joints Proceedings of a Regional Conference of the International Society for Rock Mechanics* (p. 820). Leon.

Barton, N., & Choubey, V. D. (1977). The shear strength of rock joints in theory and practice. *Rock mechanics* (Vol. 1/2, pp. 1–54). New York: Springer-Verlag and NGI-Publ. 119, 1978.

Barton, N., Lien, R., & Lunde, J. (1974). Engineering classification of rock masses for the design of tunnel support. In *Rock mechanics* (Vol. 6, No. 4, pp. 189–236). New York: Springer-Verlag.

Barton, N., Loset, F., Lien, R., & Lunde, J. (1980). Application of Q-system in design decisions concerning dimensions and appropriate support for underground installations. In *International Conference on Sub-Surface Space, Rock Store* (Vol. 2, pp. 553–561). Stockholm: Sub-Surface Space.

Bhasin, R. (2004). Personal communication with Bhawani Singh, IIT Roorkee, Uttarakhand, India.

Choubey, V. D. (1998). Landslide hazard assessment and management in Himalayas. In *International Conference on Hydro Power Development in Himalayas* (pp. 220–238). Shimla, India.

Indaratna, B., & Haque, A. (2000). *Shear behaviour of rock joints* (p. 164). Rotterdam: A. A. Balkema.

ISRM. (1978). Suggested methods for the quantitative description of discontinuities in rock masses (coordinator N. Barton). *International Journal of Rock Mechanics and Mining Sciences—Geomechanics Abstracts, 15*, 319–368.

Jaeger, J. C., & Cook, N. G. W. (1969). *Fundamentals of rock mechanics* (Article 3.4). London: Mathew and Co. Ltd.

Jain, M. (2000). Personal communication with Bhawani Singh, IIT Roorkee, Uttarakhand, India.

Lamb, T. W., & Whitman, R. V. (1979). *Soil mechanics* (Chap. 21.1, p. 533). New Delhi: Wiley Eastern Ltd.

Miller, R. P. (1965). *Engineering classification and index properties for intact rock* (p. 282). Ph.D. Thesis. Chicago: University of Illinois.

Shankar, D., Kapur, N., & Singh, B. (2002). Thrust-wedge mechanics and development of normal and reverse faults in the Himalayas. *Journal of the Geological Society, London, 159*, 273–280.

Singh, B. (1973). Continuum characterization of jointed rock mass: Part II—Significance of low shear modulus. *International Journal of Rock Mechanics and Mining Sciences—Geomechanics Abstracts, 10*, 337–349.

Singh, B., & Goel, R. K. (2002). *Software for engineering control of landslide and tunnelling hazards* (Chap. 5, p. 344). Rotterdam: A. A. Balkema (Swets & Zeitlinger).

Singh, M. (1997). *Engineering behaviour of joints model materials* (p. 339). Ph.D. Thesis. Uttarakhand, India: IIT Delhi.

Sinha, U. N., & Singh, B. (2000). Testing of rock joints filled with gouge using a triaxial apparatus. *International Journal of Rock Mechanics and Mining Sciences, 37*, 963–981.

Zhao, J. (1997). Joint surface matching and shear strength, Part B: JRC-JMC shear strength criterion. *International Journal of Rock Mechanics and Mining Sciences, 34*(2), 179–185.

Chapter 16

Shear Strength of Rock Masses in Slopes

Failure does not take place homogeneously in a material, but failure occurs by strain localization along shear bands, tension cracks in soils, rocks, concrete, masonry and necking in ductile material.

Professor G.N. Pandey (1997)

MOHR-COULOMB STRENGTH PARAMETERS

Stability analysis of a rock slope requires assessment of shear strength parameters, that is, cohesion (c) and angle of internal friction (ϕ) of the rock mass. Estimates of these parameters are usually not based on extensive field tests. Mehrotra (1992) carried out extensive block shear tests to study the shear strength parameters of rock masses. The following inferences may be drawn from this study:

1. The rock mass rating (RMR) system can be used to estimate the shear strength parameters c and ϕ of the weathered and saturated rock masses. It was observed that the cohesion (c) and the angle of internal friction (ϕ) increase when RMR increases (Figure 16.1).
2. The effect of saturation on shear strength parameters has been found to be significant. For poor saturated (wet) rock masses, a maximum reduction of 70% has been observed in cohesion (c), whereas the reduction in angle of internal friction (ϕ) is of the order of 35% when compared to those for the dry rock masses.
3. Figure 16.1 shows that there is a non-linear variation of the angle of internal friction with RMR for dry rock masses. This study also shows that ϕ values of Bieniawski (1989) are somewhat conservative.

NON-LINEAR FAILURE ENVELOPES FOR ROCK MASSES

Dilatancy in a rock mass is unconstrained near slopes as normal stress on joints is small due to weight of the wedge. Therefore, the failure of a rock mass occurs partially along joints and partially in non-jointed portions such as in solid rocks, but in massive rocks, it may occur entirely in solid rocks. Therefore, the failure of a rock mass lies within the area bounded by the failure envelope for a solid rock and a joint. The mode of failure thus depends on the quality and the type of the rock mass under investigation.

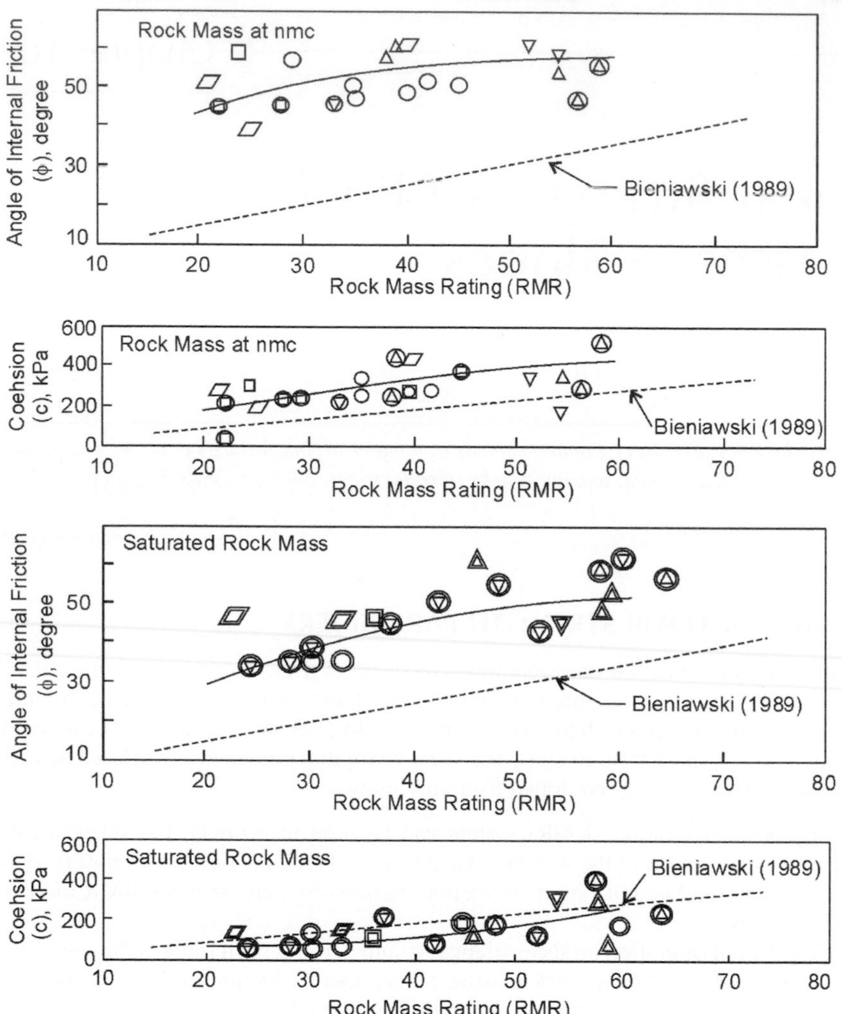

FIGURE 16.1 Relationship between rock mass rating and shear strength parameters, cohesion (c), and angle of internal friction (ϕ) (nmc: natural moisture content). *(From Mehrotra, 1992)*

For poor rock masses, the magnitude of normal stress (σ) significantly influences the shear strength; therefore a straight-line envelope is not a proper fit for such data and is likely to lead to overestimation of the angle of internal friction (ϕ) at higher normal stresses.

When the in situ rock mass is in a situation of post-peak failure of the original rock (Rao, personal communication), the failure envelopes for the rock masses generally show a non-linear trend. A straight-line criterion may be valid only when loads are small ($\sigma << q_c$), which is generally not the case in civil engineering (hydroelectric) projects where the intensity of stresses is comparatively high. The failure envelopes based on generalized empirical power law may be expressed as follows (Hoek & Brown, 1980):

$$\tau = A(\sigma + T)^B \tag{16.1}$$

where τ = shear strength of rock mass, A and B = rock mass constants, and T = tensile strength of rock mass.

For known values of power factor B, constants A and T have been worked out from a series of block shear test data. Consequently, empirical equations for the rock masses, both at natural moisture content and at saturation, have been calculated for defining failure envelopes. The values of the power factor B have been assumed to be the same as in the equations proposed by Hoek and Brown (1980) for heavily jointed rock masses.

Mehrotra (1992) plotted the Mohr envelopes for four different categories of rock masses: (1) limestones; (2) slates, xenoliths, and phyllites; (3) metabasics and traps; and (4) sandstones and quartzites. One such typical plot is shown in Figure 16.2. The constants A and T have been estimated using the results obtained from in situ block shear tests carried out on the lesser Himalayan rocks. Recommended non-linear strength envelopes (Table 16.1) can be used only for preliminary designs of dam abutments and rock slopes. There is a scope of refinement if the present data are supplemented with in situ triaxial test data. For RMR > 60, shear strength is governed by strength of rock material, because the failure plane will partly pass through solid rock.

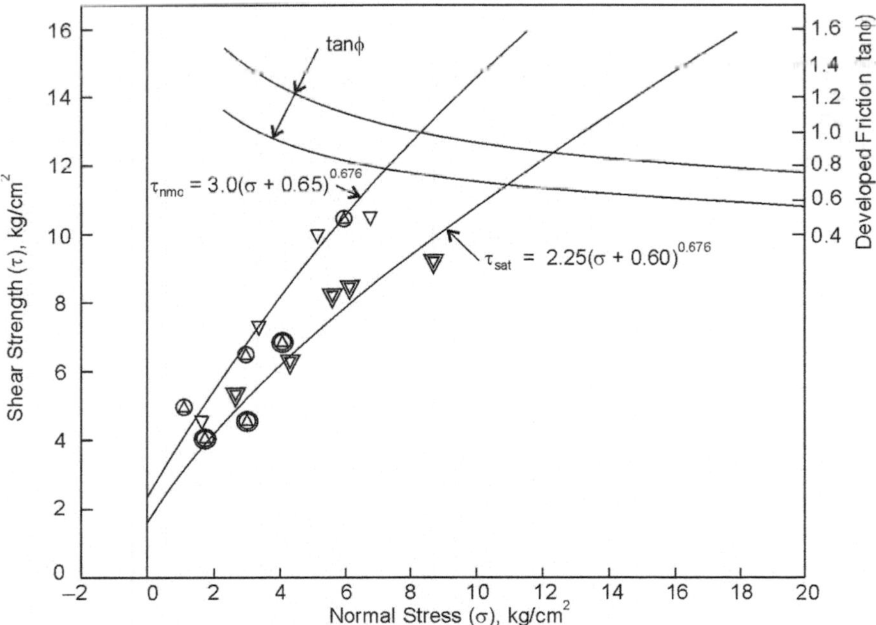

FIGURE 16.2 Failure envelopes for jointed trap and metabasic rocks at natural moisture content (nmc) and undersaturated conditions.

TABLE 16.1 Recommended Mohr Envelopes for Slopes in Jointed Rock Masses

S. No.	Rock type/quality	Limestone	Slate, xenolith, phyllite	Sandstone, quartzite	Trap, metabasics
1	Good rock mass RMR = 61–80 Q = 10–40	$\tau_n(nmc) = 0.38$ $(\sigma_n + 0.005)^{0.669}$ $\tau_n(sat) = 0.35$ $(\sigma_n + 0.004)^{0.669}$ $[S = 1]$	$\tau_n(nmc) = 0.42$ $(\sigma_n + 0.004)^{0.683}$ $\tau_n(sat) = 0.38$ $(\sigma_n + 0.003)^{0.683}$ $[S = 1]$	$\tau_n(nmc) = 0.44$ $(\sigma_n + 0.003)^{0.695}$ $\tau_n(sat) = 0.43$ $(\sigma_n + 0.002)^{0.695}$ $[S = 1]$	$\tau_n(nmc) = 0.50$ $(\sigma_n + 0.003)^{0.698}$ $\tau_n(sat) = 0.49$ $(\sigma_n + 0.002)^{0.698}$ $[S = 1]$
2	Fair rock mass RMR = 41–60 Q = 2–10	$\tau_{nmc} = 2.60$ $(\sigma + 1.25)^{0.662}$ $\tau_{sat} = 1.95$ $(\sigma + 1.20)^{0.662}$ $[S = 1]$	$\tau_{nmc} = 2.75$ $(\sigma + 1.15)^{0.675}$ $[S_{av} = 0.25]$ $\tau_{sat} = 2.15$ $(\sigma + 1.10)^{0.675}$ $[S = 1]$	$\tau_{nmc} = 2.85$ $(\sigma + 1.10)^{0.688}$ $[S_{av} = 0.15]$ $\tau_{sat} = 2.25$ $(\sigma + 1.05)^{0.688}$ $[S = 1]$	$\tau_{nmc} = 3.05$ $(\sigma + 1.00)^{0.691}$ $[S_{av} = 0.35]$ $\tau_{sat} = 2.45$ $(\sigma + 0.95)^{0.691}$ $[S = 1]$
3	Poor rock mass RMR = 21–40 Q = 0.5–2	$\tau_{nmc} = 2.50$ $(\sigma + 0.80)^{0.646}$ $[S_{av} = 0.20]$ $\tau_{sat} = 1.50$ $(\sigma + 0.75)^{0.646}$ $[S = 1]$	$\tau_{nmc} = 2.65$ $(\sigma + 0.75)^{0.655}$ $[S_{av} = 0.40]$ $\tau_{sat} = 1.75$ $(\sigma + 0.70)^{0.655}$ $[S = 1]$	$\tau_{nmc} = 2.80$ $(\sigma + 0.70)^{0.672}$ $[S_{av} = 0.25]$ $\tau_{sat} = 2.00$ $(\sigma + 0.65)^{0.672}$ $[S = 1]$	$\tau_{nmc} = 3.00$ $(\sigma + 0.65)^{0.676}$ $[S_{av} = 0.15]$ $\tau_{sat} = 2.25$ $(\sigma + 0.60)^{0.676}$ $[S = 1]$
4	Very Poor rock mass RMR < 21 Q < 0.5	$\tau_{nmc} = 2.25$ $(\sigma + 0.65)^{0.534}$ $\tau_{sat} = 0.80\ (\sigma)^{0.534}$ $[S = 1]$	$\tau_{nmc} = 2.45$ $(\sigma + 0.60)^{0.539}$ $\tau_{sat} = 0.95\ (\sigma)^{0.539}$ $[S = 1]$	$\tau_{nmc} = 2.65$ $(\sigma + 0.55)^{0.546}$ $\tau_{sat} = 1.05\ (\sigma)^{0.546}$ $[S = 1]$	$\tau_{nmc} = 2.90$ $(\sigma + 0.50)^{0.548}$ $\tau_{sat} = 1.25\ (\sigma)^{0.548}$ $[S = 1]$

$\tau_n = \tau/q_c$; $\sigma_n = \sigma/q_c$; σ is in kg/cm^2; $\tau = 0$ if $\sigma < 0$; S = degree of saturation (average value of degree of saturation is shown by S_{av}) = 1 for completely saturated rock mass.

Source: Mehrotra, 1992.

The results of Mehrotra's (1992) study for poor and fair rock masses are presented below.

Poor Rock Masses (RMR = 23 to 37)

1. It is possible to estimate the approximate shear strength from data obtained from in situ block shear tests.
2. Shear strength of the rock mass is stress dependent. The cohesion of the rock mass varies from 0.13 to 0.16 MPa for saturated and about 0.22 MPa for naturally moist rock masses.
3. Beyond the normal stress (σ) value of 2 MPa, there is no significant change in the values of tanϕ. It is observed that the angle of internal friction (ϕ) of rock mass is asymptotic at 20 degrees.

Bieniawski (1989) suggested that ϕ may decrease to zero if RMR reduces to zero. This is not borne out by field experience. Even sand has a much higher angle of internal friction. Limited direct shear tests by the University of Roorkee (now IIT, Roorkee) in India suggest that ϕ is above 15 degrees for very poor rock masses (RMR = 0–20).

Fair Rock Masses (RMR = 41 to 58)

1. It is possible to estimate approximate shear strength from in situ block shear test data.
2. Shear strength of a rock mass is stress dependent. At natural moisture content the cohesion intercept of the rock mass is about 0.3 MPa. At saturation, the cohesion intercept varies from 0.23 to 0.24 MPa.
3. Beyond a normal stress (σ) value of 2 MPa, there is no significant change in the values of tanϕ. It is observed that the angle of internal friction of a rock mass is asymptotic at 27 degrees.
4. The effect of saturation on the shear strength is found to be significant. When saturated, the reduction in the shear strength is about 25% at the normal stress (σ) of 2 MPa.

STRENGTH OF ROCK MASSES IN SLOPES

1. E_d and q_{cmass} are significantly higher in deep tunnels than those near the ground surface and rock slopes for the same value of rock mass quality except near faults and thrusts.
2. The Hoek, Wood, and Shah (1992) criterion is applicable to rock slopes and opencast mines with weathered and saturated rock masses. Block shear tests suggest q_{cmass} to be 0.38 γ $Q^{1/3}$ MPa ($Q < 10$), as joint orientation becomes a very important factor due to unconstrained dilatancy and negligible intermediate principal stress unlike in tunnels. So, block shear tests are recommended only for slopes and not for supported deep underground openings (Singh et al., 1998).
3. The angle of internal friction of rock masses with mineral-coated joint walls may be assumed as $\tan^{-1}(J_r/J_a)$ approximately for low normal stresses.
4. Rock slopes both σ_2 and σ_3 are negligible; there is insignificant or no strength enhancement. Block shear tests on rock masses give realistic results for rock slopes

and dam abutments only, because σ_2 is zero in these tests. It is most important that the blocks of rock masses are prepared with extreme care to represent the undisturbed rock mass.

5. In rock slopes, E_d is found to be lower due to complete relaxation of in situ stress, low confining pressures σ_2 and σ_3, excessive weathering, and longer length of joints. For the same Q, q_{cmass} will also be low near rock slopes.
6. Table 16.1 may be used to estimate tensile strength of rock mass (value of σ for $\tau = 0$).

BACK ANALYSIS OF DISTRESSED SLOPES

The most reliable method for estimating strength parameters along discontinuities of rock masses is by appropriate back analysis of distressed rock slopes. Software packages BASP, BASC, and BAST have been developed at IIT Roorkee in India to back calculate strength parameters for planar, circular, and debris slides, respectively (Singh & Goel, 2002). The experience of careful back analysis of rock slopes also supports Bieniawski's values of strength parameters.

REFERENCES

Bieniawski, Z. T. (1989). *Engineering rock mass classification* (p. 251). New York: John Wiley & Sons.

Hoek, E., & Brown, E. T. (1980). *Underground excavations in rocks*. Institution of Mining and Metallurgy (revised ed., p. 527). London: Maney Publishing.

Hoek, E., Wood, D., & Shah, S. (1992). A modified Hoek-Brown failure criterion for jointed rock masses. In J. A. Hudson (Ed.), *ISRM Symposium, EUROCK '92 on Rock Characterization*. London: Thomas Telford.

Mehrotra, V. K. (1992). *Estimation of engineering parameters of rock mass*. Ph.D. Thesis. Uttarakhand, India: IIT Roorkee, p. 267.

Singh, B., & Goel, R. K. (2002). *Software for engineering control of landslide and tunnelling hazards* (p. 344). Rotterdam: A. A. Balkema (Swets & Zeitlinger).

Singh, B., Goel, R. K., Mehrotra, V. K., Garg, S. K., & Allu, M. R. (1998). Effect of intermediate principal stress on strength of anisotropic rock mass. *Tunnelling and Underground Space Technology, 13*(1), 71–79.

Chapter 17

Types of Failures of Rock and Soil Slopes

I render infinite thanks to God for being so kind as to make me the first observer of marvels kept hidden in obscurity for all previous centuries.

Galileo Galilei

INTRODUCTION

The classification of rock and soil slopes is based on the mode of failure. In the majority of cases, the slope failures in rock masses are governed by joints and occur across surfaces formed by one or several joints. Some common modes of failure, which are frequently found in the field, are described in this chapter.

PLANAR (TRANSLATIONAL) FAILURE

Planar (translational) failure takes place along prevalent and/or continuous joints dipping toward the slope with strike nearly parallel ($\pm 15°$) to the slope face (Figure 17.1b). Stability condition occurs if

1. Critical joint dip is less than the slope angle
2. Mobilized joint shear strength is not enough to assure stability

Generally, a planar failure depends on joint continuity.

3D WEDGE FAILURE

Wedge failure occurs along two joints of different sets when these two discontinuities strike obliquely across the slope face and their line of intersection day-lights in the slope face, as shown in Figure 17.1c (Hoek & Bray, 1981). The wedge failure depends on joint attitude and conditions and is more frequent than planar failure. The factor of safety of a rock wedge to slide increases significantly with the decreasing wedge angle for any given dip of the intersection of its two joint planes (Hoek & Bray, 1981).

CIRCULAR (ROTATIONAL) FAILURE

Circular (rotational) failure occurs along a surface that develops only partially along joints, but mainly crosses them. This failure can only happen in heavily jointed rock masses with a very small block size and/or very weak or heavily weathered rock mass

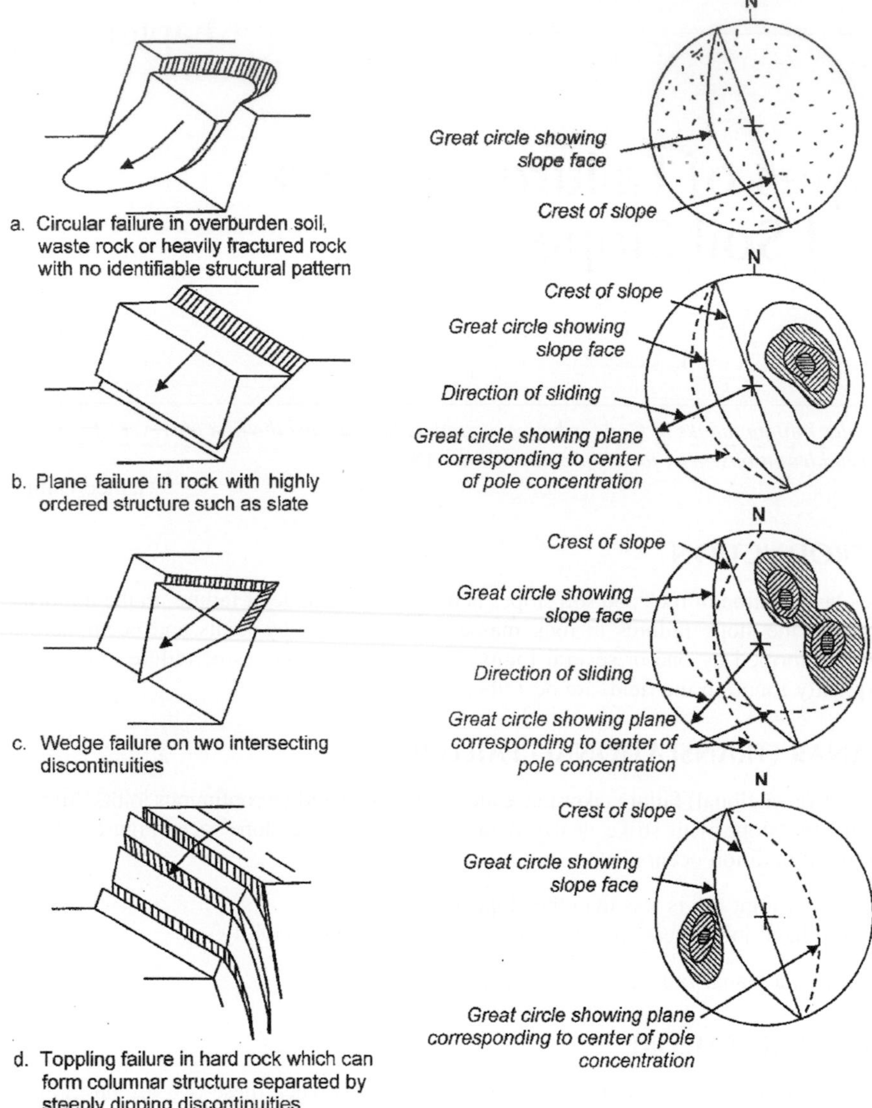

FIGURE 17.1 Main types of slope failures and stereo plots of structural conditions likely to give rise to these failures. *(From Hoek and Bray, 1981)*

(Figure 17.1a). It is essential that all the joints are oriented favorably so that planar and wedge failures or toppling is not possible.

The modes of failure discussed so far involved the movement of a mass of material upon a failure surface. An analysis of failure or a calculation of the factor of safety for these slopes requires that the shear strength of the failure surface, defined by c and ϕ, is known. There are a few types of slope failures that cannot be analyzed even if the strength of mass is known, because failure does not involve simple sliding. These cases are discussed in the next sections.

TOPPLING FAILURE (TOPPLES)

Toppling failure with its stereo plot is shown in Figure 17.1d. Consider a block of rock resting on an inclined plane as shown in Figure 17.2. Here the dimensions of the block are defined by height (h) and base length (b), and it is assumed that the force resisting the downward movement of the block is friction only, that is, cohesion is almost zero.

When the vector representing the weight of the block (W) falls within the base (b), sliding of the block occurs if the inclination of the plane (ψ) is greater than the angle of friction (ϕ). However, when the block is tall and slender (h > b), the weight vector (W) can fall outside the base (b) and, when this happens, the block will topple; that is, it will rotate about its lowest contact edge (Hoek & Bray, 1981).

The conditions for sliding and/or toppling for a rock block are defined in Figure 17.3. The four regions in this diagram are defined as follows:

Region 1: $\psi < \phi$ and $b/h > \tan\psi$, the block is stable and will neither slide nor topple

Region 2: $\psi > \phi$ and $b/h > \tan\psi$, the block will slide but will not topple

Region 3: $\psi < \phi$ and $b/h < \tan\psi$, the block will topple but will not slide

Region 4: $\psi > \phi$ and $b/h < \tan\psi$, the block can slide and topple simultaneously

Wedge toppling occurs along a rock wedge where a third joint set intersects the wedge and dips toward the hill side. Thus thin triangular rock wedges topple down successively. The process of toppling is slow during each rainy season.

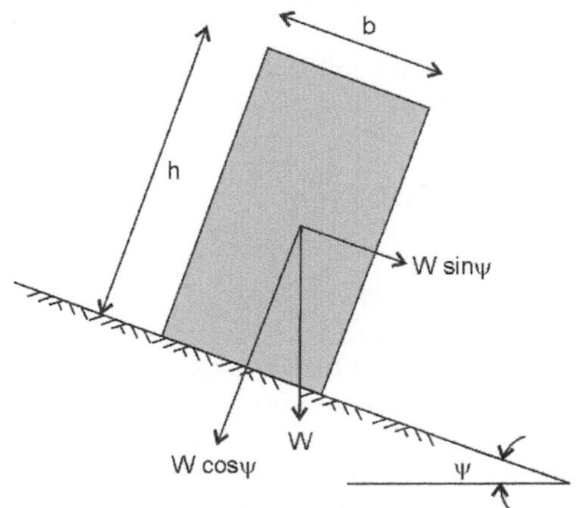

FIGURE 17.2 Geometry of block on inclined plane.

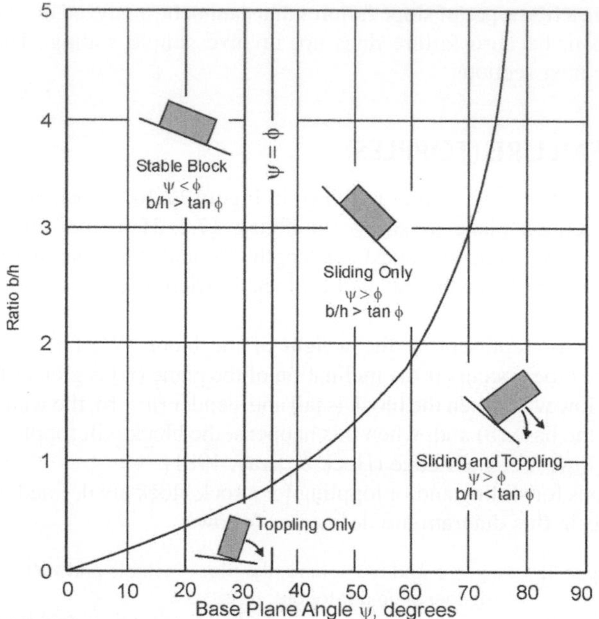

FIGURE 17.3 Conditions for sliding and toppling of a block on an inclined plane. *(From Hoek and Bray, 1981)*

RAVELING SLOPES (FALLS)

Accumulation of screes, or small pieces of rock detached from the rock mass at the base of steep slopes, and cyclic expansion and contraction associated with freezing and thawing of water in cracks and fissures in the rock mass are the principal reasons for slope raveling. A gradual deterioration of materials, which cement the individual rock blocks together, may also play a part in this type of slope failure.

Weathering or the deterioration of certain types of rock exposure also give rise to the loosening of a rock mass and the gradual accumulation of materials on the surface, which falls at the base of the slope.

It is important that the slope designer recognizes the influence of weathering on the materials for which he is designing (see the section Rock Slope Failures in this chapter).

EFFECT OF SLOPE HEIGHT AND GROUNDWATER CONDITIONS ON SAFE SLOPE ANGLE

Figure 17.4 illustrates the significant effect of slope height on stable slope angle for various modes of failure. The groundwater condition also reduces the factor of safety. IIT Roorkee developed software packages SASP, SASW/WEDGE, SARC, and SAST to analyze planar, 3D wedge, circular, and debris slides, respectively (Singh & Anbalagan, 1997; Singh & Goel, 2002). A few deep-seated landslides such as planar and rotational are more catastrophic than millions of surfacial landslides along reservoir rims of dams. Because of this, potential deep-seated landslides in the landslide hazard zonation should be identified.

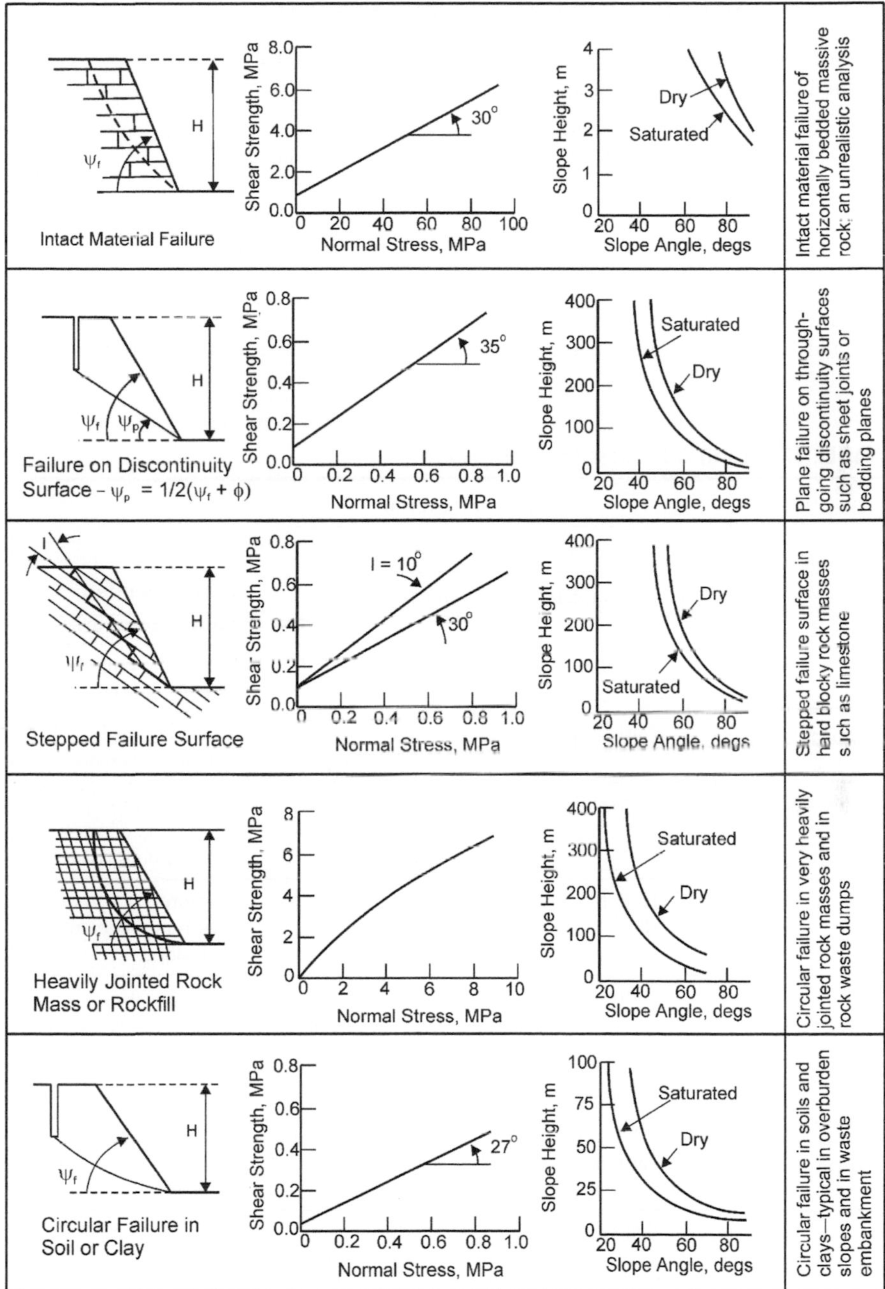

FIGURE 17.4 Slope angle versus height relationships for different materials. *(From Hoek and Bray, 1981)*

A BASIC LANDSLIDE CLASSIFICATION SYSTEM

The basic types of landslides/rockslides are summarized in Table 17.1.

The landslides are defined as follows:

Debris slide	Sliding of debris or talus on rock slopes due to a temporary groundwater table just after long rains
Debris flow	Liquid flow mixture of boulders, debris, clay, and water along a gully during rains or a cloudburst
Earth flow/ mud flow	Liquid flow mixture of soil, clay, and water along a gully during rains.

Landslide control measures may be selected from the last column of Table 17.1. Lien and Tsai (2000) showed that the slit dams have been effective in trapping big

TABLE 17.1 Landslide Classification System

Type of movement		Type of material: Soils: *Predominantly fine*	Type of material: Soils: *Predominantly coarse*	Type of material: *Bedrock*	Recommended control measures
Falls		Earth fall	Debris fall	Rock fall	Geotextile nailed on slope/spot bolting
Topples		Earth topple	Debris topple	Rock topple	Breast walls/soil nailing
Slides	Rotational	Earth slump	Debris slump	Rock slump	Flattening of slope profile and earth and rock fill buttress
	Translational	Earth block slide	Debris block slide	Rock block slide	Reinforced earth or rock reinforcement in rock slope
		Earth slide	Debris slide	Rock slide	Biotechnical measures, subsurface drainage
Lateral spreads		Earth spread	Debris spread	Rock spread	Check dams along gully
Flows		Earth flow	Debris flow	Rock flow	Series of check dams, slit dam
		Soil creep		Deep creep	Rows of deep piles
Complex		Combination of two or more principal types of movement			Combined system

Source: IS14680, 1999.

boulders and retarding the debris flow in the Himalayas in China. The slit dam is like a check dam with many slits. According to Ishikawa, Takeuchi, and Nonaka (2006), check dams of a series of triangular steel frames composed of steel pipes filled with concrete have been used to trap large rocks in debris flow and act as shock absorbers in Japan.

CAUSATIVE CLASSIFICATION

Landslides may also be classified according to their causes (Deoja et al., 1991).

1. Rainfall induced landslide: Most landslides and rock slides
2. Earthquake induced landslides: Generally rock falls and boulder jumping to long distances in hilly areas
3. Cloudburst induced landslide: Mostly mud flows and debris flows (and flash floods) along gullies in the Himalayan region
4. Landslide dam break: Resulting in flash floods and a large number of landslides due to the toe erosion along the hill rivers
5. Glacial lake outburst flood (GLOF): Common in glaciated Himalayan ridges due to melting of nearby glaciers, particularly due to global warming; such a flood causes bank undercutting, landslides, and debris flows
6. Freeze and thaw induced rock falls: Occur during sunny days in the snowbound steep Rocky Mountains

Bhandari (1987) presented strategies about landslides in the fragile Himalayas as well as very economical landslide measures there. Subsequently, Choubey (1998) highlighted the causes of rock slides in the Himalayas and stressed the need for detailed field investigations at the sites of complex landslides.

COMPREHENSIVE CLASSIFICATION SYSTEM OF LANDSLIDES

Hutchinson (1988) presented a detailed classification of landslides, which is a significant improvement over the classification by Varnes (in Schuster & Krizek, 1978). It is surprising that there are so many different types of landslides.

Table 17.2 lists a comprehensive classification system of landslides both for rocks and soils based on slope movement. Figures 17.5 through 17.12 illustrate various modes of failure of rock and soil slopes. Recommended computer programs are also mentioned with the various types of landslides. It seems that debris slides are most common along roads (Figure 17.8). Engineers generally avoid landslide or landslide-prone areas for hill development. Their interest mainly lies in developing a safe terrace system that lasts for at least 25 years; therefore, site development is the real challenge. Adjoining landslides provides a clue as to the potential mode of failure.

LANDSLIDE IN OVER-CONSOLIDATED CLAYS

Expert advice is needed when tackling landslides in over-consolidated clays. Progressive failure of slopes in clays and soft shales occurs slowly. The slope failure may take place after approximately 30 years of temporary stability. It is recommended that residual and drained shear strength parameters should be used in analyzing static stability of clay slopes. In dynamic analysis, peak undrained shear strength parameters should be used.

TABLE 17.2 Classification of Sub-Aerial Slope Movements

A. Rebound (Figure 17.5) Movements associated with:
   1. Man-made excavations
   2. Naturally eroded valleys

B. Creep:
   1. Superficial, predominantly seasonal creeps; mantle creep:
      (a) Soil creep, talus creep (non-periglacial)
      (b) Frost creep and gelifluction of granular debris (periglacial)
   2. Deep-seated, continuous creep; mass creep
   3. Pre-failure creep; progressive creep
   4. Post-failure creep

C. Sagging of mountain slopes (Figure 17.6):
   1. Single-aided sagging associated with the initial stages of landsliding:
      (a) Of rotational (essentially circular) type (R-sagging)
      (b) Of compound (markedly non-circular) type (C-sagging);
         (i) listric (CL)
         (ii) bi-planar (CB)
   2. Double-aided sagging associated with the initial stages of double landsliding, leading to ridge spreading:
      (a) Of rotational (essentially circular) type (DR-sagging)
      (b) Of compound (markedly non-circular) type (DC-sagging);
         (i) listric (DCL)
         (ii) bi-planar (DCB)
   3. Sagging associated with multiple toppling (T-sagging)

D. Landslides (Figures 17.7 and 17.8):
   1. Confined failures (Figure 17.7)
      (a) In natural slopes
      (b) In man-made slopes
   2. Rotational slips:
      (a) Single rotational slips
      (b) Successive rotational slips
      (c) Multiple rotational slips
   3. Compound slides (markedly non-circular, with listric or bi-planar slip surfaces):
      (a) Released by internal shearing toward rear
         (i) In slide mass of low to moderate brittleness
         (ii) In slide mass of high brittleness
      (b) Progressive compound slides, involving rotational slip at rear and fronted by subsequent translational slide
   4. Translational slides (Figure 17.8):
      (a) Sheet slides
      (b) Slab slides; flake slides
      (c) Peat slides
      (d) Rock slides:
         (i) Planar slides; block slides
         (ii) Stepped slides
         (iii) Wedge failures

TABLE 17.2—Cont'd

- (e) Slides of debris:
  - (i) Debris slides; debris avalanches (non-periglacial)
  - (ii) Active layer slides (periglacial)
- (f) Sudden spreading failures

E. Debris movements of flow-like form (Figure 17.9):
1. Mudslides (non-periglacial):
   - (a) Sheets
   - (b) Lobes (lobate or elongate)
2. Periglacial mudslides (gelifluction of very sensitive clays):
   - (a) Sheets
   - (b) Lobes (lobate or elongate, active and relict)
3. Flow slides:
   - (a) In loose, cohesionless materials
   - (b) In lightly cemented, high porosity silts
   - (c) In high porosity, weak rocks
4. Debris flows, very rapid to extremely rapid flows of wet debris:
   - (a) Involving weathered rock debris (except on volcanoes):
     - (i) Hillslope debris flows
     - (ii) Channeled debris flows; mud flows; mud-rock flows during heavy rains or cloudbursts
   - (b) Involving peat; bog flows, bog bursts
   - (c) Associated with volcanoes; lahars:
     - (i) Hot lahars
     - (ii) Cold lahars
5. Sturzstroms, extremely rapid flows of dry debris

F. Topples (Figure 17.10):
1. Topples bounded by pre-existing discontinuities:
   - (a) Single topples
   - (b) Multiple topples
2. Topples released by tension failure at rear of mass
3. Wedge toppling due to falling of thin triangular rock wedges slowly

G. Falls (Figure 17.10):
1. Primary, involving fresh detachment of material; rock and soil falls
2. Secondary, involving loose material, detached earlier; stone falls
3. Boulder jumping for long distances particularly just after earthquake

H. Complex slope movements (Figures 17.11 and 17.12):
1. Cambering and valley bulging (Figure 17.11)
2. Block-type slope movements (Figure 17.12)
3. Abandoned clay cliffs
4. Landslides breaking down into mudslides or flows at the toe:
   - (a) Slump-earth flows
   - (b) Multiple rotational quick-clay slides
   - (c) Thaw slumps
5. Slides caused by seepage erosion where groundwater intersects a soil slope
6. Multi-tiered slides
7. Multi-storied slides

Based principally on morphology with some account taken of mechanism, material, and rate of movement.
Source: After Hutchinson, 1988.

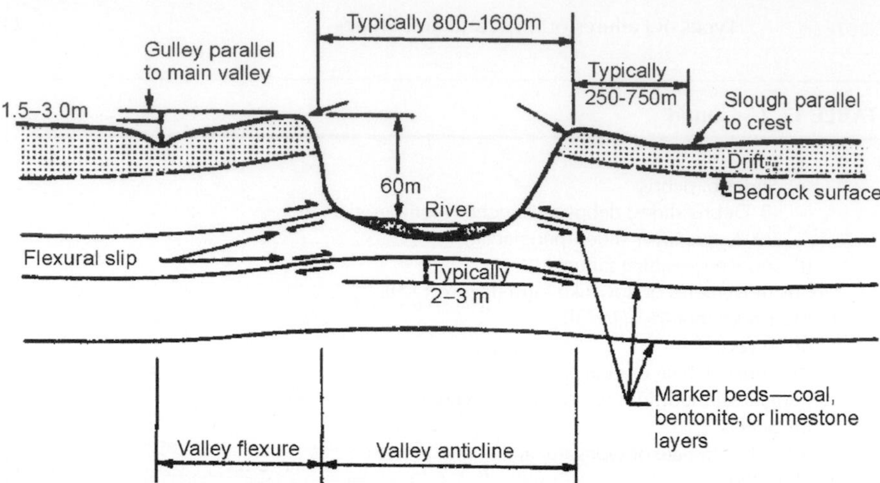

FIGURE 17.5 Valley rebound.

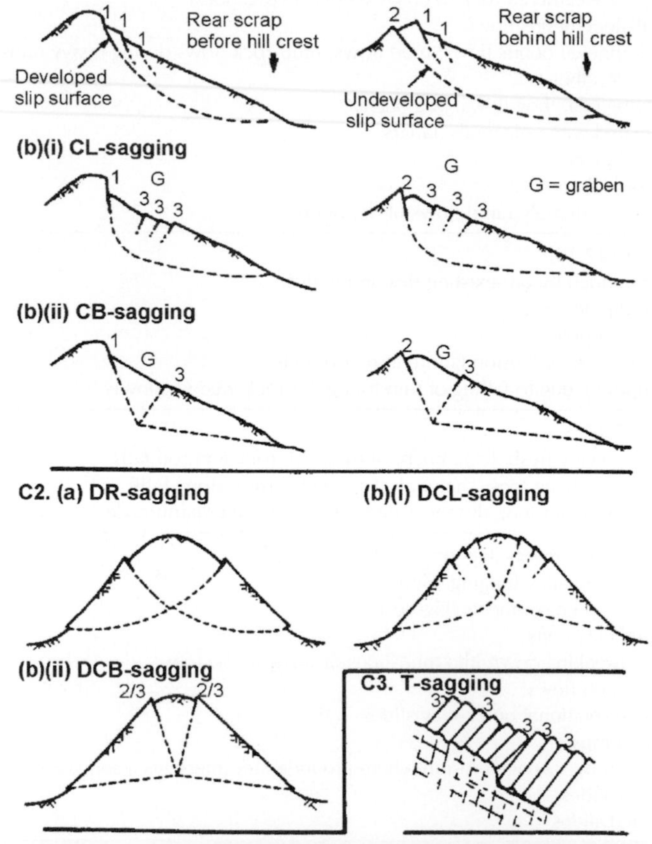

FIGURE 17.6 Main types of sagging (SANC is recommended for C1 and C2).

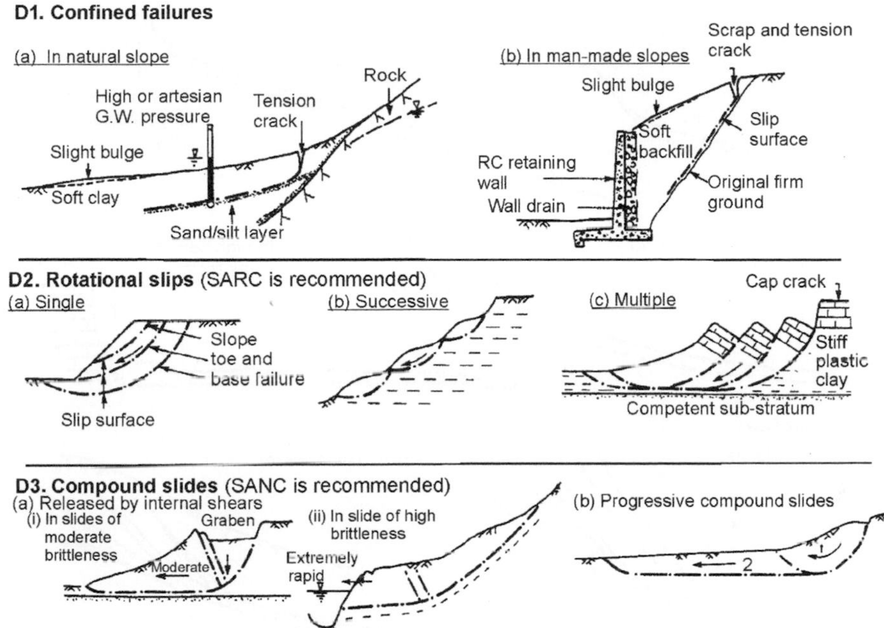

FIGURE 17.7 Main types of confined failures, rotational slips, and compound slides.

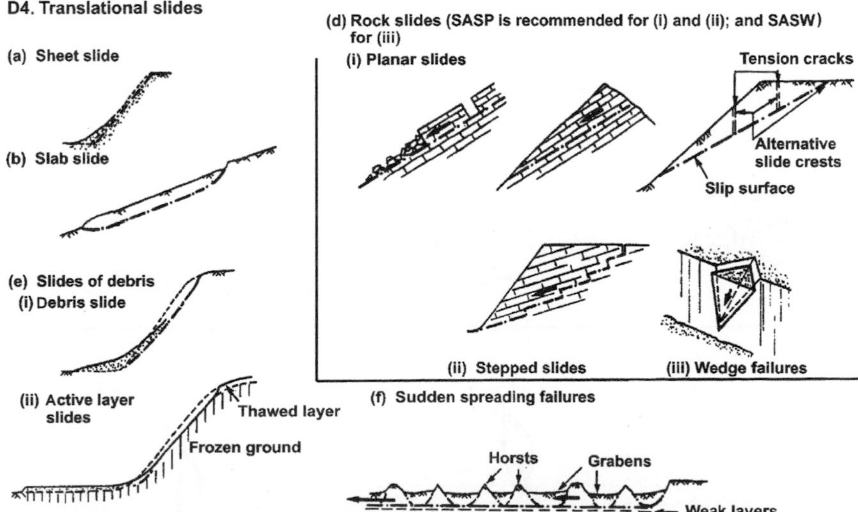

FIGURE 17.8 Main types of translational failures (SAST is recommended).

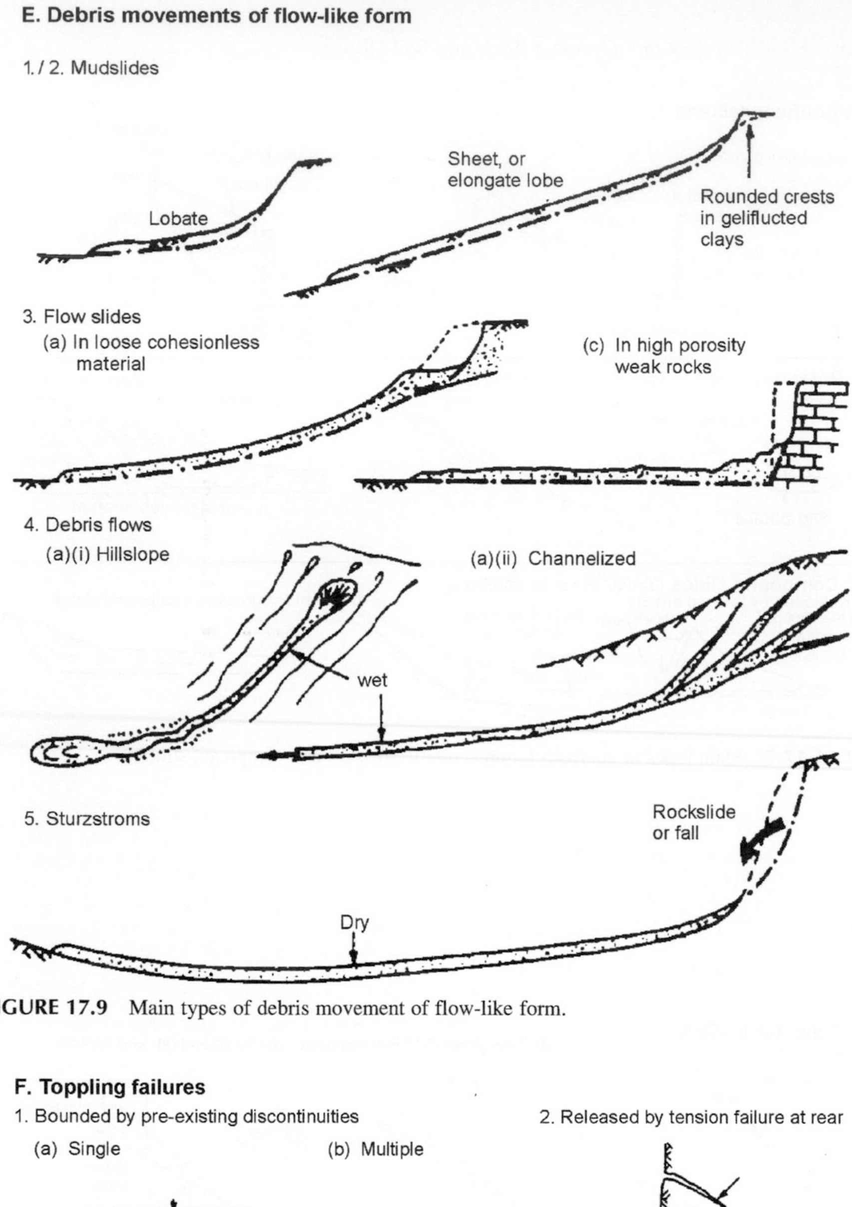

E. Debris movements of flow-like form

1./2. Mudslides

Lobate

Sheet, or elongate lobe

Rounded crests in gelifluctted clays

3. Flow slides

(a) In loose cohesionless material

(c) In high porosity weak rocks

4. Debris flows

(a)(i) Hillslope

(a)(ii) Channelized

wet

5. Sturzstroms

Rockslide or fall

Dry

FIGURE 17.9 Main types of debris movement of flow-like form.

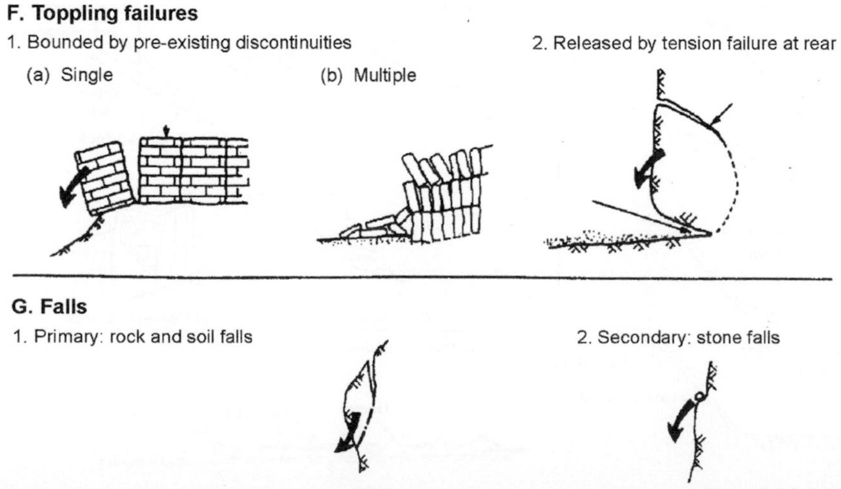

F. Toppling failures

1. Bounded by pre-existing discontinuities

(a) Single

(b) Multiple

2. Released by tension failure at rear

G. Falls

1. Primary: rock and soil falls

2. Secondary: stone falls

FIGURE 17.10 Main types of toppling failures and falls.

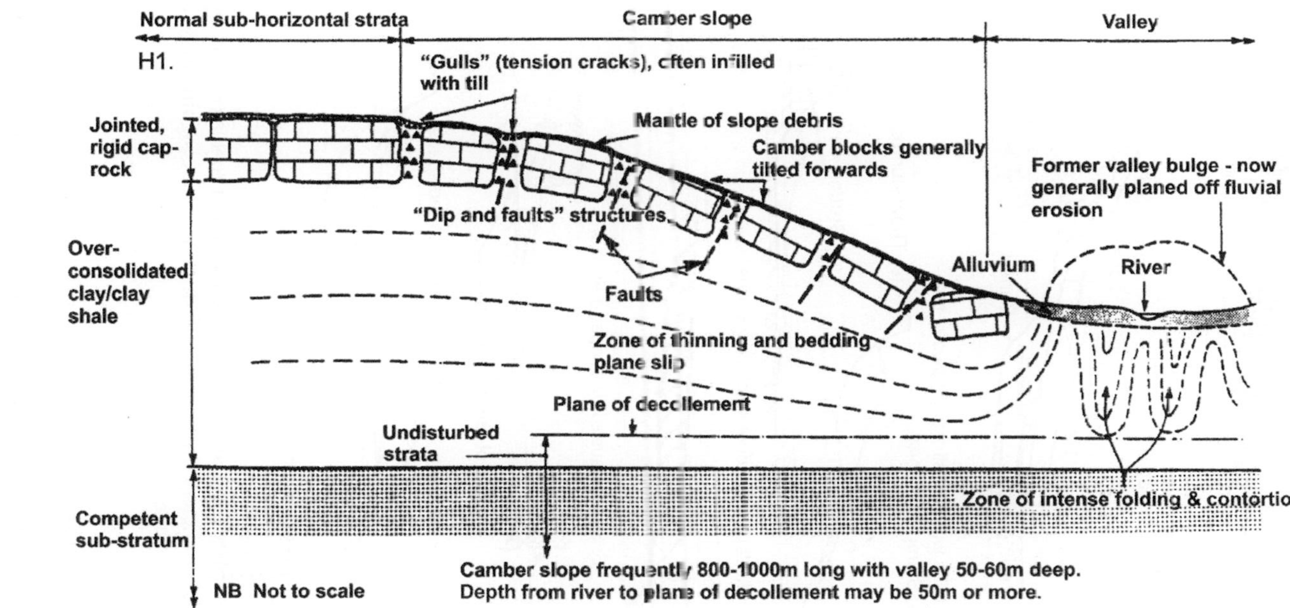
Normal sub-horizontal strata
Camber slope
Valley
H1.
"Gulls" (tension cracks), often infilled with till
Mantle of slope debris
Camber blocks generally tilted forwards
Former valley bulge - now generally planed off fluvial erosion
Jointed, rigid cap-rock
"Dip and faults" structures
Over-consolidated clay/clay shale
Alluvium
River
Faults
Zone of thinning and bedding plane slip
Plane of decollement
Undisturbed strata
Zone of intense folding & contortion
Competent sub-stratum
NB Not to scale
Camber slope frequently 800-1000m long with valley 50-60m deep.
Depth from river to plane of decollement may be 50m or more.

FIGURE 17.11 Schematic section of cambering and valley.

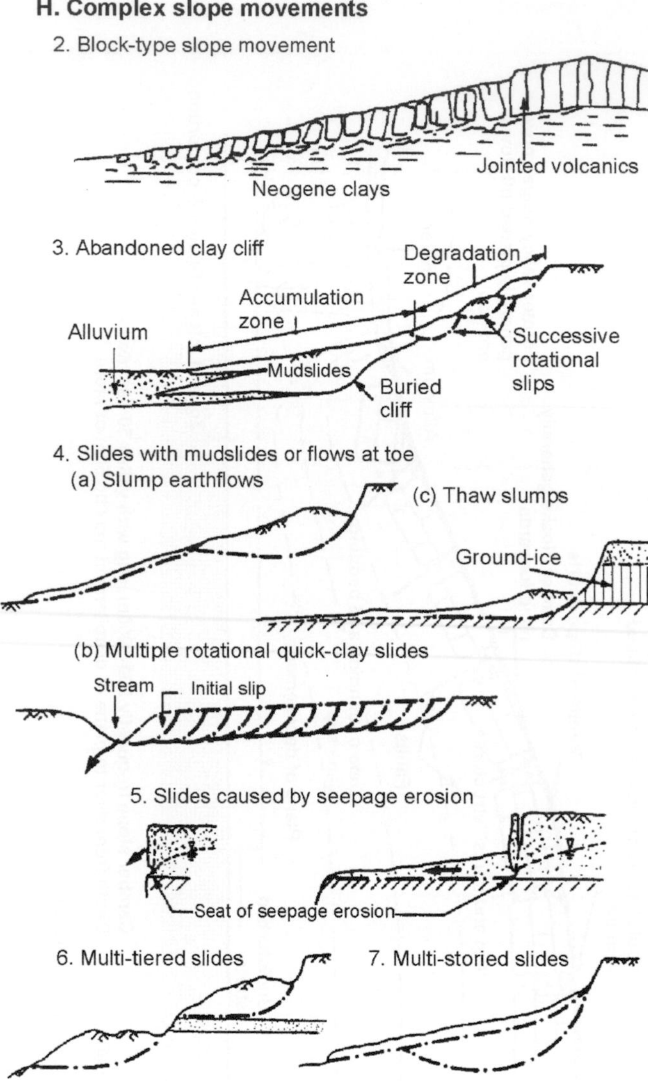

FIGURE 17.12 Some types of complex slope movements.

The orientation of platy clay particles takes place in a thin zone along the slip surface. As such, the strength parameters along the actual slip surface are significantly lower than those along any other assumed slip surface.

ROCK SLOPE FAILURES

Natural rock slopes support the foundations of dams, penstocks, buildings, abutments of bridges, and transmission towers. From a design aspect, it is essential to recognize the types of rock slides, which are often complex. Table 17.3 describes these failure modes

TABLE 17.3 Some Modes of Failure in Slopes in Rock Masses

Failure mode (1)	Description (2)	Typical materials (3)	Figure (4)
Erosion, piping	Gullies formed by action of surface or ground-water	Silty residual soils and saprolite (especially disintegrated granite), silty fault gouge, uncemented sand rocks, uncemented noncohesive pyroclastic sediments	
Raveling	Gradual erosion, particle by particle or block by block	Poorly cemented conglomerates and breccias; very high fractured hard rocks; layered rock masses being loosened by active weathering (e.g., thinly bedded sandstone/shale)	
Block sliding on a single plane	Sliding without rotation along a face; single or multiple blocks	Hard or soft rocks with well-defined discontinuities and jointing (e.g., layered sedimentary rocks, volcanic flow rocks, block-jointed granites, foliated metamorphic rocks)	
Wedge sliding	Sliding without rotation on two nonparallel planes, parallel to their line of intersection; single or multiple blocks	Blocky rock with at least two continuous and nonparallel joint sets (e.g., cross-jointed sedimentary rocks, regularly faulted rocks, block-jointed granite, and especially foliated or jointed metamorphic rocks)	
Rock slumping	Backward rotation of single or multiple blocks, moving into edge/face contact to form one or more detached beams	Hard rocks with regular, parallel joints dipping toward but not day-lighting into free space and one flat-lying joint that does day-light into free space; multiple block modes typically developed in foliated metamorphic rocks and steeply dipping sedimentary rocks; single block modes develop in block-jointed granites, sandstones, and volcanic flow rocks	17.13(a–e)
Toppling	Forward rotation about an edge — single or multiple blocks	Hard rocks with regular, parallel joints dipping away from the free space, with or without crossing joints; foliated metamorphic rocks and steeply dipping layered sedimentary rocks; also in block-jointed granites	17.13(f–h)
Slide toe toppling	Toppling at the toe of a slide in response to active loading from above	All rock types susceptible to block toppling	17.14(a, d)

Continued

TABLE 17.3 Some Modes of Failure in Slopes in Rock Masses—Cont'd

Failure mode (1)	Description (2)	Typical materials (3)	Figure (4)
Slide head toppling	Toppling behind the scarp at the top of a slide	All rock types susceptible to block toppling	17.14(b)
Slide base toppling	Toppling of beds beneath a slide mass due to shear across their tops	Typically developed in any rock type susceptible to toppling, located beneath the base of landslide (e.g., where the seat of sliding occurs along a fault surface)	17.14(c)
Block torsion	Rotary sliding in a single plane	Blocky rock where sliding on the potential slip surface is prevented by a rock bridge, asperity, or other restraint which forms a hinge	17.14(e)
Sheet failure	Tensile failure and fall or sliding of hanging sheets	Steeply dipping pre-existing sheet joints in granites and sandstone; new sheet joints in weathered rocks, friable massive sandstone, and pyroclastic sediments on steep slopes	17.14(f)
Rock bridge cracking	Failure of intact rock that restrains block motion through compressive, tensile, or flexural cracking	Weak rock forming rock bridges; hard or soft rocks with impersistent discontinuities (as in some layered sedimentary rocks, volcanic flow rocks, block-jointed granites, and foliated or jointed metamorphic rocks)	17.15(a, b)
Slide base rupture	Rupture of the rock mass beneath the slide caused by slide-transmitted shear and moment	Weak rock beneath the toe of a slide	17.15(c)
Buckling and kink band slumping	Compressive collapse of columns or slabs parallel with the rock slope face	Thinly bedded, weak sedimentary rocks inclined steeply and parallel to the slope surface; shale-sandstone and shale-chert sequences, coal measures, and foliated metamorphic rocks	17.16(a, b)
Soil-type slumping	Shearing with backward rotation, as in clay soils	Weathered or softened clay shales; thick fault-gouge; altered zones; soft tuffs; high pore pressure zones	17.16(c, d)
Rock bursting	Hard rock under breaking stress	Granite and marble quarries into high stressed rock; hard sedimentary rock at the base of deep, narrow canyons	

Source: Goodman and Kieffer, 2000.

and gives examples of typical materials in which failures occurs (Goodman & Kieffer, 2000). Adversely oriented (key) blocks move out first followed by other wedges or blocks. When sliding opportunities are inhibited, rotation of blocks may take place, causing toppling, buckling, block slumping, or torsional failures.

It is important to realize that theoretically all that is needed to stabilize a rock slope is to anchor the "key" or worst oriented wedge or block of the rock mass. Seepage erosion is also frequent enough to collapse the toe of slopes gradually in soils or soluble rock slopes. Toppling failures can be deep, large, and potentially rapid. Spillways can cause a large amount of erosion of valley slopes and slope failure (see also Figures 17.13 through 17.16).

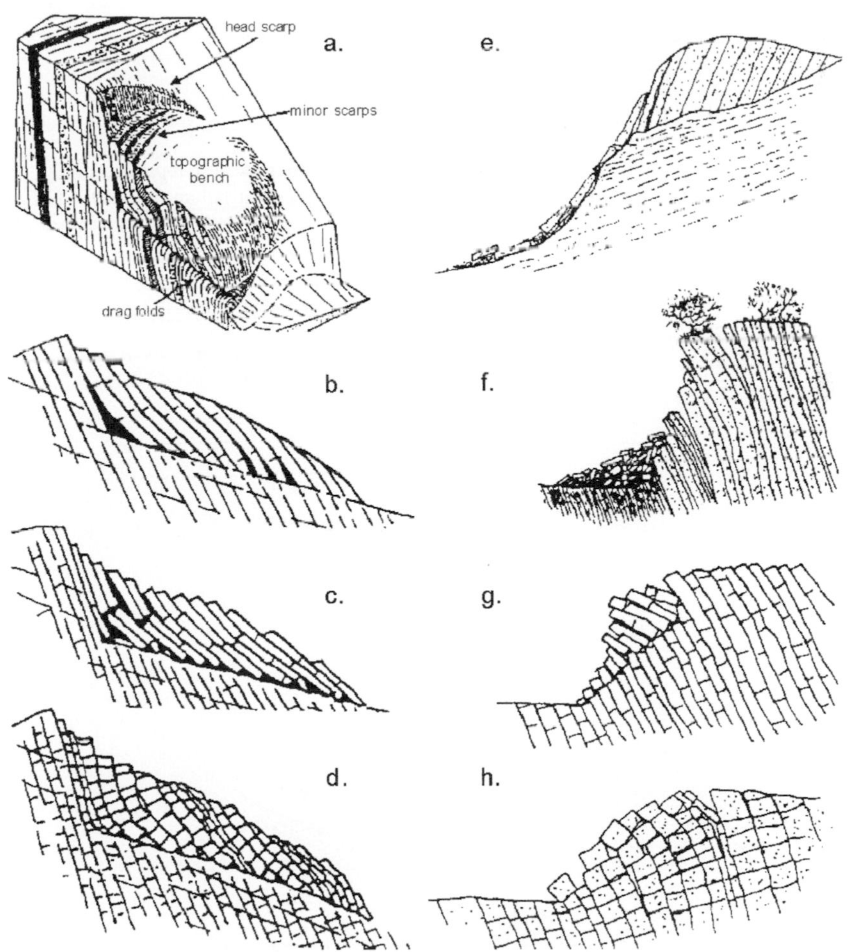

FIGURE 17.13 (a–e) Rock slumping and (f–h) toppling (use TOPPLE, UDEC, 3DEC). *(From Goodman and Kieffer, 2000)*

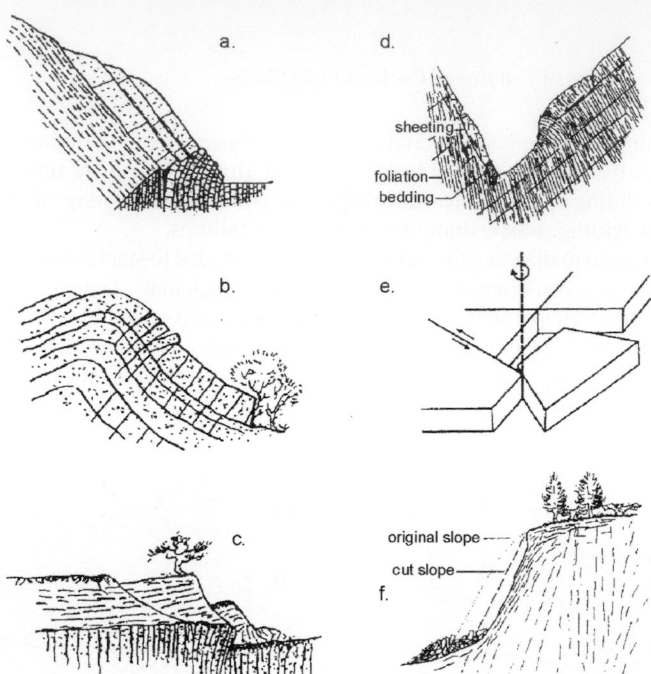

FIGURE 17.14 Secondary toppling modes: (a) slide toe toppling, (b) slide head toppling, (c) slide base toppling, (d) contrasting slide toe topples, (e) block torsion, and (f) sheet failure (use UDEC). *(From Goodman and Kieffer, 2000)*

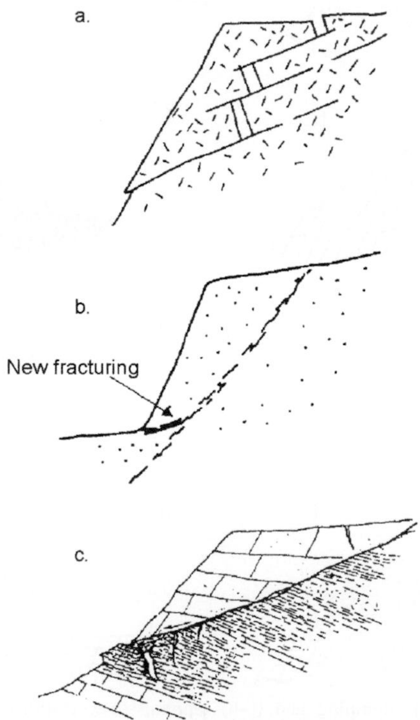

FIGURE 17.15 Additional modes involving rock fracturing: (a) rock bridge cracking in tension, (b) rock bridge failure in compression, and (c) slide base rupture (use SASP). *(From Goodman and Kieffer, 2000)*

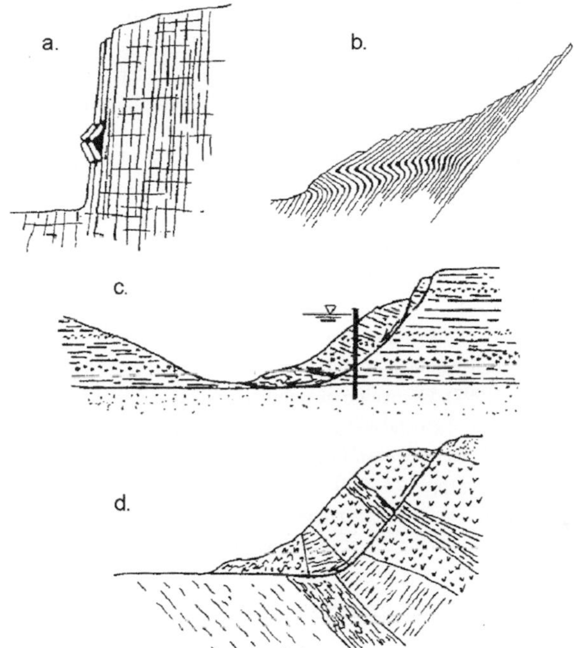

FIGURE 17.16 (a) Buckling, (b) kink band slumping, and (c, d) soil type slumping (use SANC, FLAC). *(From Goodman and Kieffer, 2000)*

LANDSLIDE DAMS

Landslide dams are formed in steep valleys due to a deep-seated landslide in deforested hills. They are also created by huge deposits of debris, which are brought about by a network of gullies during cloudbursts. The dam–river water quickly forms a reservoir submerging roads and houses (Bhandari, 1987; Choubey, 1998). The reservoir water may also enter in the tail race tunnels of nearby hydroelectric projects. This back water has caused immense damage to two underground powerhouses in the Himalayas in India. The silt content of rivers has increased nearly three times due to landslides that are caused by recent deforestation. Silting may be checked by building a new dam upstream of the proposed dam site.

REFERENCES

Bhandari, R. K. (1987). Slope instability in the fragile Himalaya and strategy for development. Ninth IGS Annual Lecture. *Indian Geotechnical Journal*, *17*(1), 1–78.

Choubey, V. D. (1998). Landslide hazard assessment and management in Himalayas. In *International Conference On Hydro Power Development in Himalayas* (pp. 220–238). Shimla, India.

Deoja, B., Dhital, M., Thapa, B., & Wagner, A. (1991). *Mountain risk engineering handbook Part I and II*. Kathmandu, Nepal: International Centre of Integrated Mountain Development.

Goodman, R. E., & Kieffer, D. S. (2000). Behaviour of rock in slopes. *Geotechnical and Geoenvironmental Engineering*, *126*(8), 675–684.

Hoek, E., & Bray, J. W. (1981). *Rock slope engineering* (revised 3rd ed., p. 358). The Institution of Mining and Metallurgy. London: Maney Publishing.

Hutchinson, J. N. (1988). Morphological and geotechnical parameters of landslides in relation to geology and hydrology, general report. In *Proceedings of the 5th International Symposium on Landslide* (Vol. 1, pp. 3–35). Oxford, UK: Taylor & Francis.

IS14680. (1999). *Landslide control guidelines.* New Delhi: Bureau of Indian Standards.

Ishikawa, N., Takeuchi, D., & Nonaka, T. (2006). Impact behaviour of new steel check dams under debris flow. In C. Majorana, et al. (Eds.), *3rd International Conference on Protection of Structures Against Hazards* (pp. 31–42), Venice. Singapore: CI-Premier.

Lien, H. P., & Tsai, F. U. (2000). Debris flow control by using slit dams. *International Journal of Sediment Research, 15*(4), 391–409.

Schuster, R. L., & Krizek, R. J. (1978). *Landslides — Analysis and control* (Special Report 176, p. 234). Washington, D.C.: National Academy of Sciences.

Singh, B., & Anbalagan, R. (1997). Evaluation of stability of dam and reservoir slopes—Mechanics of landslide, seismic behaviour of ground and geotechnical structures. In *Proceedings on Special Technical Session on Earthquake Geotechnical Engineering, XIV International Conference on Soil Mech. and Foundation Engineering* (pp. 323–339). Hamburg, Germany.

Singh, B., & Goel, R. K. (2002). *Software for engineering control of landslide and tunnelling hazards* (p. 344). Rotterdam: A. A. Balkema (Swets & Zeitlinger).

Chapter 18

Slope Mass Rating

The Mother Nature is Motherly!

Veda, Gita, and Durgasaptashati

THE SLOPE MASS RATING

For evaluating the stability of rock slopes, Romana (1985) proposed a classification system called the "slope mass rating" (SMR) system. SMR is obtained from Bieniawski's rock mass rating (RMR) by subtracting adjustment factors of the joint–slope relationship and adding a factor depending on method of excavation

$$SMR = RMR_{basic} + (F_1 \cdot F_2 \cdot F_3) + F_4 \tag{18.1}$$

where RMR_{basic} is evaluated according to Bieniawski (1979, 1989) by adding the ratings of five parameters (see Chapter 6). F_1, F_2, and F_3 are adjustment factors related to joint orientation with respect to slope orientation, and F_4 is the correction factor for method of excavation.

F_1 depends upon parallelism between joints and slope face strikes. It ranges from 0.15 to 1.0. It is 0.15 when the angle between the critical joint plane and the slope face is more than 30° and the failure probability is very low; it is 1.0 when both are near parallel.

The value of F_1 was initially established empirically. Subsequently, it was found to approximately match the following relationship:

$$F_1 = (1 - \sin A)^2 \tag{18.2}$$

where A denotes the angle between the strikes of the slope face (α_s) and that of the joints (α_j), that is, ($\alpha_s - \alpha_j$).

F_2 refers to joint dip angle (β_j) in the planar failure mode. Its values also vary from 0.15 to 1.0. It is 0.15 when the dip of the critical joint is less than 20 degrees and 1.0 for joints with dips greater than 45 degrees. For the toppling mode of failure, F_2 remains equal to 1. So

$$F_2 = \tan \beta_j \tag{18.3}$$

F_3 refers to the relationship between the slope face and joint dips.

In planar failure (Figure 18.1), F_3 refers to a probability of joints "day-lighting" in the slope face. Conditions are called "fair" when the slope face and the joints are parallel. If the slope dips 10 degrees more than the joints, the condition is termed "very

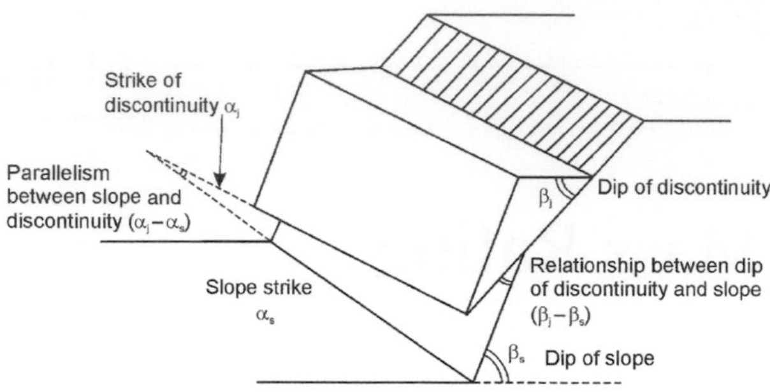

FIGURE 18.1 Planar failure.

unfavorable." For the toppling failure, unfavorable conditions depend upon the sum of the dips of joints and the slope $\beta_j + \beta_s$.

Values of adjustment factors F_1, F_2, and F_3 for different joint orientations are given in Table 18.1.

F_4 pertains to the adjustment for the method of excavation. It includes the natural slope, or the cut slope excavated by pre-splitting, smooth blasting, normal blasting, poor blasting, and mechanical excavation (see Table 18.2 for adjustment rating F_4 for different excavation methods).

- *Natural slopes* are more stable, because of long-time erosion and built-in protection mechanisms (vegetation, crust desiccation), so $\mathbf{F_4 = +15}$.

TABLE 18.1 Values of Adjustment Factors for Different Joint Orientations

Case of slope failure		Very favorable	Favorable	Fair	Unfavorable	Very unfavorable
P T W	$\lvert\alpha_j - \alpha_s\rvert$ $\lvert\alpha_j - \alpha_s - 180°\rvert$ $\lvert\alpha_i - \alpha_s\rvert$	>30°	30–20°	20–10°	10–5°	<5°
P/W/T	$\mathbf{F_1}$	0.15	0.40	0.70	0.85	1.00
P W	$\lvert\beta_j\rvert$ $\lvert\beta_i\rvert$	<20°	20–30°	30–35°	35–45°	>45°
P/W	$\mathbf{F_2}$	0.15	0.40	0.70	0.85	1.00
T	$\mathbf{F_2}$	1.0	1.0	1.0	1.0	1.0
P W	$\lvert\beta_j - \beta_s\rvert$ $\lvert\beta_i - \beta_s\rvert$	>10°	10–0°	0°	0 – (–10°)	<–10°
T	$\lvert\beta_j + \beta_s\rvert$	<110°	110–120°	>120°	—	—
P/W/T	$\mathbf{F_3}$	0	–6	–25	–50	–60

P, planar failure; T, toppling failure; W, wedge failure; α_s, slope strike; α_j, joint strike; α_i, plunge direction of line of intersection; β_s, slope dip; β_j, joint dip (see Figure 18.1); β_i, plunge of line of intersection.

Source: Romana, 1985.

TABLE 18.2 Values of Adjustment Factor F_4 for Method of Excavation

Method of excavation	Value of F_4
Natural slope	+15
Pre-splitting	+10
Smooth blasting	+8
Normal blasting or mechanical excavation	0
Poor blasting	−8

Source: Romana, 1985.

- *Normal blasting* applied with sound methods does not change slope stability conditions, so $\mathbf{F_4 = 0}$.
- *Deficient blasting* or *poor blasting* damages the slope stability, so $\mathbf{F_4 = -8.0}$.
- *Mechanical excavation* of slopes, usually by ripping, can be done only in soft and/or very fractured rock and is often combined with some preliminary blasting. The plane of slope is difficult to finish. The method neither increases nor decreases slope stability, so $\mathbf{F_4 = 0}$.

The minimum and maximum values of SMR from Eq. (18.1) are 0 and 100, respectively. It is needless to mention that the slope stability problem is not found in areas where the discontinuities are steeper than the slope; therefore, this condition is not considered in the empirical approach.

Romana (1985) used planar and toppling failures for his analysis. The wedge failures have been considered as a special case of plane failures and analyzed in forms of individual planes, and the minimum value of SMR is taken for assessing the rock slopes. Dip β_i and dip direction α_i of the intersection of these planes should be taken as β_j and α_j, respectively; that is, $\beta_j = \beta_i$ and $\alpha_j = \alpha_i$ where wedge failure is likely to occur (Figure 18.2).

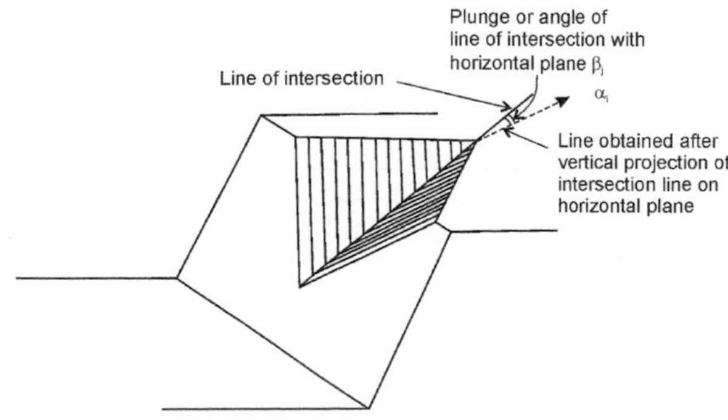

FIGURE 18.2 Wide angle wedge failure.

The effect of future weathering on the slope stability cannot be assessed with rock mass classification, because it is a process that depends mostly on the mineralogical conditions of rock and the climate. In certain rock masses (e.g., some marls, clays, and shales), slopes are stable when excavated but fail sometime afterward — usually one to two years later. In such conditions, the classification should be applied twice: initially and afterward for weathered conditions. It is always prudent to check SMR against adjoining stable rock slopes before applying it to rock slopes in distress.

In some cases, the SMR may be more than the RMR, as F_4 is $+15$ for the natural slopes where all the joint sets are oriented favorably.

Hack (1998) developed the slope stability probability classification (SSPC) system for weathered and unweathered soil and rock slopes under European climatic conditions. He developed a chart to assess the probability of failure of a slope. He also found correlations for the sliding angle of friction (ϕ) along joints. SSPC (slope stability probability classification) is enjoying popularity in hilly regions of Europe, but it needs to be tested in the Himalayas in India and in other climatic conditions.

Water conditions govern the stability of many slopes, which are stable in summer and fail in winter because of heavy raining or freezing. The worst possible water conditions must be assumed for analysis.

SSPC technique is not applicable to mountains that are covered by snow most of the time. Moreover, freezing and thawing of water in rock joints cause rock slides in these regions.

SLOPE STABILITY CLASSES

According to the SMR values, Romana (1985) defined five stability classes. These are described in Table 18.3.

It is inferred from Table 18.3 that the slopes with an SMR value below 20 may fail very quickly. No slope has been registered with an SMR value below 10, because such slopes would not physically exist.

TABLE 18.3 Various Stability Classes as per SMR Values

Class No.	V	IV	III	II	I
SMR value	0–20	21–40	41–60	61–80	81–100
Rock mass description	Very bad	Bad	Normal	Good	Very good
Stability	Completely unstable	Unstable	Partially stable	Stable	Completely stable
Failures	Big planar or soil-like or circular	Planar or big wedges	Planar along some joints and many wedges	Some block failure	No failure
Probability of failure	0.9	0.6	0.4	0.2	0

Source: Romana, 1985.

The stability of slope also depends upon length of joints along the slope. Table 18.3 is found to overestimate SMR where length of joint along the slope is less than 5% of the affected height of the landslide. SMR is also not found to be applicable to opencast mines, because heavy blasting creates new fractures in the rock slope and the depth of cut slope is also large.

SMR is successfully used for landslide zonation in rocky and hilly areas in the Himalayas in India. Detailed studies should be carried out where SMR is less than 40, because life and property are in danger and slopes should be stabilized accordingly. Otherwise, a safe cut slope angle should be determined to increase SMR to 60 (see the section Portal and Cut Slopes in this chapter).

SUPPORT MEASURES

Many remedial measures can be taken to support a slope. Both detailed study and good engineering sense are necessary to stabilize a slope. Classification systems can only try to point out the normal techniques for each different class of supports as given in Table 18.4.

In a broader sense, the SMR range for each group of support measures is as follows:

SMR	65–100	None, scaling
SMR	30–75	Bolting, anchoring
SMR	20–60	Shotcrete, concrete
SMR	10–30	Wall erection, re-excavation

TABLE 18.4 Suggested Supports for Various SMR Classes

SMR classes	SMR values	Suggested supports
Ia	91–100	None
Ib	81–90	None, scaling is required
IIa	71–80	(None, toe ditch, or fence), spot bolting
IIb	61–70	(Toe ditch or fence nets), spot or systematic bolting
IIIa	51–60	(Toe ditch and/or nets), spot or systematic bolting, spot shotcrete
IIIb	41–50	(Toe ditch and/or nets), systematic bolting/ anchors, systematic shotcrete, toe wall and/or dental concrete
IVa	31–40	Anchors, systematic shotcrete, toe wall and/or concrete (or re-excavation), drainage
IVb	21–30	Systematic reinforced shotcrete, toe wall and/or concrete, re-excavation, deep drainage
Va	11–20	Gravity or anchored wall, re-excavation

Less popular support measures are given in brackets.

As pointed out by Romana (1985), wedge failure has not been discussed separately in his SMR classification system. To overcome this problem, Anbalagan, Sharma, and Raghuvanshi (1992) modified SMR to also make it applicable for the wedge mode of failure. This modification is presented in the next section.

MODIFIED SMR APPROACH

Although SMR accounts for planar and toppling failures in rock slopes, it also takes into consideration different planes forming the wedges and analyzing the different planes individually in wedge failure. The unstable wedge is a result of the combined effect of the intersection of various joints (Figure 18.2). Anbalagan et al. (1992) considered plane and wedge failures as different cases and presented a modified SMR approach for slope stability analysis.

In the modified SMR approach, the same method is applicable for planar failures, and the strike and the dip of the plane are used for the analysis. For wedge failures, the plunge and the direction of the line of intersection of the unstable wedge are used. Thin wedges with low angles are likely to be stable and should not be considered. In Table 18.5, adjustment ratings for F_1, F_2, and F_3 are also given in wedge failure as suggested by Anbalagan et al. (1992).

TABLE 18.5 Calculations for Adjustment Factors F_1, F_2, and F_3

A. Details of geological discontinuities		
	Dip direction	*Dip*
Joint J_1	N 60°	45°
Joint J_2	N 325°	35°
Slope	N 10°	50°

B. Details of line of intersection of J_1 and J_2	
Trend = 4°	See Figure 18.3
Plunge = 28°	

C. Adjustment factors F_1, F_2, and F_3 for different conditions

No.	*Condition*	F_1	F_2	F_3	*Adjustment factor* ($F_1 \cdot F_2 \cdot F_3$)
1.	Considering joint J_1 and slope	0.15	0.85	−50	−6.4
2.	Considering joint J_2 and slope	0.15	0.70	−60	−6.3
3.	Considering the plunge and trend of line of intersection of J_1 and J_2 and the slope (modified SMR approach)	0.85	0.40	−60	−20.4

Example 18.1

Consider two joint sets having dips of 45 and 35 degrees and dip directions of 66 and 325 degrees, respectively. The inclination of slope is N10°/50°. The plunge and the trend of line of intersection of these two joints forming the wedge are 28 and 4 degrees, respectively (Figure 18.3).

According to the SMR approach, the SMR value for the previously mentioned two joint sets are worked out separately, and the critical value of SMR is adopted for classification purposes, and the adjustment factor ($F_1 \cdot F_2 \cdot F_3$) for the first joint set and the slope works out to be −6.4 (Table 18.5). Similarly, considering the second joint set and slope, the adjustment factor works out to be −6.3 (Table 18.5).

Now, if we consider the plunge and the trend of the wedge formed by the two joint sets and the slope, the adjustment factor works out to be −20.4. This clearly shows that the SMR calculated for the third case is more critical than the first and second cases. Therefore, it is more logical and realistic to use the plunge and the trend of line of intersection for potential wedge failure.

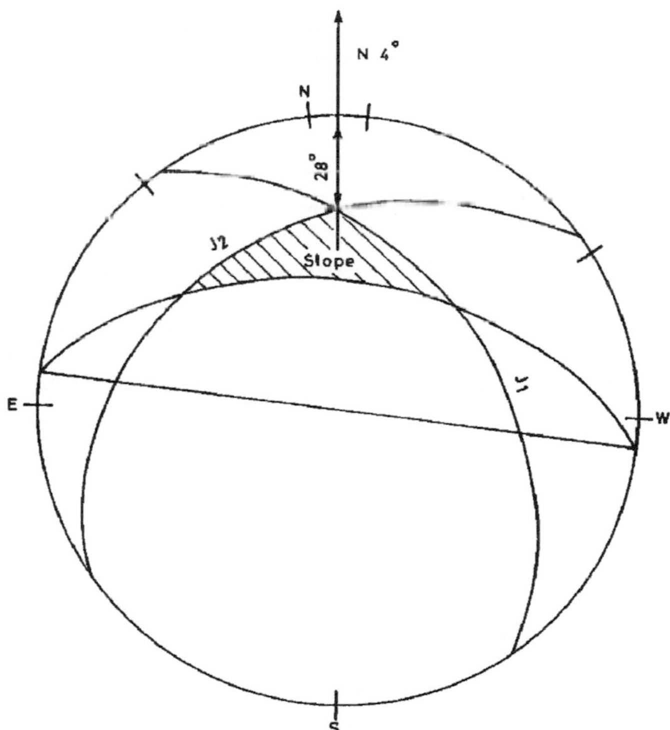

FIGURE 18.3 Usage of stereo plot for identifying the wedge.

CASE STUDY OF STABILITY ANALYSIS USING MODIFIED SMR APPROACH

Anbalagan et al. (1992) analyzed 20 different slopes using the modified SMR approach along the Lakshmanjhula-Shivpuri road in the lesser Himalayas in India.

Geology

The Lakshmanjhula-Shivpuri road section area forms the northern part of the Garhwal syncline. The road section has encountered Infra-Krol formation (Krol A, Krol B, and Krol C + D formations; lower Tal formation; upper Tal formation; and Blaini formation). The rocks are folded in the form of a syncline called the "Narendra Nagar syncline." The axis of the syncline is aligned in a NE-SW direction so that the sequence of the Blaini and Tal formations from Lakshmanjhula are repeated again to the north of the syncline axis.

The Infra-Krol formation mainly consists of dark gray shales, whereas Krol A consists of shaly limestones and Krol B includes red shales. The Krol C + D formation is comprised of gypsiferous limestones. The lower Tal formation consists of shales, whereas the upper Tal is comprised of quartzites. The rocks of the Blaini formation are exposed near Shivpuri and include laminated shales.

Rock Slope Analysis

Twenty rock slopes along the road were chosen because they cover different rock types (Figure 18.4). The RMR_{basic} for different rock types were estimated (Table 18.6). The graphical analysis was performed for the joints to deduce the mode of failure. Using this method, the poles of discontinuities were plotted on an equal area stereonet and contours were drawn to get the maxima pole concentrations. The probable failure patterns were determined by studying the orientation of various joints and the intersection and comparing them with the slope. The graphical analysis of the individual slope is shown in Figures 18.5 and 18.6. The results of the SMR approach are listed in Table 18.7.

The modified approach for wide angle wedge failure appears to be valid as SMR predictions match with the observed failure modes. However, for identifying potentially unstable wedges, just use your judgment.

PORTAL AND CUT SLOPES

It is better to locate the tunnel portals deeper into the ground or mountain where rock cover at least equal to the width of the tunnel is available. The slope of the portal should be stable with an SMR > 60. Otherwise, the tunnel should be reinforced properly with rock anchors. Alternatively, a thick breast wall (i.e., 1 m) of concrete should be constructed to ensure the stability of portals. Singh and Goel (2002) presented several methods and software for slope stabilization according to the precise mode of failure.

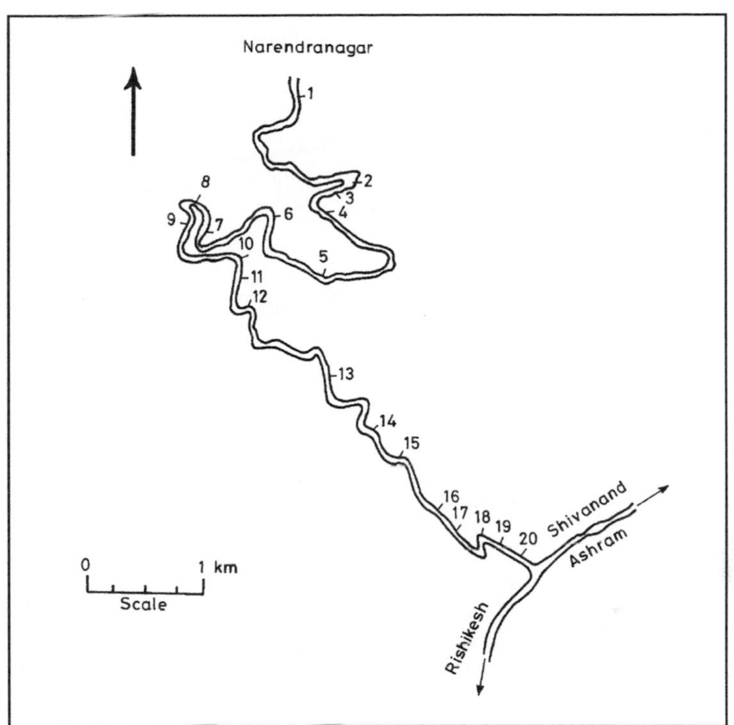

FIGURE 18.4 Location map of slope stability study showing locations in Table 18.7.

TABLE 18.6 Rock Mass Rating (RMR) for Various Rock Types of Lakshmanjhula-Shivpuri Area

Rock type	Uniaxial compressive strength	RQD from J_v	Joint spacing	Joint condition	Ground-water condition	RMR_{basic}
Infra-Krol shales	7	13	8	22	15	65
Krol A shaly limestones	12	13	8	22	15	70
Krol B shales	12	13	8	22	15	70
Krol C + D limestones	12	13	8	22	15	70
Lower Tal shales	7	13	8	22	15	65
Upper Tal quartzites	12	17	10	22	15	76
Blaini shales	7	13	8	22	15	65

Source: Anbalagan et al., 1992.

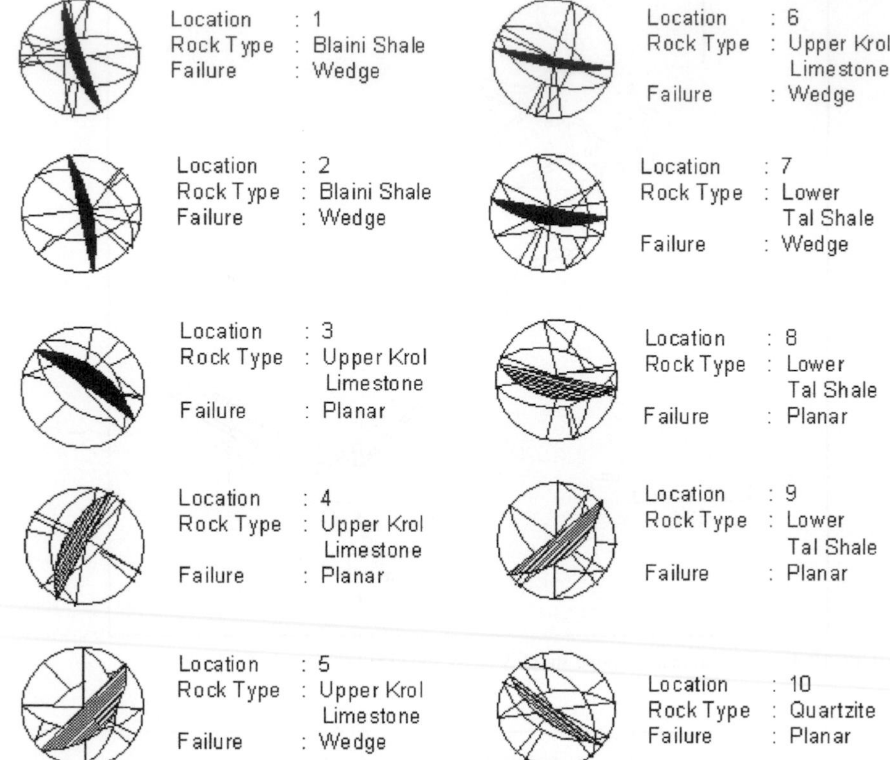

FIGURE 18.5 Stability analysis of wedge/planar failure. *(From Anbalagan et al., 1992)*

It is needless to mention that the side slopes of open trenches should be stable. Deoja et al. (1991) showed the dip of safe cut slopes with and without protective measures for both rocks and soils (Table 18.8). Rail lines are also being built in hilly terrains and Table 18.8 is recommended for deciding safe cut slope angles in those hills. This is very important because landslides/rock falls have suddenly taken place near portals after heavy rains causing very serious train accidents. Table 6.11 also lists safe cut slope angles according to RMR.

The approach to a road/rail line tunnel should be widened sufficiently. Catch drains of proper depth and width should be made on both sides of the track according to the heights and slopes of cuts and sizes of boulders on the slope. A fence of about 3.5 m in height should be erected along both drains and tied to steel poles at about 2 m center to center with horizontal bracings at 1 m center to center. Poles should then be anchored in the slopes. This is a valuable approach if the wire net (4 mm diameter wires welded at 10 $\times$10 cm or alternative) withstands the impact of rock fall jumping. The wire net should then be replaced as soon as required (Hoek, 2000). Wyllie and Mah (2004) described the rock fall hazard rating system on an exponential scale. They also presented the remedial measures for rock fall.

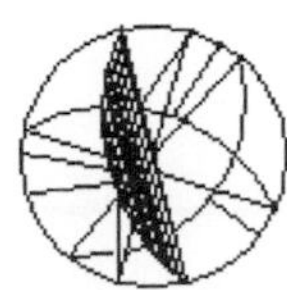

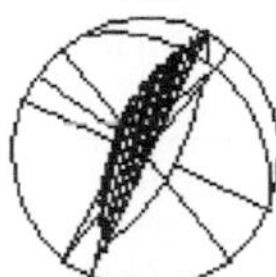

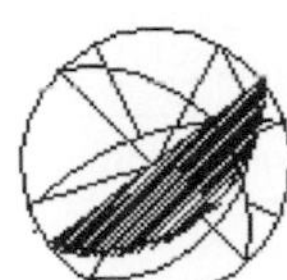

FIGURE 18.6 Stability analysis of wedge/planar failure. *(From Anbalagan et al., 1992)*

TABLE 18.7 Slope Stability Analysis along Lakshmanjhula-Shivpuri Area

Location No. (Figure 18.4)	SMR value	Class No.	Slope description	Stability	Observed failure
1	44.2	III	Normal	Partially stable	Wedge failure
2	47.8	III	Normal	Partially stable	Wedge failure
3	36.3	IV	Bad	Unstable	Planar failure
4	32.4	IV	Bad	Unstable	Planar failure
5	18.0	V	Very bad	Completely unstable	Big wedge failure

Continued

TABLE 18.7 Slope Stability Analysis along Lakshmanjhula-Shivpuri Area—Cont'd

Location No. (Figure 18.4)	SMR value	Class No.	Slope description	Stability	Observed failure
6	24.0	IV	Bad	Unstable	Planar or big wedge failure
7	26.0	IV	Bad	Unstable	Wedge failure
8	40.6	III	Normal	Partially stable	Planar failure
9	56.8	III	Normal	Partially stable	Planar failure
10	30.0	IV	Bad	Unstable	Planar failure
11	69.6	II	Good	Stable	Some block failure
12	55.2	III	Normal	Partially stable	Planar failure
13	51.6	III	Normal	Partially stable	Planar failure
14	36.6	IV	Bad	Unstable	Wedge failure
15	60.9	II	Good	Stable	Some block failure
16	24.0	IV	Bad	Unstable	Planar failure
17	61.8	II	Good	Stable	Some block failure
18	57.0	III	Normal	Partially stable	Wedge failure
19	22.65	IV	Bad	Unstable	Planar failure
20	18.5	V	Very Bad	Completely unstable	Big planar failure

Source: Anbalagan et al., 1992.

TABLE 18.8 Preliminary Design of Cut Slopes for Height of Cut Less Than 10 m

S. No.	Type of soil/rock protection work	Stable cut slope without any breast wall or minor protection work (vertical: horizontal)	Stable cut slope with breast wall (vertical: horizontal)
1	Soil or mixed with boulders	1:1	n:1*
	(a) Disturbed vegetation (b) Disturbed vegetation overlaid on firm rock	Vertical for rock portion and 1:1 for soil portion	Vertical for rock portion and n:1 for soil portion

TABLE 18.8—Cont'd

S. No.	Type of soil/rock protection work	Stable cut slope without any breast wall or minor protection work (vertical: horizontal)	Stable cut slope with breast wall (vertical: horizontal)
2	Same as above but with dense vegetation forests, medium rock, and shales	1:0.5	5:1
3	Hard rock, shale, or harder rocks with inward dip	1:0.25 to 1:0.10 and vertical or overhanging	Breast wall is not needed
4	Same as above but with outward dip or badly fractured rock/shale	At dip angle or 1:0.5 or dip of intersection of joint planes	5:1
5	Conglomerates/very soft shale/sandrock, which erode easily	Vertical cut to reduce erosion	5:1

n is 5 for $H < 3$ m; 4 for $H = 3–4$ m, and 3 for $H = 4–6$ m.

Source: Deoja et al., 1991.

REFERENCES

Anbalagan, R., Sharma, S., & Raghuvanshi, T. K. (1992). Rock mass stability evaluation using modified SMR approach. In *Proceedings of the 6th National Symposium on Rock Mechanics* (pp. 258–268). Bangalore, India.

Bieniawski, Z. T. (1979). The geomechanics classification in rock engineering applications. (Reprinted from *Proceedings of the 4th Congress of the International Society for Rock Mechanics/ Comptes-rendus/Berichte-Montreux*, Suisse, 2–8 Sept. (p. 2208, 3 Vols.). Rotterdam: A. A. Balkema.

Bieniawski, Z. T. (1989). *Engineering rock mass classifications* (p. 251). New York: John Wiley.

Deoja, B. B., Dhittal, M., Thapa, B., & Wagner, A. (1991). *Mountain risk engineering handbook* (Part II, Table 22.15). Kathmandu, Nepal: International Centre of Integrated Mountain Development.

Hack, R. (1998). *Slope stability probability classification—SSPC*, International Institute for Aerospace Survey and Earth Sciences (2nd ed., No. 43, p. 258). Delft, Netherlands: ITC Publication.

Hoek, E. (2000). *Practical rock engineering*. http://www.rocscience.com.

Romana, M. (1985). New adjustment ratings for application of Bieniawski classification to slopes. In *International Symposium on the Role of Rock Mechanics* (pp. 49–53). Zacatecas, Mexico.

Singh, B., & Goel, R. K. (2002). *Software for engineering control of landslide and tunnelling hazards* (p. 344). Rotterdam: A. A. Balkema (Swets & Zeitlinger).

Wyllie, D. C., & Mah, C. W. (2004). *Rock slope engineering—Civil and mining* (4th ed., p. 431). Based on the third edition by E. Hoek and J. Bray. London: Spon Press, Taylor & Francis Group.

Chapter 19

Landslide Hazard Zonation

Landslide is a mountain cancer. It is cheaper to cure than to endure it.

Anonymous

INTRODUCTION

The landslide hazard zonation (LHZ) map is an important tool for designers, field engineers, and geologists to classify the land surface into zones of varying degree of hazards based on the estimated significance of causative factors that influence stability (Anbalagan, 1992). The LHZ map is a rapid technique for hazard assessment of the land surface (Anbalagan & Singh, 2001; Gupta & Anbalagan, 1995; Gupta et al., 2000). It is useful for the following purposes:

1. To help planners and field engineers identify the hazard-prone areas, therefore enabling them to choose favorable locations for site development schemes. If the site cannot be changed and it is hazardous, zonation before construction helps with adopting proper precautionary measures to tackle the hazard problems.
2. To identify and delineate the hazardous area of instability for adopting proper remedial measures to check further environmental degradation of the area.
3. Geotechnical monitoring of structures on the hills should be done in the hazardous areas by preparing a contour map of displacement rates. Landslide control measures and construction controls can be identified accordingly for the safety of buildings on the hilly areas.
4. To realign tunnels to avoid regions of deep-seated major landslides to eliminate risks of high displacement rates. The tunnel portals should be relocated in the stable rock slope (see the section Portal and Cut Slopes in Chapter 18). The outlet of the tailrace tunnel of a hydroelectric project should be much higher than the flood level in the deep gorges, which are prone to landslide dams.

There are three categories of scale on LHZ maps:

1. Mega–regional: Scale of 1:50,000 or more
2. Macro-zonation and risk zonation: 1:25,000 to 1:50,000
3. Meso–zonation: Scale of 1:2000 to 1:10,000

How to prepare an LHZ map is described in the following sections along with an example to show how to apply the LHZ mapping technique in the field for demarcating the landslide-prone areas.

LANDSLIDE HAZARD ZONATION MAPS—THE METHODOLOGY

Factors

LHZ was developed by Anbalagan (1992). Many researchers have developed various methods of landslide zonation, but they are not based on causative factors. The main merit of Anbalagan's method is that it considers causative factors in a simple way. His method has become very popular in India, Italy, Nepal, and other countries. The technique, in a broader sense, classifies the area into five zones on the basis of the following six major causative factors:

1. Lithology: Characteristics of rock and land type
2. Structure: Relationship of structural discontinuities with slopes
3. Slope morphometry
4. Relative relief: Height of slope
5. Land use and land cover
6. Groundwater condition

These factors are called the "landslide hazard evaluation factors" (LHEF). Ratings of all of the LHEFs are listed in Table 19.1, whereas the maximum assigned rating for each LHEF is given in Table 19.2. The basis of assigning ratings in Table 19.1 is discussed by parameter in the following sections. There have been minor changes in ratings for the lithology (S. No. 1) and depth of soil cover in the land use and land cover (S. No. 5). These changes were suggested by the Geological Survey of India and Bureau of Indian Standards, New Delhi, India, in 2006, with the kind agreement of Professor R. Anbalagan, IIT Roorkee.

Lithology

The erodibility or the response of rocks to the processes of weathering and erosion should be the main criterion for awarding ratings for lithology. Rock types such as unweathered quartzites, limestones, and granites are generally hard and massive and more resistant to weathering, therefore forming steep slopes, but ferruginous sedimentary rocks are more vulnerable to weathering and erosion. The phyllites and schists are generally more weathered close to the surface. Accordingly, a higher LHEF rating should be awarded (Table 19.1).

With soil-like materials, genesis and age are the main considerations when awarding ratings. The older alluvium is generally well compacted and has high strength, whereas slide debris is generally loose and has low shearing resistance. Nearly vertical slopes of interlocked sand are stable for several decades in the lesser Himalayas in India. Gupta et al. (2000) observed in Garhwal in the Himalayas, that maximum landslides were found in rocks with large amounts of talc minerals.

Structure

This includes primary and secondary rock discontinuities, such as bedding planes, foliations, faults, and thrusts. The discontinuities in relation to slope direction have greater influence on slope stability, and these three types of relationships are important:

1. Extent of parallelism between the directions of discontinuity or the line of intersection of two discontinuities and the slope
2. Steepness of the dip of discontinuity or plunge of the line of intersection of two discontinuities
3. Difference in the dip of discontinuity or plunge of the line of intersection of two discontinuities of the slope

TABLE 19.1 LHEF Rating Scheme

S. No.	Contributory factor	Category	Rating	Remarks
1	**Lithology**			
	(a) Rock type	*Type I*		*Correction factor for weathering:*
		- Quartzite and limestone, banded hematite quartzite	0.2	(a) Highly weathered — rock discolored joints open with weathering products, rock fabric altered to a large extent; correction factor C_1
		- Granite, gabbro, basalt, charnokite	0.3	
		- Gneiss	0.4	
		Type II		
		- Well-cemented ferruginous sedimentary rocks, dominantly sandstone with minor beds of claystone	1.0	(b) Moderately weathered—rock discolored with fresh rock patches, weathering more around joint planes but rock intact in nature; correction factor C_2
		- Poorly cemented ferruginous sedimentary rocks, dominantly sandstone with minor clay shale beds	1.3	
		Type III		
		- Slate and phyllite	1.2	(c) Slightly weathered—rock slightly discolored along joint planes, which may be moderately tight to open, intact rock; correction factor C_3
		- Schist	1.3	
		- Shale with interbedded clayey and non clayey rocks	1.8	
		- Highly weathered shale, phyllite, and schist; any rock with talc mineral	2.0	
	(b) Soil type	- Older well-compacted fluvial fill material/RBM (alluvial)	0.8	The correction factor for weathering should be a multiple with the fresh rock rating to get the corrected rating
		- Clayey soil with naturally formed surface (alluvial)	1.0	
		- Sandy soil with naturally formed surface (alluvial)	1.4	
		- Debris comprising mostly rock pieces mixed with clayey/sandy soil (colluvial)		*For rock type I* $C_1 = 4, C_2 = 3, C_3 = 2$ *For rock type II* $C_1 = 1.5, C_2 = 1.25, C_3 = 1.0$
		I. Older well compacted material	1.2	
		II. Younger loose material	2.0	
2	**Structure**			
	*(a) Parallelism between the slope and discontinuity** PLANAR ($\alpha_j - \alpha_s$) WEDGE ($\alpha_i - \alpha_s$)	I. >30°	0.2	α_j = *dip direction of joint*
		II. 21–30°	0.25	α_i = *direction of line of intersection of two discontinuities*
		III. 11–20°	0.3	
		IV. 6–10°	0.4	
		V. <5°	0.5	α_s = *direction of slope inclination*

Continued

TABLE 19.1 LHEF Rating Scheme—Cont'd

S. No.	Contributory factor	Category	Rating	Remarks
	(b) Relationship of dip of discontinuity and inclination PLANAR ($\beta_j - \beta_s$) WEDGE ($\beta_I - \beta_s$)	I. >10°	0.3	β_j = *dip of joint*
		II. 0–10°	0.5	β_i = *plunge of line of intersection*
		III. 0°	0.7	β_s = *inclination of slope*
		IV. 0–(−10°)	0.8	
		V. <−10°	1.0	
	(c) Dip of discontinuity PLANAR (β_j) WEDGE (β_i)	I. <15°	0.2	*Category* I = very favorable
		II. 16–25°	0.25	II = favorable
		III. 26–35°	0.3	III = fair
		IV. 36–45°	0.4	IV = unfavorable
		V. >45°	0.5	V = very unfavorable
3	**Slope Morphometry**			
	- *Escarpment/cliff*	>45°	2.0	
	- *Steep slope*	36–45°	1.7	
	- *Moderately steep slope*	26–35°	1.2	
	- *Gentle slope*	16–25°	0.8	
	- *Very gentle slope*	≤15°	0.5	
4	**Relative Relief**			
	Low	<100 m	0.3	
	Medium	101–300 m	0.6	
	High	>300 m	1.0	
5	**Land Use and Land Cover**			
	- *Agricultural land/ populated flatlands*		0.60	
	- *Thickly vegetated forest area*		0.80	
	- *Moderately vegetated*		1.2	
	- *Sparsely vegetated with lesser ground cover*		1.2	
	- *Barren land*		2.0	
	- *Depth of soil cover*	<5 m	0.65	
		6–10 m	0.85	
		11–15 m	1.2	
		16–20 m	1.5	
		>20 m	2.0	
6	**Groundwater Condition**	Flowing	1.0	
		Dripping	0.8	
		Wet	0.5	
		Damp	0.2	
		Dry	0.0	

In regions of low seismicity (1, 2, and 3 zones), the maximum rating for relative relief may be reduced to 0.5 times and hydrogeological conditions increased to 1.5 times. For high seismicity (4 and 5 zones), no corrections are required.

**Discontinuity refers to the planar discontinuity or the line of intersection of two planar discontinuities, whichever is important concerning instabilities.*

Source: Gupta and Anbalagan, 1995

TABLE 19.2 Proposed Maximum LHEF Rating for Different Causative Factors for LHZ Mapping

Contributory factor	Maximum LHEF rating
Lithology	2
Structure — relationship of structural discontinuities with slopes	2
Slope morphometry	2
Relative relief	1
Land use and land cover	2
Groundwater condition	1
Total	10

Source: Gupta and Anbalagan, 1995.

These three relationships are the same as F_1, F_2, and F_3 of Romana (1985) and are discussed in Chapter 18. Various subclasses of the previously discussed conditions are also similar to Romana (1985).

It may be noted that the inferred depth of soil should be considered when rating rock types.

Slope Morphometry

Slope morphometry defines slope categories based on the frequency of a particular slope angle occurrence. Five categories representing the slopes (escarpment/cliff, steep slope, moderately steep slope, gentle slope, and very gentle slope) are used in preparing slope morphometry maps. Regionally, the angle can be obtained from topographic sheets for initial study.

Relative Relief

Relative relief maps represent the local relief of maximum height between the ridgetop and the valley floor within an individual facet. Three categories of slopes of relative relief—low, medium, and high—should be used for hazard evaluation purposes. The peak ground acceleration is maximum at the hilltop during a major earthquake.

A facet is part of a hill slope that has a consistent slope direction and inclination. In thickly populated areas, smaller facets of rock slopes may be taken into consideration.

Land Use and Land Cover

The nature of land cover is an indirect indication of hill slope stability. Forest cover, for instance, protects slopes from the effects of weathering and erosion. A well-developed and spread root system increases the shearing resistance of the slope material. Barren and sparsely vegetated areas show faster erosion and greater instability. Ratings are awarded based on the vegetation cover and its intensity for this parameter. (Review of the literature shows that extra cohesion due to root reinforcement is seldom more than 5 T/m^2.) Thus, continuous vegetation and grass cover on an entire hill slope is not fully responsible in landslide control because of root reinforcement, but drastic decrease in the infiltration rate of rainwater through a thin humus layer because of grass cover is more beneficial.

Groundwater Conditions

Since the groundwater in hilly terrain is generally channeled along structural discontinuities of rocks, it does not have a uniform flow pattern. The observational evaluation of the groundwater on hill slopes is not possible over large areas. Therefore, for quick appraisal, surface indications of water such as damp, wet, dripping, and flowing are used for rating purposes (Table 6.6). It is suggested that studies should be carried out soon after the monsoon season.

Other Factors

A 100–200 m wide strip on either side of major faults and thrusts and intra-thrust zones may be awarded an extra rating of 1.0 to consider higher landslide susceptibility depending upon intensity of fracturing.

Experiences in Garhwal, in the Himalayas, and in Indonesia show that extensive landslides are likely to occur during the heavy rainy season soon after a major earthquake in that area, which cracks and loosens the slope mass extensively near the surface. (See the legend of Table 19.1.)

Landslide Hazard Zonation

Ratings of all of the parameters are added to obtain a total estimated hazard rating (TEHR). Various zones of landslide hazard have subsequently been classified on the basis of TEHR as seen in Table 19.3.

TABLE 19.3 Classification of LHZ

Zone	Value of TEHR	Description of LHZ	Practical significance
I	<3.5	Very low hazard (VLH)	Safe for development schemes
II	3.5–5.0	Low hazard (LH)	
III	5.1–6.0	Moderate hazard (MH)	Local vulnerable zones of instabilities
IV	6.1–7.5	High hazard (HH)	Unsafe for development schemes
V	>7.5	Very high hazard (VHH)	

Source: Gupta and Anbalagan, 1995.

Presentation of LHZ Maps

The results should be presented in the form of maps. Terrain evaluation maps are prepared in the first stage showing the nature of facet-wise distribution of parameters. The terrain evaluation maps are superimposed and TEHR is estimated for individual facets. Subsequently, LHZ maps are prepared based on facet-wise distribution of TEHR values. For this exercise, two types of studies are performed: a desk or laboratory study and a field study. The general procedures of LHZ mapping techniques have been outlined in the form of a flow chart (Figure 19.1). This method has been adopted by the Bureau of Indian Standards (IS 14496 Part 2, 1998).

Caution: This technique is not applicable to mountains that are most often covered by snow. Moreover, freezing and thawing of water in rock joints causes rock slides in these regions. LHZ is also not suitable for areas of cloudbursts (rainfall $>$ 500 mm per day).

In the next section, a case history has been presented to clarify the LHZ methodology and to develop confidence among users.

A CASE HISTORY (GUPTA AND ANBALAGAN, 1995)

This investigation covers the Tehri-Pratapnagar area, which lies between latitude (30°22′15″–30°30′5″) and longitude (78°25′–78°30′) (Figure 19.2).

Geology of the Area

The study area lies in the Tehri District of Uttar Pradesh in India. The rock masses of the area belong to Damtha, Tejam, and Jaunsar groups. The stratigraphic sequence of the area and its vicinity is as follows (Valdiya, 1980):

Nagthat-Berinag Formation

Chandpur formation	Jaunsar group
Deoban formation	Tejam group
Rautgara formation	Damtha group

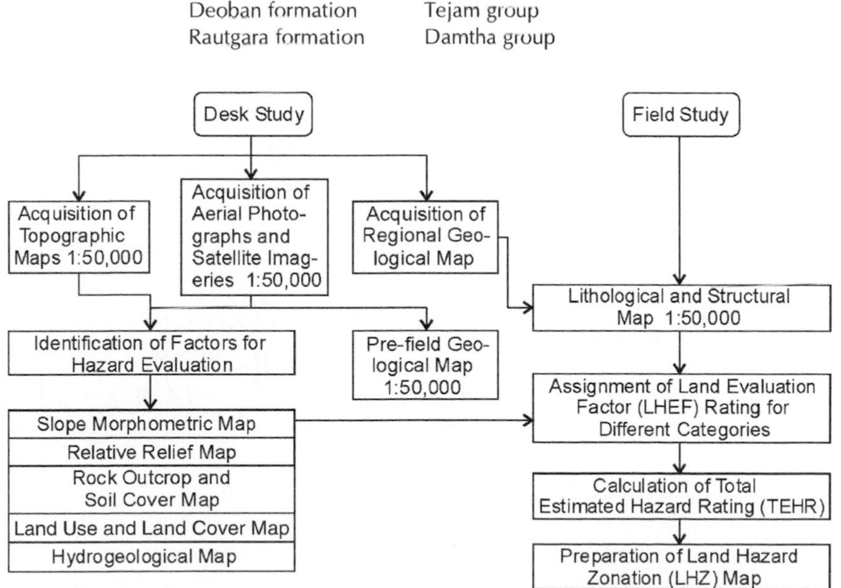

FIGURE 19.1 Procedure for macro-regional LHZ mapping.

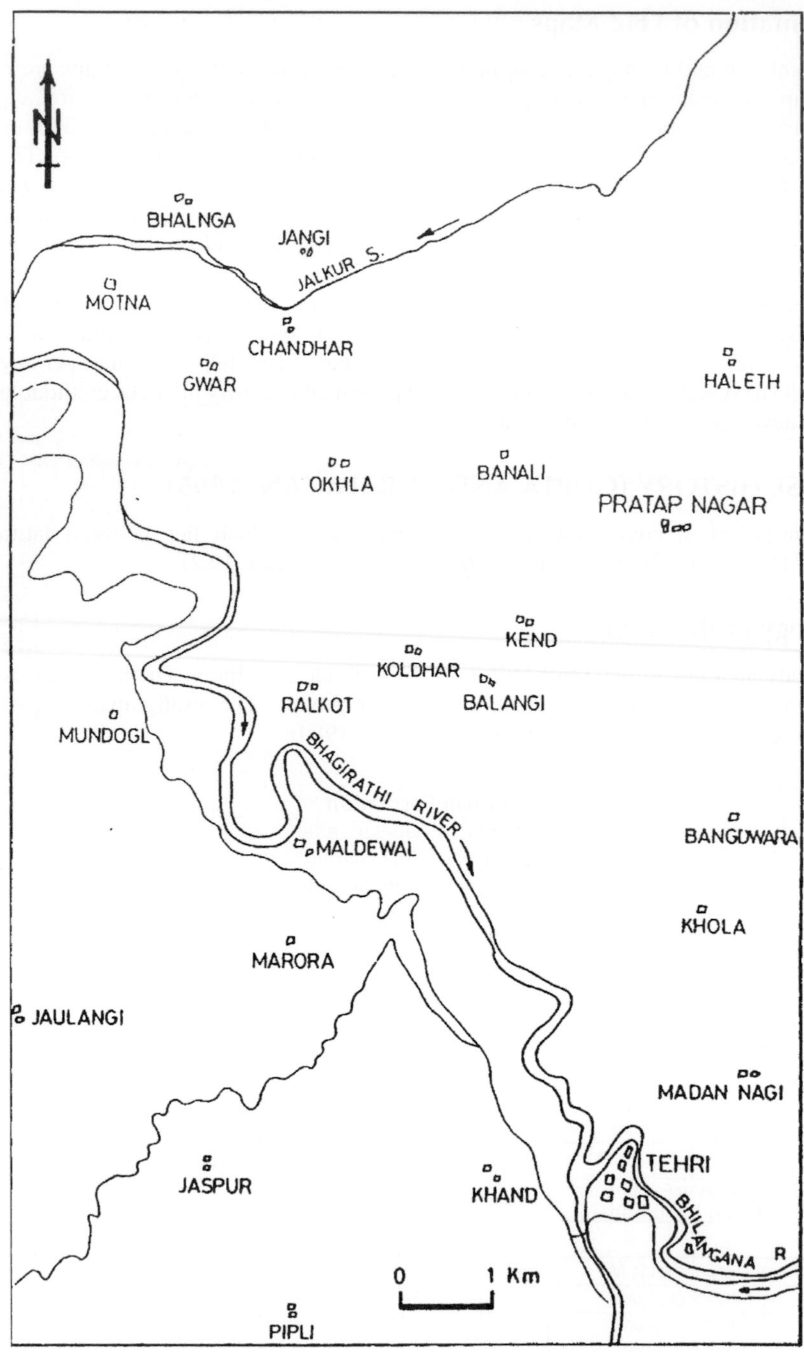

FIGURE 19.2 Location map of the study area. *(From Gupta and Anbalagan, 1995)*

This area has been mapped on a 1:50,000 scale for studying its lithology and structure. The rocks exposed in the area include phyllites of Chandpur formation interbedded with sublitharenites of Rautgara formation, dolomitic limestone of Deoban formation, and quartzites of Nagthat-Berinag formation. The phyllites are gray and olive green interbedded with metasiltstones and quartzitic phyllites. The Rautgara formation is comprised of purple, pink, and white medium-grained quartzites interbedded with medium-grained gray and dark green sublitharenites and slates as well as metavolcanics. The Deoban formation consists of dense, fine-grained white and light pink dolomites with minor phyllitic intercalations. They occupy topographically higher ridges. The Nagthat-Berinag formation includes purple, white, and green quartzites interbedded with greenish and gray slates as well as gray phyllites.

The Chandpur formation is delimited toward the north by a well-defined thrust called "North Almora thrust" trending roughly northwest-southeast and dipping southwest. Moreover, the Deoban and the Nagthat-Berinag formations have a thrusted contact with the thrust trending parallel to the North Almora thrust and dipping northeast. This is called the "Pratapnagar thrust". The rocks are badly crushed in the thrust zones.

LHZ Mapping

The LHZ map of this area has been prepared on a 1:50,000 scale using the LHEF rating scheme for which a facet map of the area has been prepared (Figure 19.3). A facet is a part of a hill slope that shows consistent slope direction and inclination. The thematic maps of the area—including the lithological map (Figure 19.4), structural map (Figure 19.5), slope morphometry map (Figure 19.6), land use and land cover map (Figure 19.7), relative relief map (Figure 19.8), and groundwater condition map (Figure 19.9)—have been prepared using the detailed LHEF rating scheme (Table 19.1).

Lithology

Lithology (see Figure 19.4) is one of the major causative factors for slope instability. The major rock types observed in the area include phyllites, quartzites, and dolomitic limestones. In addition, fluvial terrace materials are abundant to the right of the Bhagirathi River, all along its course.

Phyllites are exposed on either bank close to the Bhagirathi River. Although older terrace materials are present at lower levels, thick alluvial and colluvial soil cover are present in the upper levels on the right bank. On the left bank, the phyllites are generally weathered close to the surface and support thin soil cover. The thickness of soil cover increases up to 5 m in some places.

The North Almora thrust separates the Chandpur phyllites on the south from the quartzites of the Rautgara formation. The Rautgara quartzites interbedded with minor slates and metavolcanics are pink, purple, and white, well-jointed, and medium grained. The rocks and soil types in the area have the following distribution: phyllites, 44.17%; quartzites, 27.41%; marl/limestones, 12.48%; metabasics, 0.25%; river terrace material, 6.11%; phyllites with thin alluvial soil cover, 6.16%; and quartzites with thin soil cover, 3.41%.

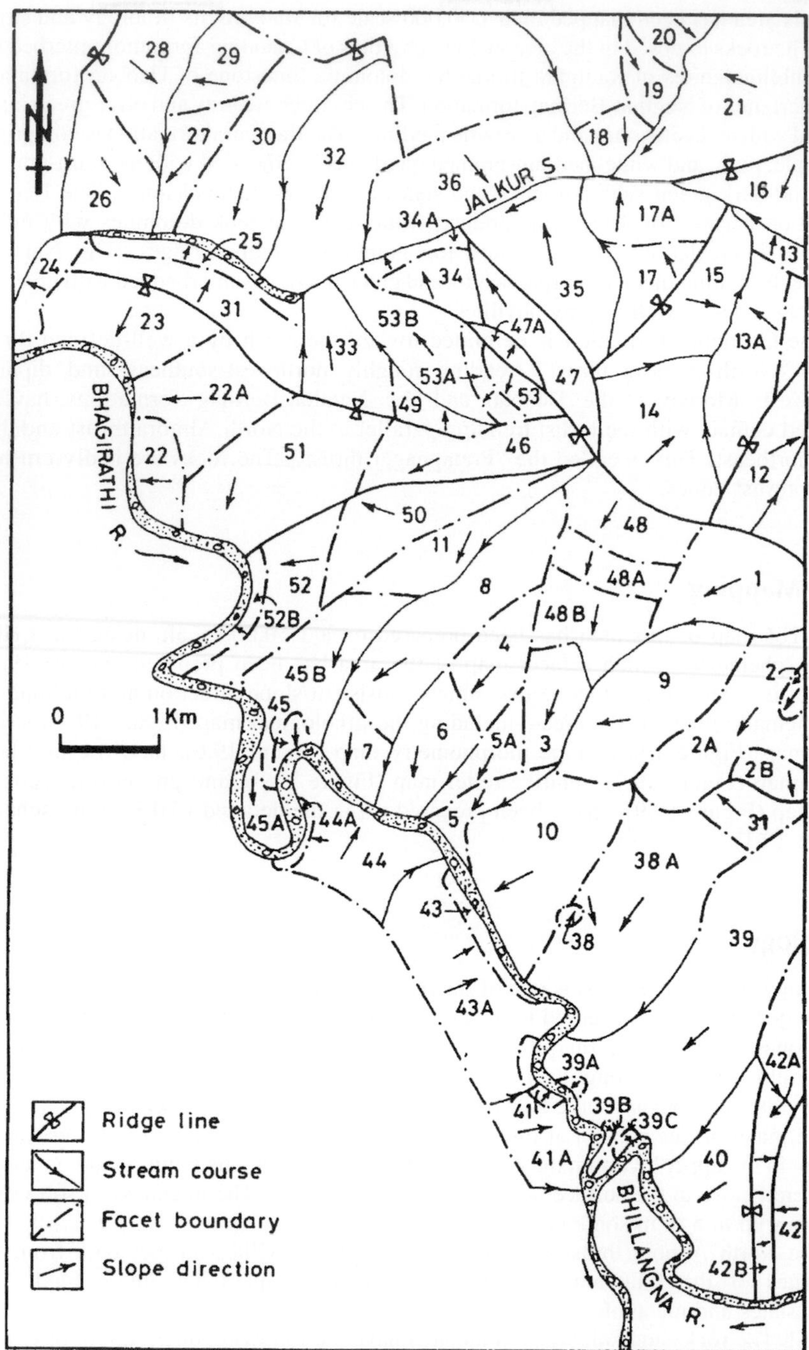

FIGURE 19.3 Facet map of the study area. *(From Gupta and Anbalagan, 1995)*

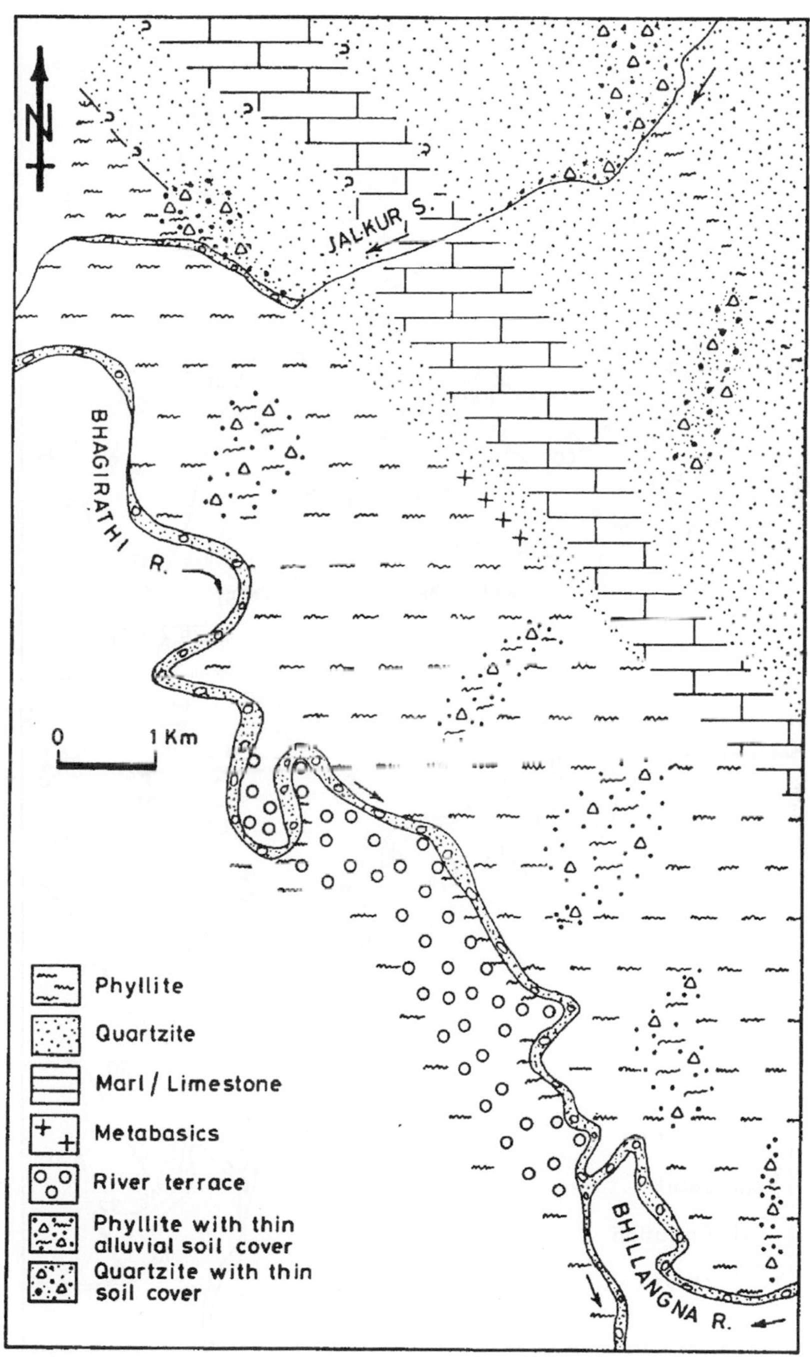

FIGURE 19.4 Lithological map. *(From Gupta and Anbalagan, 1995)*

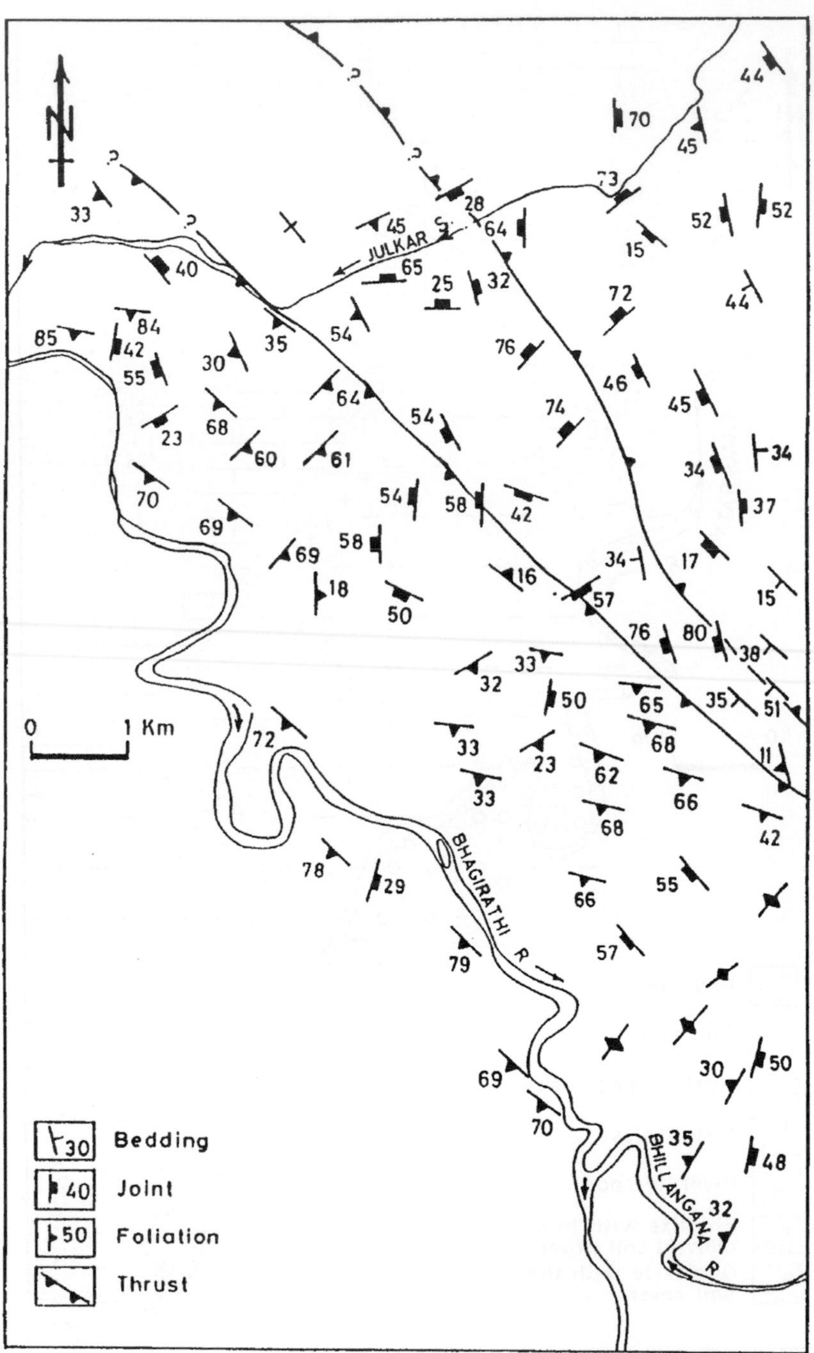

FIGURE 19.5 Structural map. *(From Gupta and Anbalagan, 1995)*

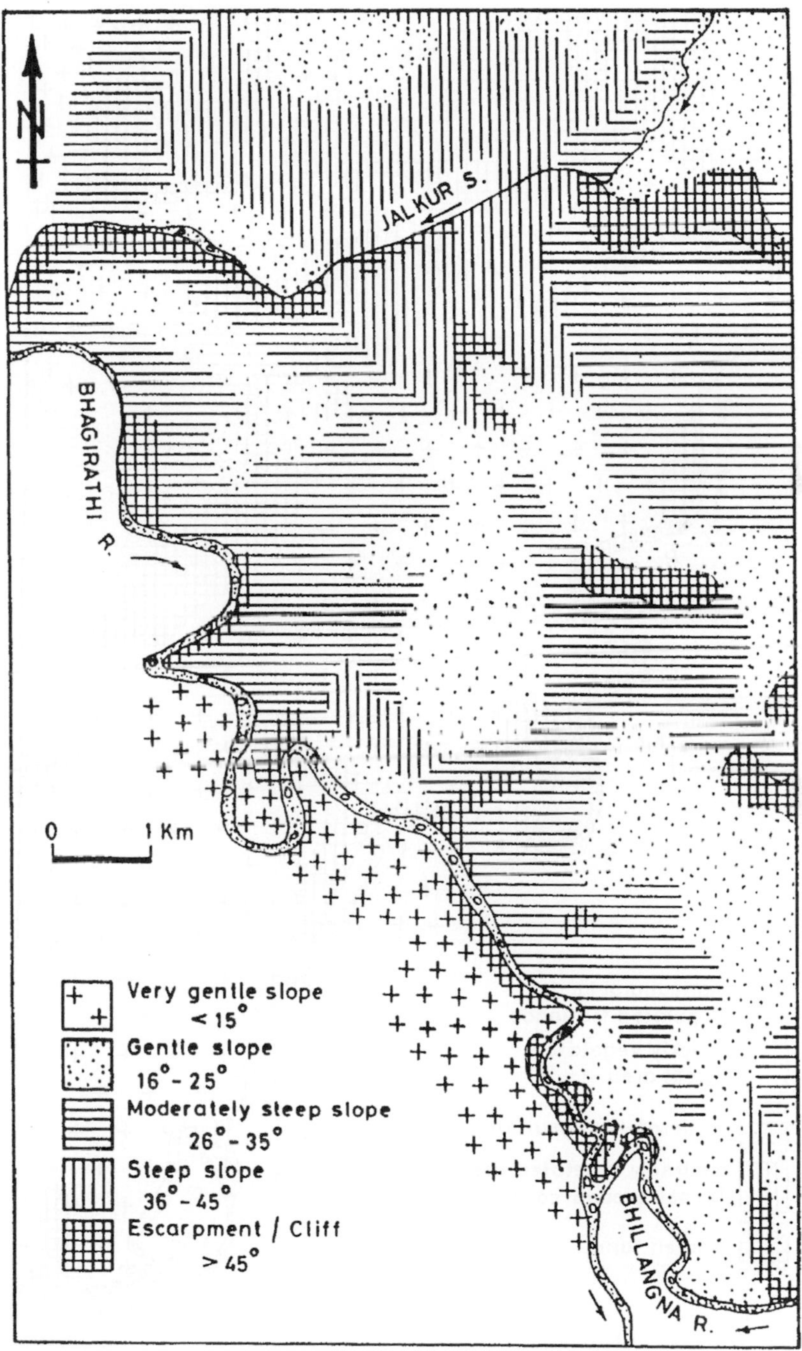

FIGURE 19.6 Slope morphometry map. *(From Gupta and Anbalagan, 1995)*

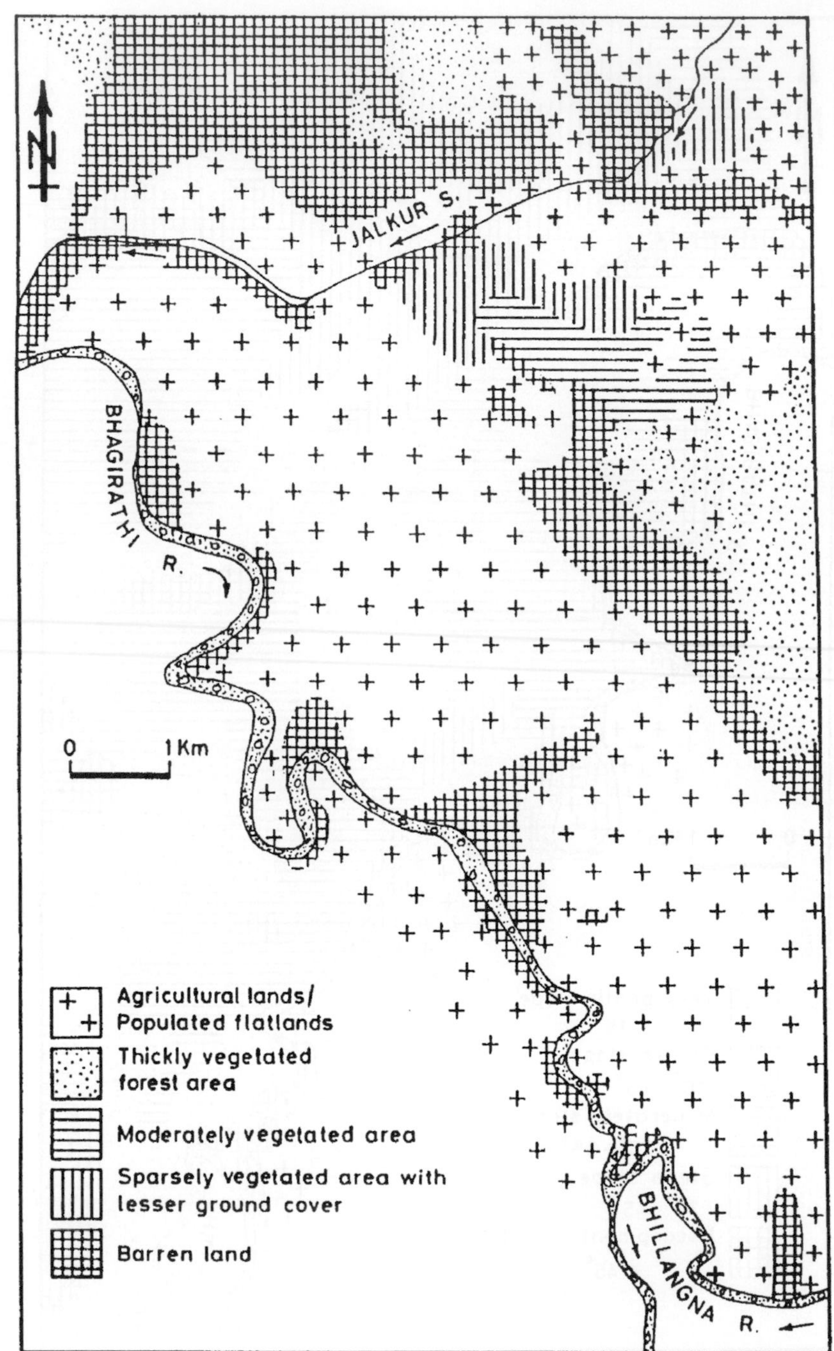

FIGURE 19.7 Land use and land cover map. *(From Gupta and Anbalagan, 1995)*

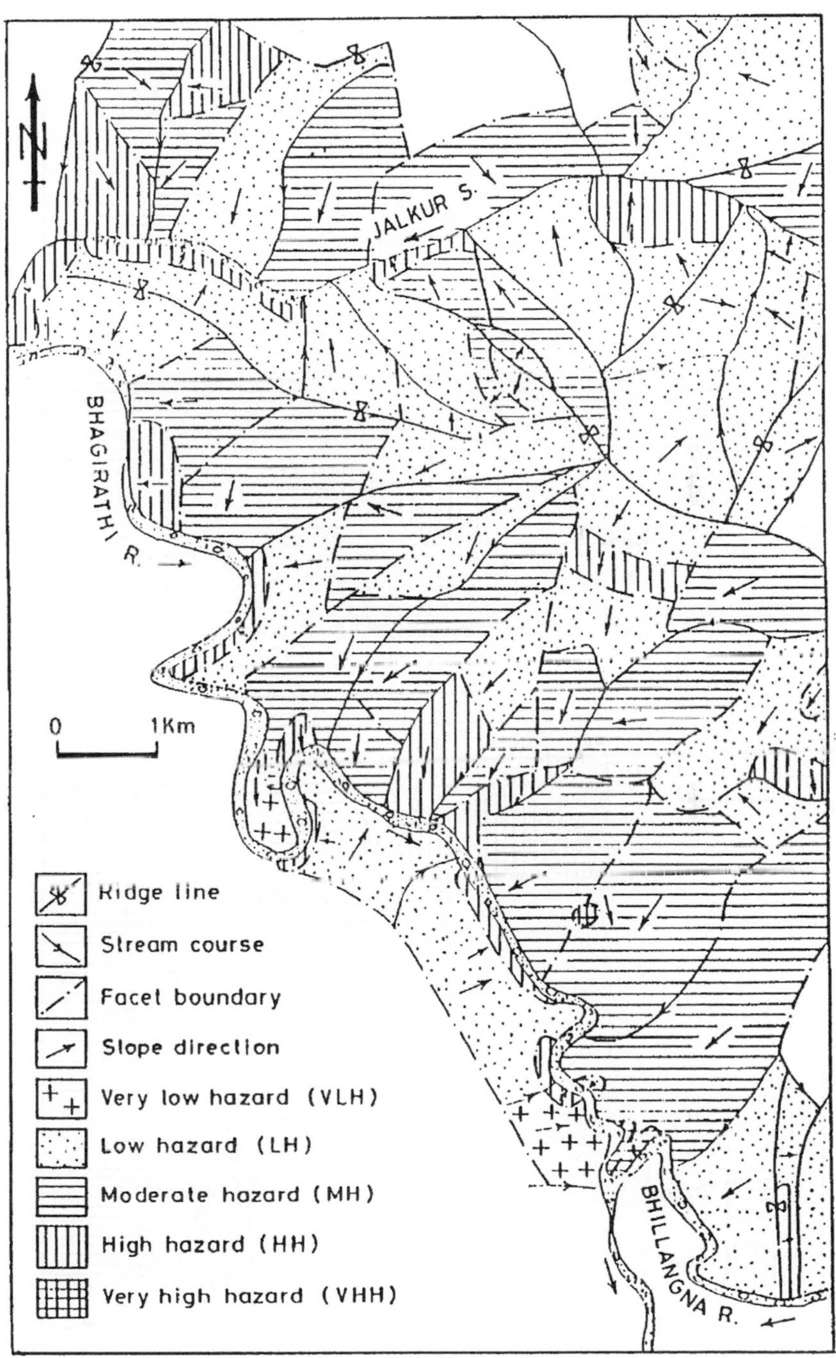

FIGURE 19.8 Relative relief map. *(From Gupta and Anbalagan, 1995)*

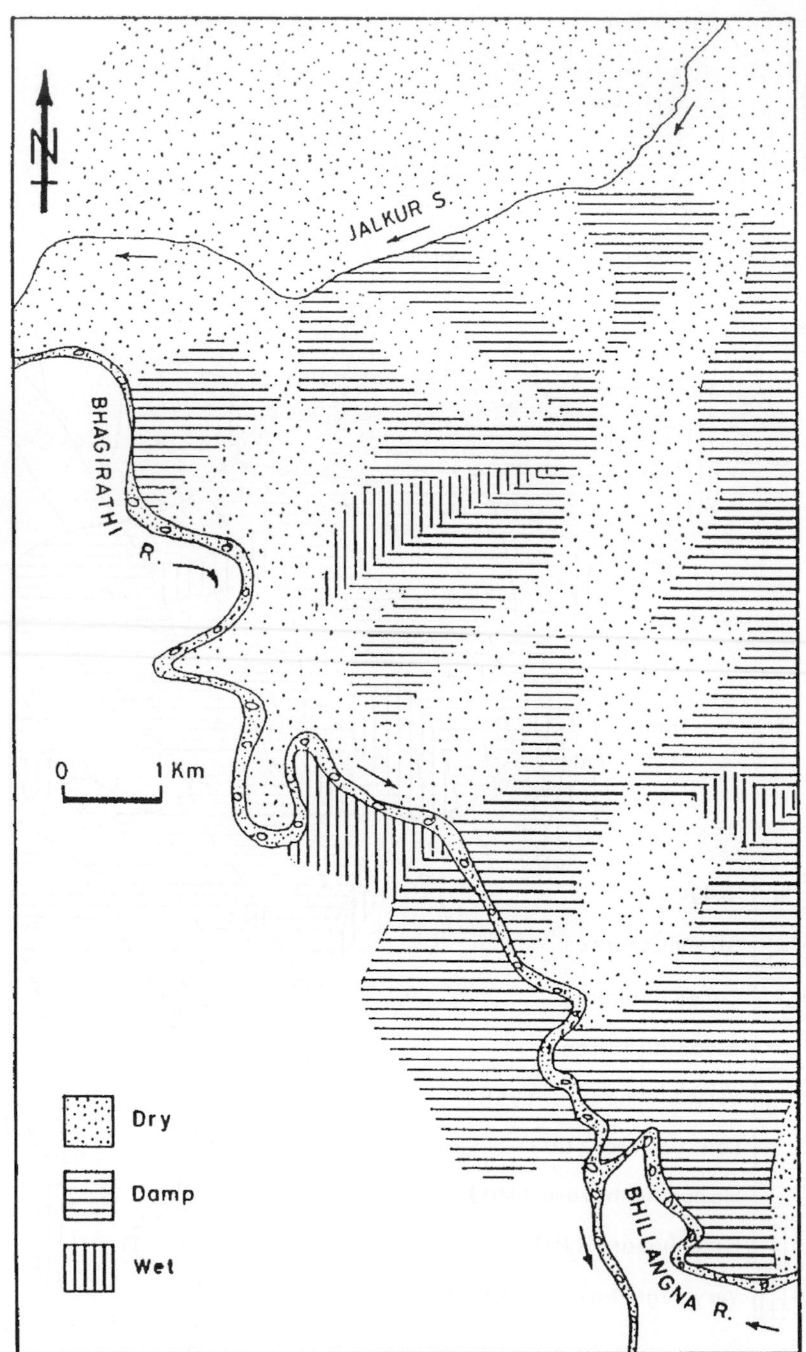

FIGURE 19.9 Groundwater condition map. *(From Gupta and Anbalagan, 1995)*

Structure

Major structural features (see Figure 19.5) seen in the area include the North Almora and the Pratapnagar thrusts, which form part of the Berinag thrust. The structures used for LHZ mapping include beddings, joints, and foliations. The dispositions of the structures have been plotted in a stereonet for individual facets. The interrelation of the structural discontinuity with the slope is studied carefully before ratings are awarded.

Slope Morphometry

A slope morphometry map (see Figure 19.6) represents the zones of different slopes, which have a specific range of inclination. The area of study has a good distribution of slope categories. The area to the west of the Bhagirathi River, mainly occupied by terrace deposits, falls in the category of a very gentle slope and these are mainly confined to agricultural fields. Moderately steep slopes mainly occur in the central and eastern parts of the area, steep slopes mainly occur in the central and the eastern parts of the area, and very steep slopes occur in the northern part of the study area adjoining the Jalkur stream.

The Jalkur stream flows through a tight, narrow, V-shaped gorge. Very steep slopes/escarpments occur in small patches, mainly close to the watercourses, possibly because of toe erosion. This area has the following distribution: very gentle slope, 6.14%; gentle slope, 31.92%; moderately steep slope, 42.32%; steep slope, 11.37%; and very steep slope/escarpment, 8.27%.

Land Use and Land Cover

Vegetation cover generally smoothes the action of climatic agents and protects the slope from weathering and erosion (see Figure 19.7). The nature of land cover may indirectly indicate the stability of hill slopes. Agricultural lands/populated flatlands are extensively present in the central, southeastern, and southern areas, and in parts of the northeastern areas. Thickly vegetated forest areas are seen in the Pratapnagar-Bangdwara area. Moderately vegetated areas are mainly present in small patches to the west of thickly vegetated areas. Sparsely vegetated and barren lands are mainly confined to quartzitic and dolomitic limestone terrain where steep to very steep slopes are present. These types of slopes are seen along the Bhagirathi Valley adjoining the river course, generally on steep slopes (Figure 19.2). The five categories of land use and land cover include agricultural lands/populated flatlands, thickly vegetated forest areas, moderately vegetated areas, sparsely vegetated areas, and barren areas, with the distribution of 65.44%, 5.94%, 1.73%, 3.78%, and 23.10%, respectively, in the study area.

Relative Relief

Relative relief (see Figure 19.8) is the maximum height between the ridgetop and the valley floor within an individual facet. The three categories of relative relief include high relief, medium relief, and low relief and occupy 75.53%, 15.96%, and 8.74% of the study area, respectively.

Groundwater Condition

The surface manifestation of groundwater (see Figure 19.9), such as wet, damp, and dry, has been observed in the study area. This area predominantly shows dry conditions in about 54.86% of it, damp conditions in about 40.96% of it, and 4.8% of the study area is covered by wet groundwater conditions. The dry condition is mainly observed in the northern part and is well distributed in the remaining study area, and damp and wet conditions are present in a number of facets in the southern, eastern, and central parts.

LHZ Map

The sum of all causative factors (see Figure 19.10) within an individual facet gives the TEHR for a facet. The TEHR indicates the net probability of instability within an individual facet. Based on the TEHR value, facets are divided into different categories of hazard zones (Anbalagan, 1992).

The five categories of hazards are very low hazard (VLH), low hazard (LH), moderate hazard (MH), high hazard (HH), and very high hazard (VHH) and are present in the study area. The areas showing VLH and LH constitute about 2.33% and 43.27% of the study area, respectively. They are well distributed within the area. MH zones are mostly present in the immediate vicinities to the east of the Bhagirathi River. HH and VHH zones occur as small patches, mostly close to the watercourses. They represent areas of greater instability where detailed investigations should be carried out.

Some difficulty was experienced in zonation at the boundary lines. The visual inspection of existing landslides matched with Figure 19.10 for more than 85% of the area. As such, Anbalagan's technique may be adopted in all mountainous terrains with minor adjustments in his ratings. For rocky hill areas, the slope mass rating is preferred (see Chapter 18).

PROPOSITION FOR TEA GARDENS

Tea gardens are recommended in medium and high hazard zones because of suitable soil and climatic conditions in these areas. Rains are nature's boon to plants and forests. The tea gardens significantly reduce infiltration of rainwater into the debris, stabilizing landslide-prone areas. Tea gardens also provide job opportunities to local people and ease their poverty. Herbal farming should also be adopted.

GEOGRAPHIC INFORMATION SYSTEM

Geographic Information Systems (GIS) are software tools used to store, analyze, process, manipulate, and update information in layers where geographic location is an important characteristic or critical to the analysis (Aronoff, 1989).

LHZ mapping, as described in the section Landslide Hazard Zonation Maps — The Methodology in this chapter, can be done efficiently by using GIS. LHEF can be used as layers of information to the GIS using various input devices. For example, Figures 19.2–19.9 can be used as the layers of information to a GIS using input devices such as a digitizer, scanner, and so forth to carry out LHZ mapping of the considered area providing an output similar to Figure 19.10. Amin et al. (2001) developed a software package called "GLANN" using GIS, neural network analysis, and genetic algorithms for automatic landslide zonation. They also successfully used Anbalagan's LHZ system.

Handling and analyzing data referenced to a geographic location are key capabilities of a GIS, but the power of the system is most apparent when the quantity of data involved in mapping LHZ is too large to be handled manually. Over and above the main causative factors mentioned earlier, there may be many other features considered for

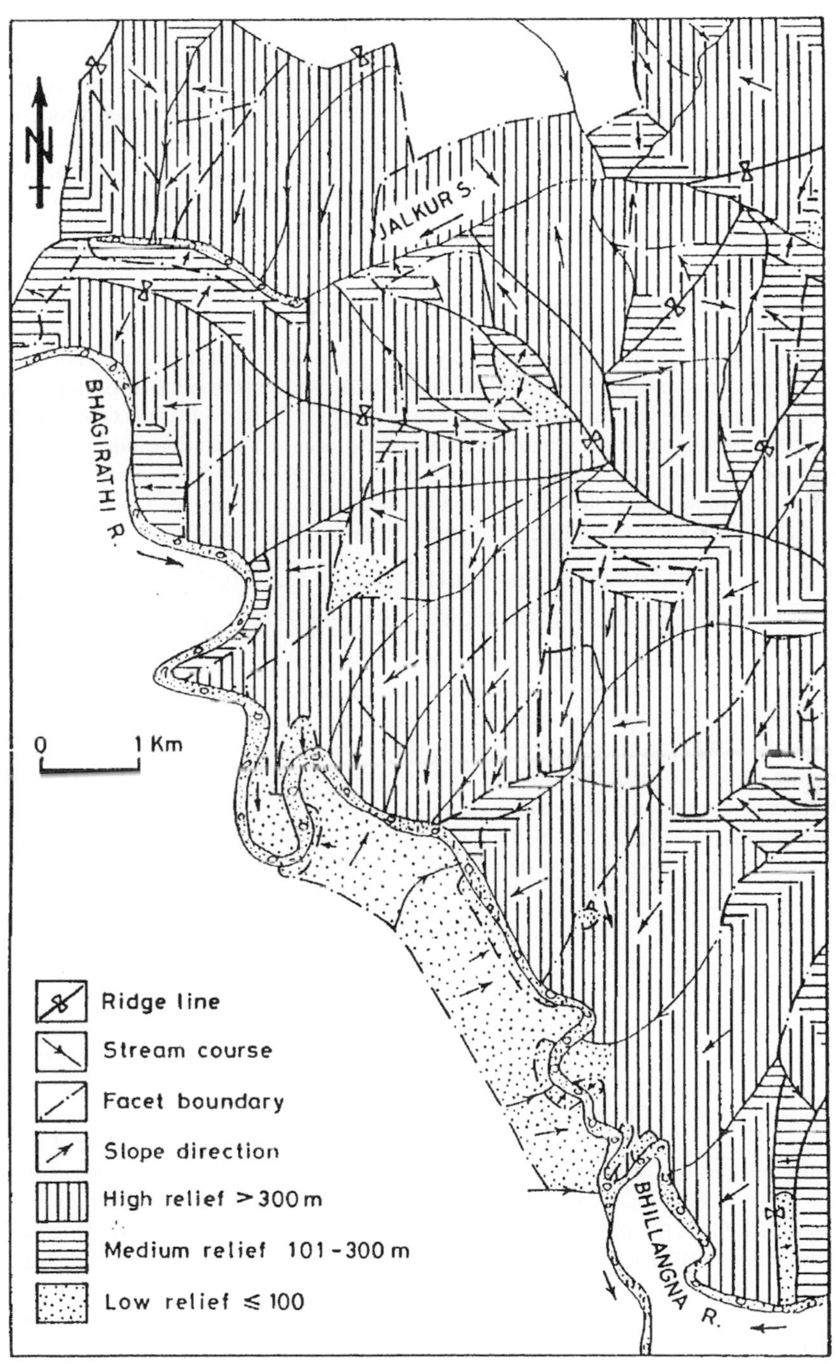

FIGURE 19.10 LHZ map. *(From Gupta and Anbalagan, 1995)*

LHZ based on site-specific conditions and many factors associated with each feature or location. These data may exist as maps, tables of data, or even lists of names (Figure 19.2). Such large volumes of data cannot be efficiently handled by manual methods. However, when those data are input into a GIS, they can be easily processed and analyzed efficiently and economically.

Geographical position systems (GPS) have been used successfully in monitoring landslides with an accuracy of nearly 2 mm (Brunner, Hartinger, & Richter, 2000). This practical method of LHZ plots contours of rates of displacement per year.

MEGA-REGIONAL LANDSLIDE ZONATION

The first law of geomorphology is that everything is related to everything else, but closer things are related more. In this method of zonation the region is divided into square grid cells (e.g., 1 × 1 km). The various geomorphological parameters such as maximum slope, aspect ratio, density of vegetation or normalized difference vegetation index (NDVT), positive or negative curvature of slope, and relative height of hill are considered. Erener (2005) found the following relation between the probability of landslide occurrence (L_n) and the previously discussed parameters (scale 1:250,000):

$$\begin{aligned} L_n(\text{Odds}) = 5.08 + 0.11\psi_f - 0.0015A_s - 0.82VD \\ -0.0035NC + 0.019PC - 0.013H \end{aligned} \quad (19.1)$$

where ψ_f = maximum slope near a grid point, A_s = aspect ratio of slope, VD = density of vegetation (NDVT or normalized difference vegetation index), NC = negative curvature of slope surface, PC = positive curvature of slope surface, and H = relative relief.

Equation (19.1) clearly shows that the density of vegetation is extremely effective in stabilizing slopes, as expected. Erener (2005) further analyzed that the density of roads, river density, and lineaments (smaller faults) do not significantly improve the model. It appears that Eq. (19.1) may be modified for different regions and used for mega-regional landslide zonation using GIS. Further research is needed.

REFERENCES

Amin, S., Gupta, R., Saha, A. K., Arora, M. K., & Gupta, R. P. (2001). Genetic algorithm based neural network for landslide hazard zonation: Some preliminary results. In *Workshop on Application of Rock Engineering in Nation's Development* (In honor of Professor Bhawani Singh) (pp. 203–216). Uttarakhand, India: IIT Roorkee.

Anbalagan, R. (1992). Terrain evaluation and landslide hazard zonation for environmental regeneration and land use planning in mountainous terrain. In *International Symposium on Landslides* (pp. 861–868). Christchurch, New Zealand.

Anbalagan, R., & Singh, B. (2001). Landslide hazard and risk mapping in the Himalayas. In L. Tianchi et al. (Eds.), *An ICIMOD Publication on Landslide Hazard Mitigation in the Hindu Kush Himalayas* (pp. 163–188), Kathmandu, Nepal.

Aronoff, S. (1989). *Geographic information systems: A management perspective*. Ottawa, Canada: WDL Publications.

Brunner, F. K., Hartinger, H., & Richter, B. (2000). Continuous monitoring of landslides using GPS: A progress report. In *Proceedings of the Physical Aspects of Mass Movements,* S. J. Bauer & F. K. Weber (Eds.), Vienna: Austrian Academy of Sciences.

Erener, A. (2005). Landslide hazard assessment by using spatial regression techniques (p. 57), Project Report, Ankara, Turkey: Middle East Technical University.

Gupta, P., & Anbalagan, R. (1995). Landslide hazard zonation, mapping of Tehri-Pratapnagar Area, Garhwal Himalayas. *Journal of Rock Mechanics and Tunnelling Technology*, *1*(1), 41–58.

Gupta, P., Jain, N., Anbalagan, R., & Sikdar, P. K. (2000). Landslide hazard evaluation and geostatistical studies in Garhwal Himalaya. *Journal of Rock Mechanics and Tunnelling Technology*, *6*(1), 41–60.

IS14496 (Part 2). (1998). Indian Standard Code on Preparation of Landslide Hazard Zonation Maps in Mountainous Terrains: Guidelines—Part 2: Macro-zonation, New Delhi, India.

Romana, M. (1985). New adjustment ratings for application of Bieniawski classification to slopes. In *International Symposium on the Role of Rock Mechanics* (pp. 49–53). Zacatecas, Mexico.

Valdiya, K. S. (1980). *Geology of Kumaon Lesser Himalaya* (p. 291). Dehradun, India: Wadia Institute of Himalayan Geology.

Chapter 20

Allowable Bearing Pressure for Shallow Foundations

If A is success in life, then A equals x + y + z. Work is x; y is play (sports); and z is keeping your mouth shut (inner joy).

Albert Einstein

INTRODUCTION

Foundations on weak, faulted, and highly undulating rock surfaces may pose serious problems. Because rocks can be more heterogeneous than soil, the problem of differential settlement may be serious in heterogeneous subsurface rocks. The design of a foundation depends upon the subsurface strata and its bearing capacity. Where the foundation rests on rocks, the bearing pressure can be obtained from the available classification tables as described in this chapter. If a site is covered partly by rocks and partly by talus deposits or soil, the heterogeneity in deformability of soil and rocks must be taken into account. It is generally suggested that plate load tests be conducted on talus or soil with a bearing pressure of 12 mm settlement criterion, which is the same for rock masses. Ramamurthy (2007) presented a detailed state-of-the-art report on foundations on rock masses.

A CLASSIFICATION FOR NET SAFE BEARING PRESSURE

Pressure acting on a rock bed from a building foundation should not be more than the safe bearing capacity of the rock foundation system. Both the effect of eccentricity and the effect of interference of different foundations should also be considered.

Universally applicable values of safe bearing pressure for rocks cannot be given since many factors influence it, and it is frequently controlled by settlement criterion. Nevertheless, it is often useful to estimate the safe bearing pressure for preliminary designs based on the classification approach, although such values should be checked or treated with caution for the final design.

Orientation of joints plays a dominant role in stress distribution below strip footings due to low shear modulus as shown in Figure 20.1 (Singh, 1973). Bray (1977) derived a simple solution for radial stress (σ_r) below a line load in the layered rock mass and proved that $\tau_{r\theta} = 0$ and $\sigma_\theta = 0$, which leads to similar pressure bulbs as shown in Figure 20.1. Model tests of Bindlish (2007) confirmed these twin pressure bulbs below the strip footing on rock mass with two joint sets. The bearing capacity of rocks is drastically low for nearly vertical joints with strike parallel to the footing length as the pressure bulb extends

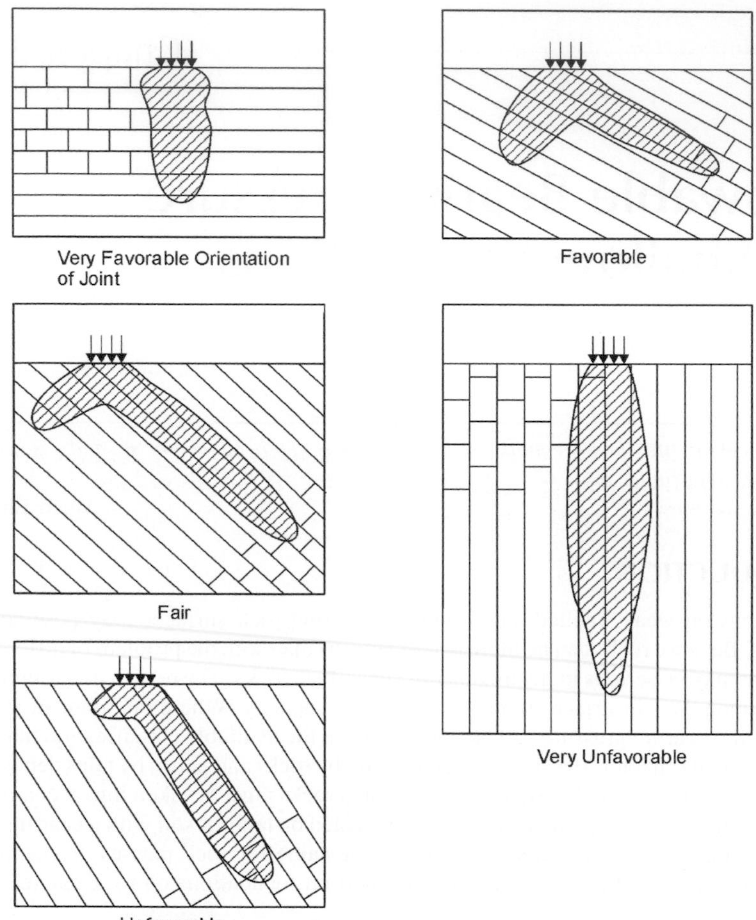

FIGURE 20.1 Theoretical pressure bulbs (10% intensity) below strip load on a medium of rock mass having low shear modulus. *(From IS 12070, 1987; Singh, 1973)*

deep into the strata. Several uniaxial jacking tests with MPBX by Choudhary (2007) in sandstones showed that displacement does not decrease with depth below the plate as rapidly as predicted by Boussinesque's theory for isotropic elastic medium. Anisotropic rock mass causes tilt of a uniformly loaded foundation where joints are asymmetrically inclined (Figure 20.1). Shear zones and clay seams, if present below foundation level, need to be treated to improve bearing capacity and reduce differential settlement as discussed in Chapter 2 and IS 12070 (1987).

A rock mass classification for assessing net safe bearing pressure is presented in Table 20.1 (Peck, Hansen, & Thornburn, 1974). The net safe bearing pressure and the allowable bearing pressure are terms that can be used interchangeably, but the net safe bearing pressure here means the ultimate safe bearing pressure, whereas the allowable bearing pressure means the bearing pressure considered for the designs (i.e., allowable bearing pressure) after accounting for the safety factor. The allowable bearing pressure is also safe for settlement criterion.

TABLE 20.1 Net Safe Bearing Pressure (q_{ns}) for Various Rock Types

S. No.	Rock type/material	Safe bearing pressure, q_{ns} (t/m^2)
1	Massive crystalline bedrock including granite, diorite, gneiss, trap rock, hard limestone, and dolomite	1000
2	Foliated rocks such as schist or slate in sound condition	400
3	Bedded limestone in sound condition	400
4	Sedimentary rock, including hard shales and sandstones	250
5	Soft or broken bedrock (excluding shale) and soft limestone	100
6	Soft shales	30

Source: Peck et al., 1974.

ALLOWABLE BEARING PRESSURE

Using Rock Mass Rating

Bieniawski's rock mass rating (RMR; Chapter 6) may also be used to obtain net allowable bearing pressure as per Table 20.2 (Singh, 1991; Mehrotra, 1992). Engineering classifications listed in Table 20.2 were developed based on plate load tests at about 60 sites and calculating the allowable bearing pressure for a 6 m wide raft foundation with settlement of 12 mm. Figure 20.2 shows the observed trend between allowable bearing pressure and RMR (Mehrotra, 1992), which is similar to the curve from plate test data from IIT Roorkee (Singh, 1991). The permissible settlement is reduced as failure strain of a geological material decreases such as in rock mass. The plate load test is the most reliable method for determining the allowable bearing pressure of both rock mass and soil.

TABLE 20.2 Net Allowable Bearing Pressure (q_a) Based on RMR

Class No.	I	II	III	IV	V
Description of rock	Very good	Good	Fair	Poor	Very poor
RMR	100–81	80–61	60–41	40–21	20–0
q_a (t/m^2)	600–440	440–280	280–135	135–45	45–30

The RMR should be obtained below the foundation at depth equal to the width of the foundation, provided RMR does not change with depth. If the upper part of the rock, within a depth of about one-fourth of foundation width, is of lower quality the value of this part should be used or the inferior rock should be replaced with concrete. Since the values here are based on limiting the settlement, they should not be increased if the foundation is embedded into rock.

During earthquake loading, the values of allowable bearing pressure from Table 20.2 may be increased by 50% in view of rheological behavior of rock masses.

Source: Mehrotra, 1992.

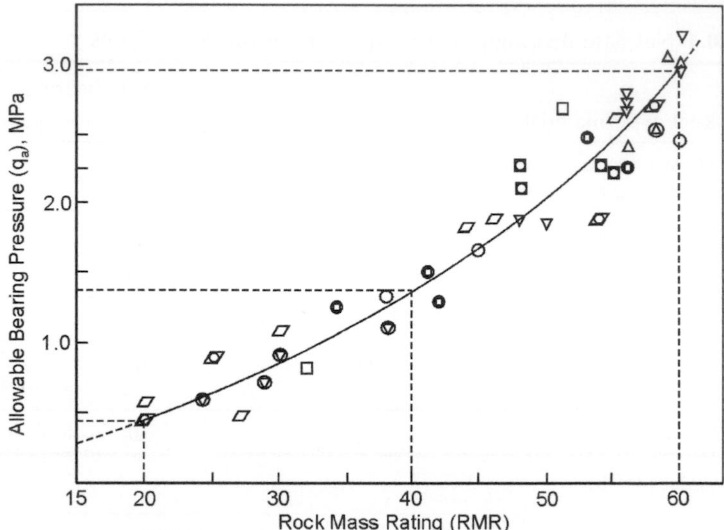

FIGURE 20.2 Allowable bearing pressure on the basis of rock mass rating and natural moisture content (nmc = 0.60–6.50%). *(From Mehrotra, 1992)*

Sinha et al. (2003) reported that contamination of rock mass by seepage of caustic soda not only reduces the bearing capacity of foundation by about 33% in comparison to that of uncontaminated rock mass, but it also causes swelling and heaving of the concrete floors. Because of this, the alkaline soda was neutralized by injecting acidic compound and grouting the rock mass with cement grout.

Bearing Pressure Using RQD

Peck et al. (1974) used the RQD directly to assess the allowable bearing stress (q_a), provided that the applied stress did not exceed the uniaxial compressive strength (UCS) of the intact rock ($q_a < q_c$). The RQD relationship is shown in Figure 20.3. These values appear to be too high (see Table 20.2).

Classification for Bearing Pressure

Another classification of rock masses for allowable bearing pressure is given in Table 20.3.

The Canadian practice for socketed piles and shallow foundations (Gill, 1980) results in the following simple formula for safe bearing pressure.

$$q_a = q_c \cdot N_j \cdot N_d \tag{20.1}$$

where q_a = net allowable safe bearing pressure, q_c = average laboratory UCS of rock material (Table 8.13), and N_j = empirical coefficient depending on the spacing of discontinuities (see Table 20.4).

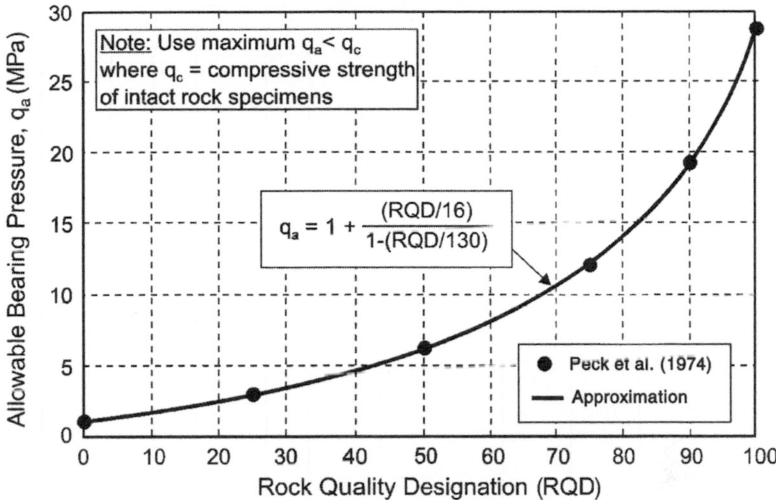

FIGURE 20.3 Allowable bearing stress for foundations on fractured rock from RQD. *(From Peck et al., 1974)*

$$N_j = \frac{3 + (s/B)}{10 \cdot \sqrt{1 + (300 \cdot \delta/s)}} \tag{20.2}$$

where s = spacing of joints in centimeters, B = footing width in centimeters, and δ = opening of joints in centimeters.

$$N_d = 0.8 + 0.2\, h/D < 2 \tag{20.3}$$

$N_d \geq 1.0$ and 1.0 for shallow foundations of buildings, h = depth of socket (embedment) in rock, and D = diameter of socket (embedment) of pile or pier.

Equation (20.1) may also be applied to shallow foundations considering $N_d = 1$; however, the previous correlation (Eq. 20.1) does not account for orientation of joints. The socket of piers should be deeper in the cloudburst-prone hills in erodible rocks. The results of plate load tests show that the settlement consideration of 12 mm generally results in a lower allowable bearing pressure than the strength consideration (Eq. 20.1). It is safer, therefore, to use settlement considerations in heterogeneous rocks.

It is debatable which correction should be applied if a rock mass is submerged. It is suggested that the bearing pressure be reduced by 25 to 50% depending upon the clay content of the gouge and its thickness. A correction must also be applied if the dip of the joints is unfavorable—that is, slopes with steeply inclined joints in flat ground and joints dipping toward a valley (IS 12070, 1987).

Because of this, it is recommended that plate load tests should be conducted on poor rocks where allowable bearing pressure is likely to be less than 100 t/m^2. Rock mass is more heterogeneous compared to soil; therefore, a large number of observation pits should be made at about a rate of at least three per important structure. These tests should be conducted in the pit representing the poorest rock qualities. The allowable bearing pressure is frequently found to decrease with the number of observation pits and tests.

The safe depth of a shallow foundation is at least 50 cm below the top level of the surface. With solution cavities in soluble rocks, the shallow foundation must rest on 80% of the area of excavated rock mass or dental concreting should be done up to this

TABLE 20.3 Net Allowable Pressure q_a (t/m^2) of Various Rock Types under Different Weathering Conditions

Rock type	Highly weathered structure unfavorable for stability*	Fairly weathered structure unfavorable for stability	Highly weathered structure favorable for stability	Fairly weathered structure favorable for stability	Unweathered rock structure unfavorable for stability	Unweathered rock structure favorable for stability
Marls and marls interbedded with sandstone	15	30	35	50	60	110
Calc-schist and calc-schist interbedded with quartzites	15	30	45	65	100	200
Slates, phyllites, and schists interbedded with hard sandstones and/or quartzite or gneiss	20	35	60	75	90	130
Limestone, dolomites, and marbles	50	80	90	130	150	200
Sandstone	40–60 (massive)	90	120	150	170	220
Calcareous conglomerates (massive)	60	100	120	200	200	330
Quartzite (massive)	50–70	150	120	180	200	330
Gneiss (massive)	30–60	150	120	180	200	330
Granite and leucocratic plutonic rocks	20	250	>330	—	—	—

The values reported above have a minimum factor of safety of 3.
**This column indicates sites with highly weathered rock and unfavorable geological structures, subjected to instability*

Source: Krahenbuhl and Wagner, 1983.

TABLE 20.4 Value of N_j

Spacing of discontinuities (cm)	N_j
300	0.4
100–300	0.25
30–100	0.1

depth (to follow IS code). The depth of subsurface exploration in a rocky area should be more than two times the width of concerned footing.

Faults and shear zones are seen often in the Himalayas, so the sites of tall structures are changed repeatedly until a safe site is discovered. The natural frequency of vibrations of tall structures, high silos, and large bridges is so low that seismic forces are insignificant. Instead, wind forces govern the design of tall structures. Foundations should be robust structures, embedded all along into the rock mass. They are restrained from all sides to prevent excessive displacements during vibrations. A safe edge distance from the slope (e.g., 10 m in high hills) should be planned due to possible surveying errors in steep hilly terrain. (An error of 1 mm in contour lines means an error of 50 m horizontally on a map on a scale of 1:50,000.) Stability of slopes, together with heavily loaded foundations, is of critical importance. A minimum factor of safety of 1.2 is recommended in the static case and 1.0 in the dynamic case. Block shear tests and uniaxial jacking tests should be conducted carefully on the undisturbed rock mass inside the drifts or pits up to the foundation level to get realistic strength parameters and moduli of deformations for the detailed design of tall and very costly structures on rocks.

COEFFICIENT OF ELASTIC UNIFORM COMPRESSION FOR MACHINE FOUNDATIONS

The coefficient of uniform compression (C_u) is defined as the ratio between pressure and corresponding settlement of block foundation. Typical values of coefficient of elastic uniform compression (C_u) for machine foundations on a rock mass are listed in Table 20.5 (Ranjan et al., 1982). The coefficient of uniform shear is generally $C_u/2$. It may be noted that C_u is less than 10 kg/cm^3 in very poor rocks.

The elastic modulus of rock mass E_e (Eq. 8.19) may be used for calculating C_u. Cyclic plate load tests are more reliable for this purpose.

SCOUR DEPTH AROUND BRIDGE PIERS

Approximately 400 bridge sites were surveyed by Hopkins and Beckham (1999). They observed insignificant rock scour around exposed bridge piers and abutments. Only a few sites experienced rock scour holes. Figure 20.4 shows the trend of correlation between depth of scour holes and RQD. It is observed that the depth of scour is less than about 30 cm below bed level where RQD is more than 10%. Scour depth was maximum up to 1 m at zero RQD, and the minimum depth of socketing of well foundations or bridge piers is 50 cm (Peck et al., 1974). The design scour depth is twice the actual scour

TABLE 20.5 Coefficient of Elastic Uniform Compression (C_u) for Rock Masses

S. No.	Rock type	Allowable bearing pressure (t/m^2)	C_u (kg/cm^2/cm)
1	Weathered granites	—	17
2	Massive limestones	160	25
3	Flaky limestones	75	12
4	Shaly limestones	50	7
5	Soft shales	45	7
6	Saturated soft shales	33	1.5
7	Saturated non-plastic shales	27	2.6

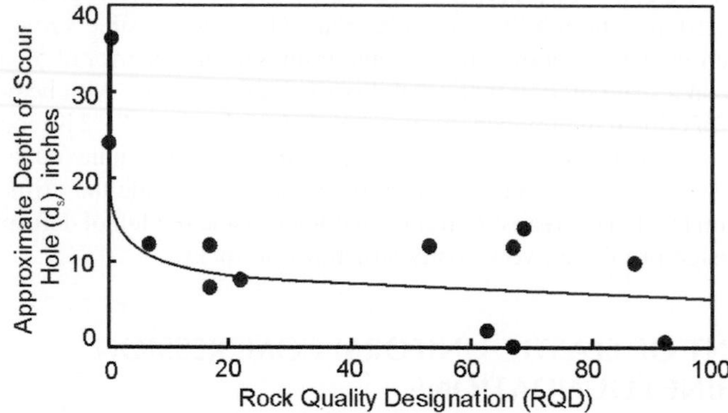

FIGURE 20.4 Approximate depth of scour and RQD. *(From Hopkins and Beckham, 1999)*

depth at the nose of piers. Figure 20.4 was recently used for finding the depth of socketing of well foundations for a bridge in New Delhi, India. The well or socket foundations were anchored by the pre-stressed cable anchors to prevent overturning during a major earthquake or storms. A key is also provided at the base of the well foundation to resist horizontal forces on the same.

ROCK PARAMETERS TO SELECT TYPE OF DAM

The type of dam built depends on the value of the modulus of deformation and shear parameters. The guidelines outlined in Tables 20.6 and 20.7 are useful for dam construction (Kilkuchi, Saito, & Kusonoki, 1982).

Table 20.7 indicates the type of dam considered suitable for different grades of rocks as defined by their physical characteristics from Table 20.6. The section Treatment for Dam Foundations in Chapter 2 describes dental treatment of the shear zone below the concrete dam foundation.

TABLE 20.6 Range of Physical Values of Rocks

Rock grade	Static modulus of elasticity (GPa)	Modulus of deformation (GPa)	Cohesion of rock material (MPa)	Angle of internal friction, ϕ (degrees)	Velocity of elastic P-wave (km/sec)	Rebound of the rock test hammer
	Grade A values very much higher than those of grade B					
B	8 or more	5 or more	4 or more	55–65	3.7 or more	36 or more
C	8–4	5–2	4–2	40–55	3.7–3	36–27
D	4–1.5	2–0.5	2–1	30–45	3–1.5	27–15
E–F	1.5 or less	0.5 or less	1 or less	15–38	1.5 or less	15 or less

Source: Kilkuchi et al., 1982.

TABLE 20.7 Criteria for Property of Rocks as Dam Foundations for Various Types of Dams

Rock grade	Property as concrete arch dam foundation	Property as concrete dam foundation (dam height 60 m or more)	Property as rockfill core foundation (dam height 60 m or more)
A	Very good	Very good	Very good
B	Very good	Very good	Very good
C	Almost good	Almost good	Almost good
D	Bad Hard and medium hard rocks close to grade C may be considered but soft rocks are not proper as dam foundations	Bad Hard and medium hard rocks may be improved; soft rocks are not so proper as high dam foundations	Almost good concerning bearing resistance
E	Very bad	Bad But near the portion affected by small acting force such as the area near the dam; foundation treatment may be applied to use the rocks as dam foundation	Generally rocks of this grade are not good as foundation, but use of rocks capable of being improved—close to grade D—and watertight is not impossible
F	Very bad	Very bad	Very bad

Source: Kilkuchi et al., 1982.

TABLE 20.8 Shear Strength Parameters from Different Projects in India and Bhutan

S. No.	Name of the project	Rock type	Shear strength parameters							
			Rock to rock interface				*Concrete to rock interface*			
			Peak values		*Residual values*		*Peak values*		*Residual values*	
			c (MPa)	$\phi°$	*c_r (MPa)*	$\phi_r°$	*c (MPa)*	$\phi°$	*c_r (MPa)*	$\phi_r°$
	Hydroelectric projects in India									
1	Lakhwar Dam	Trap	0.68	42.0	0.58	40.0	—	—	—	—
2	Chamera Dam project	Phyllites	—	—	—	—	0.13	53.3	0.00	49.5
3	Hibra hydroelectric project	Phyllitic quartzite	—	—	—	—	0.10	56.5	0.00	55.1
4	Srinagar Dam (left bank)	Quartzite								
	Main drift		0.50	68.0	0.15	60.0	0.25	58.5	0.22	45.0
	T-section		1.20	59.0	0.60	58.0	—	—	—	—
5	Srinagar Dam (right bank)	Quartzite								
	Main drift		—	—	—	—	0.44	59.0	0.00	51.0
	T-section		—	—	—	—	0.40	64.0	0.00	64.0
6	Srinagar Dam (right bank) Main drift	Metabasic	0.76	46.0	—	—	0.16	41.0	0.10	40.0
7	Greater Shillong Dam	Phyllites	0.27	70.0	0.01	69.0	0.25	66.0	0.01	62.0
8	Nathpa Jhakri Dam	Mica schist	0.25	57.5	0.13	50.3	—	—	—	—
9	Kalpong Dam	Ultrabasic	—	—	—	—	0.12	59.0	0.01	46.0

10	Rupsiabagar Khasipara (left bank)	Quartz biotite schist	0.99	58.8	0.76	40.5	0.32	55.8	0.23	53.5
11	Rupsiabagar Khasipara (right bank)	Quartz biotite schist	—	—	—	—	0.37	60.6	0.35	59.4
12	Vishnugad Pipalkothi (left bank)	Quartzite	—	—	—	—	0.79	54.8	0.48	49.1
13	Vishnugad Pipalkothi (right bank)	Quartzite	—	—	—	—	0.30	58.8	0.22	57.1
14	Kotlibhel Dam (left bank)	Quartzitic sandstones	0.35	67.9	0.22	65.4	0.42	54.3	0.26	53.5
15	Kotlibhel Dam (right bank)	Quartzitic sandstones	0.31	65.0	0.24	52.0	0.34	54.0	0.13	50.0
	Hydroelectric projects in Bhutan									
16	Bunakha Dam	Biotite gneiss	—	–	—	—	0.65	62.0	0.38	61.0
17	Sankosh Main Dam	Phyllites	0.17	60.0	0.00	57.0	—	—	—	—
18	Sankosh Lift Dam	Sandstone	0.11	38.0	0.00	37.5	0.13	52.0	0.00	48.0
19	Tala Dam (right bank)	Biotite gneiss	0.37	62.9	0.025	57.1	0.50	49.0	0.08	46.0
20	Tala Dam (left bank)	Biotite gneiss	0.35	63.8	0.14	57.4	0.54	46.0	0.16	45.0
Variations in values			0.11–1.20	38–70	0.00–0.76	37.5–69	0.12–0.79	41–64	0.00–0.48	40–64
Average values			0.48	58.4	0.22	49.6	0.35	55.8	0.15	52.0

Source: Singh, 2009.

Figure 23.2 shows which criterion to ascertain when grouting is needed in the dam foundation. The foundation of a concrete dam should go deeper than debris into a good grouted-rock mass. Its depth in rock mass should be more than twice the scour depth (Figure 20.4). The foundation should be undulating for seismic stability by increasing the joint wall roughness coefficient (JRC) of the dam foundation surface.

Singh (2009) proved that the sliding angle of friction between concrete and rock masses is higher at low normal stresses than previously believed. Table 20.8 provides the shear strength parameters from in situ direct shear tests at 20 different hydroelectric project sites in the Himalayas in India. The residual strength parameters are likely to reduce with more sliding of the concrete blocks. Shear strength of rock mass is anisotropic and the least in the direction of tectonic movement, so the dam axis should be inclined to the direction of tectonic motion for better strength of rock mass.

Example 20.1

A clear water reservoir is to be built by cutting the top of the hill of a highly weathered rock mass of about 35 m height. It is igneous boulder rock mass with no boulder bigger than 60 to 75 cm on average in size. Suggest the allowable bearing pressure. It is not practicable to do plate load tests at the site. The rock mass is classified as poor according to RMR after rating adjustment of the slope all around the site.

Table 20.2 suggests a least bearing pressure of 30 t/m^2 (0.3 MPa). In view of the steep slopes, the recommended allowable bearing pressure is 15 t/m^2 (0.15 MPa), which is sufficient to take the pressure due to water and the tank. The minimum distance of foundation from the edge of the natural slope is 2 m (beyond the filled up soil). The depth of foundation is 0.5 m below the plane of excavation of the hilltop. The raft foundation is provided to prevent its cracking due to possible differential settlement as well as penetration of water toward the slope, which may cause distress to these slopes. Suitable drainage measures for the surface water should also be implemented.

Example 20.2

A 270 m high chimney is to be built for a thermal powerhouse. The rock mass is granite beneath a soil cover of 25 m. The average UCS of rock material is 85 MPa. A core loss of 40% was observed during drilling, but all pieces of rock core were longer than 20 cm. The site is located in a no earthquake zone. Design the foundation.

A raft foundation is suggested with 26 m long cast in situ concrete piles of 60 cm diameter in because the structure is very tall. The piles should be socketed into the rock mass up to 1 m in depth (1D–2D in strong rocks). The minimum spacing of piles should be 1.8 m c/c (3D).

The safe bearing pressure of the rock mass according to Eq. (20.1) is

$$q_a = q_c\ N_j\ N_d$$

$$N_j = \frac{3 + (s/B)}{10\sqrt{1 + (300\ \delta/s)}},\quad s \cong 0.2\text{ m},\quad B = 0.6\text{ m},\quad \delta/s \cong \text{core loss}/100 \quad \text{and} \quad \cong \ 0.4,$$

$N_j = \dfrac{3 + (0.2/0.6)}{10\sqrt{1 + (300 \times 0.4)}} = 0.03$, $N_d = 0.8 + 0.2\ h/D$ (between 1 and 2), $N_d = 0.8 + 0.2 \times 1/0.6 = 1.1$, and $q_a = 85 \times 0.03 \times 1.1 = 2.8$ MPa (t/m^2) and is < safe compressive stress in concrete.

The allowable vertical end bearing capacity of a pile is as shown next (neglecting the side shearing resisting and overburden pressure of soil cover).

$$Q_a = A_p\, q_a$$

$$= \frac{\pi(0.6)^2}{4} \times 280 = 79 \text{ tonnes}$$

The initial test pile should be tested up to two-and-a-half times the estimated load or up to failure load. The safe load on a pile will be the least of the following:

1. Fifty percent of the load at 12 mm vertical settlement or the load corresponding to the 6 mm vertical settlement of the rock mass
2. One-third of the ultimate failure load

The two initial test piles should also be tested up to failure by the lateral load. The lateral safe load should be the least of following:

1. Fifty percent of the final lateral load, which corresponds to 8 mm total lateral displacement
2. Final load, which corresponds to 4 mm total lateral displacement

A precaution should be taken to grout the slush of broken rocks first at the bottom of the borehole in the rock mass before concreting the piles.

REFERENCES

Bindlish, A. (2007). *Bearing capacity of strip footings on jointed rocks* (p. 329). Ph.D. Thesis. Department of Civil Engineering, IIT Roorkee, Uttarakhand, India.

Bray, J. (1977). Unpublished notes. Imperial College, London: Royal School of Mines (see Ramamurthy, 2007).

Choudhary, J. S. (2007). Personal Communication with Bhawani Singh. India: IIT Roorkee.

Gill, S. A. (1980). Design and construction of rock cassions. In *International Conference of Structural Foundations on Rock, Sydney* (pp. 241–252). Rotterdam: A. A. Balkema.

Hopkins, T. C., & Beckham, T. L. (1999). *Correlation of rock quality designation and rock scour around bridge piers and abutments founded on rock*. Research Report KTC-99-57. College of Engineering, University of Kentucky. *www.ktc.uky.edu*.

IS12070. (1987). *Indian standard code of practice for design and construction of shallow foundations on rocks*. New Delhi: Bureau of Indian Standards.

Kilkuchi, K., Saito, K., & Kusonoki, K. (1982). Geotechnically integrated evaluation on the stability of dam foundation on rocks. In *14th ICOLD, Q53R4* (pp. 49–74), Rio de Janeiro.

Krahenbuhl, J. K., & Wagner, A. (1983). *Survey design and construction of trail suspension bridges for remote area* (Vol. B, p. 325). Survey. SKAT, Swiss Centre for Technical Assistance, Zurich, Switzerland.

Mehrotra, V. K. (1992). *Estimation of engineering properties of rock mass* (p. 267). Ph. D. Thesis. Uttarakhand, India: IIT Roorkee.

Peck, R. B., Hansen, W. E., & Thornburn, T. H. (1974). *Foundation engineering* (2nd ed., Chap. 22, p. 512). New York: John Wiley.

Ramamurthy, T. (2007). *Engineering in rocks for slopes, foundations and tunnels* (Chap. 16, p. 731). New Delhi: Prentice-Hall of India Pvt. Ltd.

Ranjan, G., Agarwal, K. B., Singh, B., & Saran, S. (1982). Testing of rock parameters in foundation design. In *IVth Congress of International Association of Engineering Geology, New Delhi* (Vol. III, pp. 273–287).

Singh, B. (1973). Continuum characterization of jointed rock mass Part II—Significance of low shear modulus. *International Journal of Rock Mechanics and Mining Sciences—Geomechanics Abstracts*, *10*, 337–349.

Singh, B. (1991). Application of rock classification methods for underground construction in river valley projects. In *Proceedings of the Workshop on Rock Mechanics Problems of Tunnels, Mine Roadways, and Caverns* (pp. IV-1–IV-41). Ooty, India.

Singh, R. (2009). Measurement of in situ shear strength of rock mass. *Journal of Rock Mechanics and Tunnelling Technology, ISRMTT, India*, *15*(2), 131–142.

Sinha, U. N., Sharma, A. K., Bhargva, S. N., Minocha, A. K., & Kumar, P. (2003). Effect of seepage of caustic soda on foundation and remedial measure in alumina plant. In M. N. Viladkar (Ed.), *IGC 2003 on Geotechnical Engineering for Infrastructural Development, IGS* (Vol. 1, pp. 229–234). IIT Roorkee, Uttarakhand, India.

Chapter 21

Method of Excavation

Blasting for underground construction purposes is a cutting tool, not a bombing operation.
Svanholm, Persson, and Larsson (1977)

EXCAVATION TECHNIQUES

Excavation of rock or soil is an important aspect of a civil engineering project. The excavation techniques or the methods of excavation of rocks differ from those in soil. Similarly, these also change with the type of project.

Broadly, methods of excavation can be classified according to their purpose, that is, whether the excavation is for foundations, slopes, or underground openings. Methods of excavation in a broader sense can be divided into three types:

1. Digging
2. Ripping
3. Blasting

A system was proposed by Franklin, Broch, and Walton (1972) to classify methods of excavation based on rock material strength (Figure 21.1). Figure 21.1a shows a plot between the point load strength index of rocks and fracture spacing, whereas Figure 21.1b is drawn between point load strength index and rock quality. Using these figures, we can select a method of excavation for a particular rock; for instance, a rock of medium strength and medium fracture spacing is classified as medium rock (Figure 21.1a) and should be excavated by ripping (Figure 21.1b). There is too much confusion on the soil–rock boundary line. ISO 14689 (2003) defined a geological (rock) material having a uniaxial compressive strength (UCS) less than 1 MPa as soil.

This classification is useful when estimating the cost of excavation, which should be paid to a contractor who may not prefer to change the method of excavation according to rock condition.

ASSESSING THE RIPPABILITY

Assessing the rippability is also an important aspect of excavation. Even stronger rocks such as limestones and sandstones, when closely jointed or bedded, are removed by heavy rippers to at least the limit of weathering and surfacial stress relief.

Sedimentary rocks are usually easily ripped. Rippability of metamorphic rocks, such as gneisses, quartzites, schists, and slates, depends on their degree of lamination and mica content. Igneous rocks are often not possible to rip, unless they are very thinly laminated as in some volcanic lava flows.

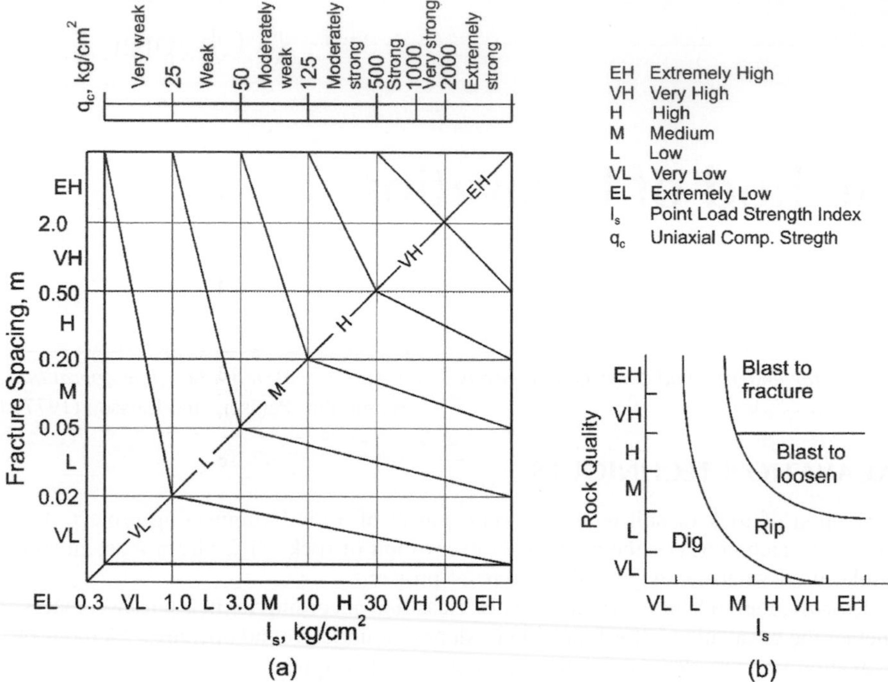

FIGURE 21.1 Rock mass classification for excavation. *(From Franklin et al., 1972)*

Ripping is comparatively easier in open excavations. In confined areas or in a narrow trench, however, the same rock often requires blasting due to confinement effect and difficulties in using a ripper in a confined space. Rippability can also be assessed by using the seismic refraction survey and knowing the seismic velocities.

ROCK MASS CLASSIFICATION ACCORDING TO EASE OF RIPPING

Based on the combined effects of the following five parameters, a rippability index classification (RIC) was developed by Singh et al. (1986, 1987) and is presented in Table 21.1.

1. Uniaxial tensile strength of rock material, determined by Brazilian disc test or derived from point load index values
2. Degree of weathering, determined by visual observations
3. Seismic wave velocity, determined by surface or cross-hole seismic surveys; the velocity may be as high as 6 km/sec for a strong, dense, and unweathered rock mass or as low as 300 m/sec for loose unsaturated soil
4. Abrasiveness of rock material, the abrasiveness index classification based on the Cerchar index value, and the examination of physical and mineralogical properties of rock, given by Singh et al. (1986)
5. Spacing of discontinuities, measured by the scan line survey

The RIC is the result of a broad examination of existing rippability classifications and experience gained from a number of opencast sites in the United Kingdom and

TABLE 21.1 Classification of Rock Mass According to Rippability Index

Parameter	Class 1	Class 2	Class 3	Class 4	Class 5
Uniaxial Tensile Strength (MPa)	<2	2–6	6–10	10–15	>15
Rating	**0–3**	**3–7**	**7–11**	**11–14**	**14–17**
Weathering	Completely	Highly	Moderately	Slightly	Unweathered
Rating	**0–2**	**2–6**	**6–10**	**10–14**	**14–18**
Sound vel. (m/s)	400–1100	1100–1600	1600–1900	1900–2500	>2500
Rating	**0–6**	**6–10**	**10–14**	**14–18**	**18–25**
Abrasiveness	Very low	Low	Moderately	Highly	Extremely
Rating	**0–5**	**5–9**	**9–13**	**13–18**	**18–22**
Discontinuity spacing (m)	<0.06	0.06–0.3	0.3–1	1–2	>2
Rating	**0–7**	**7–15**	**15–22**	**22–28**	**28–33**
Total rating	**<30**	**30–50**	**50–70**	**70–90**	**>90**
Ripping assessment	Easy	Moderate	Difficult	Marginal	Blast
Recommended dozer	Light duty	Medium duty	Heavy duty	Very heavy duty	

Source: Singh et al., 1986.

Turkey (Singh et al., 1986). The rippability index is the algebraic sum of the values of the weighted parameters given in Table 21.1. Subsequently, it has been used to indicate the quality of rock mass with respect to its rippability.

Abdullatif and Cruden (1983) compared three other systems: the Franklin (1974), the Norwegian Q, and the South African RMR systems. These are all based on block size and rock strength. They conducted excavation trials with rock mass quality measurements in limestone, sandstone, shale, and several igneous rocks at 23 sites in the United Kingdom and found that the RMR system (Chapter 6) gave the best predictions. They offered the following guidelines for selecting a method of excavation (Table 21.2).

TABLE 21.2 Selection of Method of Excavation Based on RMR

RMR value	Excavation method
<30	Digging
31–60	Ripping
61–100	Blasting

EMPIRICAL METHODS IN BLASTING

The study of Ibarra at the Aguamilpa hydropower tunnels in Mexico presented by Franklin (1993) showed the application of empirical methods for optimization of blast designs. Based on 92 measured tunnel sections, overbreak was shown to correlate with rock mass quality Q. As expected, overbreaks were found to be inversely proportional to rock mass quality (Figure 21.2).

In addition, Ibarra found that for any given rock quality Q, the overbreak increases in proportion to the perimeter powder factor, which is defined as the weight of explosives in the perimeter blastholes divided by the volume of rock removed (perimeter length × drillhole depth × burden). Using the results of Figure 21.3, the optimum perimeter powder factor can be determined for the given quality of a rock mass.

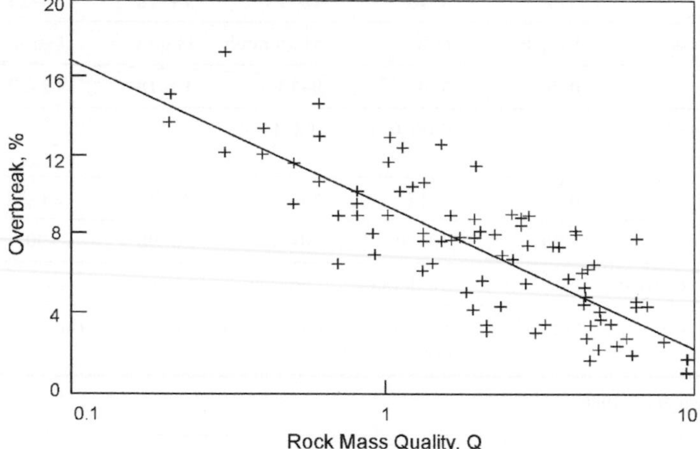

FIGURE 21.2 Overbreak as a function of rock mass quality Q. *(From Franklin, 1993)*

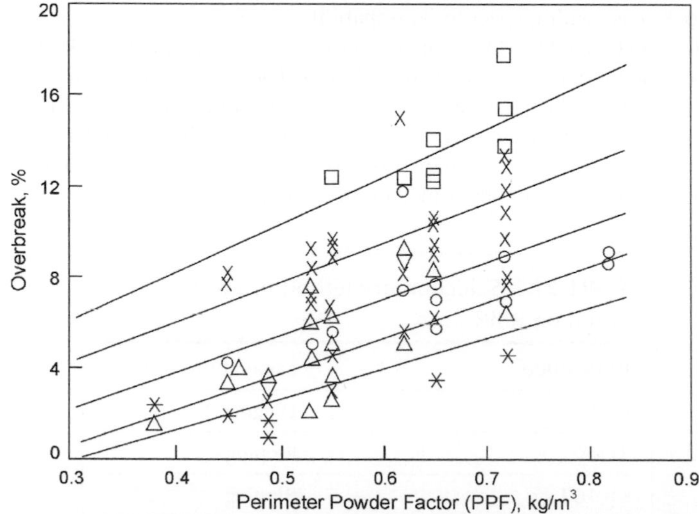

FIGURE 21.3 Overbreak as a function of perimeter powder factor. *(From Franklin, 1993)*

Chakraborty, Jethwa, and Dhar (1997) found the following trend between average powder factor, p_f (weight of explosive divided by volume of broken rock), and weighted average of rock mass quality Q in tunnels within massive basalts:

$$p_f = 1.02 + 0.0005\,Q\ \mathrm{kg/m^3} \tag{21.1}$$

The coefficient of correlation is 0.82. Chakraborty et al. (1997) also inferred that p_f increases directly with UCS (q_c). They have used these correlations to suggest the tunnel rock blasting index (TBI) for reliable prediction of powder factor. Further research may give specific classifications for rock blasting in tunnels.

REFERENCES

Abdullatif, O. M., & Cruden, D. M. (1983). The relationship between rock mass quality and ease of excavation. *Bulletin of the International Association of Engineering Geology*, *28*, 183–187.

Chakraborty, A. K., Jethwa, J. L., & Dhar, B. B. (1997). Predicting powder factor in mixed-face condition: Development of a correlation based on investigations in a tunnel through basaltic flows. *Engineering Geology*, *47*, 31–41.

Franklin, J. A. (1974). Rock quality in relation to the quarrying and performance of rock construction materials. In *Procedings 2nd International Congress, International Association of Engineering Geology* 11pp, Paper IV- PC-2, Sao Paulo, Brazil.

Franklin, J. A. (1993). Empirical design and rock mass characterisation. In J. A. Hudson (Ed.), *Comprehensive rock engineering* (Vol. 2, Chap. 32, pp. 759–806). New York: Pergamon.

Franklin, J. A., Broch, E., & Walton, G. (1972). Logging the mechanical character of rock. *Transactions of the Institute of Mining Metallurgy*, A80, A1–A9 and Discussion A81, A34, A51.

ISO 14689-1, (2003). (E) *Geotechnical investigation and testing—Identification and classification of rock—Part 1: Identification and description* (pp. 1–16). Geneva, Switzerland: International Organization for Standardization.

Singh, R. N., Denby, B., & Egretli, I. (1987). Development of a new rippability index for coal measures excavations. In *Proceedings of the 28th U.S. Symposium on Rock Mechanics* (pp. 935–943), Tucson. (Reprinted from I. W. Farmer, J. J. K. Daeman, C. S. Desai, C. E. Glass, & S. P. Neuman (Eds.), *Rock mechanics: Proceedings of the 28th U.S. Symposium* (pp. 12–64), Tucson, June 29–1 July 1. Rotterdam: A. A. Balkema.

Singh, R. N., Denby, B., Egretli, I., & Pathan, A. G. (1986). Assessment of ground rippability in open cast mining operations, Nottingham University. *Mining Department Magazine*, *38*, 21–34.

Svanholm, B. O., Persson, P. A., & Larsson, B. (1977). Smooth blasting for reliable underground openings. In *Proceedings of the 1st International Symposium on Storage in Excavated Rock Caverns* (Vol. 3, pp. 37–43), Stockholm. London: Pergamon.

Chapter 22

Rock Drillability

Get happiness out of your work or you may never know what happiness is.

Elbert Hubbard

DRILLABILITY AND AFFECTING PARAMETERS

The rock drillability or speed of drilling for a blasthole and rock bolting needs to be estimated to assess the cycle time of tunneling for a given setup of tunneling machines. Construction time for back grouting and consolidation grouting also depends on the same.

Rock drillability is defined as the ease of drilling a hole in the rock mass. Studies have shown that the drillability of rock and the penetration rate of a drill are affected by

1. Rock hardness
2. Rock texture and density
3. Rock fracture pattern
4. General structure of the formation/rock mass

These parameters do not account for the drilling equipment characteristics. Each of the listed properties affecting drillability is considered separately. An experienced driller can tell how a rock will drill. The important thing to know is how fast it will drill. Considering these four properties, rock drillability may be classified into five conditions: fast, fast average, average, slow average, and slow. Various properties can be determined as follows.

Hardness

Hardness of a mineral may be obtained by the Mohs scale of hardness shown in Table 22.1. The number for each mineral in Table 22.1 indicates its hardness. A higher number means the mineral is harder than the next lower number. Minerals with a higher number can scratch the ones with the same or a lower number. Rocks may contain more than one mineral, so tests should be made at several places on a piece of rock to determine the average hardness. Mohs' hardness kit for testing minerals can also be used in the field.

Texture

Texture may be determined by visual inspection of the grain structure of the rock and then classified for the drilling condition as shown in Table 22.2 (Wilbur, 1982).

TABLE 22.1 Mohs' Hardness Scale

1	Talc	6	Feldspar
2	Gypsum	7	Quartz
3	Calcite	8	Topaz
4	Fluorite	9	Corundum
5	Apatite	10	Diamond

Source: Nast, 1955.

TABLE 22.2 Texture

Drilling condition	Type of rock and texture
Fast	Porous (cellular or filled with cavities)
Fast average	Fragmental (fragments, loose or semi-consolidated)
Average	Granitoid (grains large enough to be readily recognized — average grained granite)
Slow average	Porphyritic (large crystals in fine-grained granite)
Slow	Dense (grain structure too small to identify with the naked eye)

Source: Wilbur, 1982.

Fracture

Fracture in drillability refers to how a rock breaks apart when struck by a blow with a hammer. Five drilling conditions are correlated with type of rock and its fracture pattern in Table 22.3.

Formation

Formation describes the condition of rock mass structure. Various formations facilitating the five drilling conditions are shown in Table 22.4. A high drilling rate is possible in massive rocks, whereas slow drilling is obtained in blocky and seamy rock masses.

The rock chart in Figure 22.1 shows drilling characteristics for the five drilling conditions (Nast, 1955).

CLASSIFICATION FOR DRILLING CONDITION

When the characteristics of a rock fall into different conditions, which is usually the case, it is necessary to compute final drilling conditions. This may be done by using the point system chart shown in Table 22.5. The chart may be used as explained in the next paragraph.

TABLE 22.3 Fracture

Drilling condition	Type of rock and fracture pattern
Fast	Crumbly (crumbles into small pieces when struck lightly)
Fast average	Brittle (rock breaks with ease when struck lightly)
Average	Sectile (when slices can be shaved or split off and rock crumbles when hammered)
Slow average	Tough (rock resists breaking when struck with heavy blow)
Slow	Malleable (rock that tends to flatten under blow of hammer)

Source: Wilbur, 1982.

TABLE 22.4 Formation

Drilling condition	Type of rock with respect to formation
Fast	Massive (solid or dense with practically no seams)
Fast average	Sheets (layers or beds 4–8 feet (1.2–2.4 m) thick with thin horizontal seams)
Average	Laminated (thin layers 1–3 feet (0.3–0.9 m) thick with horizontal seams with little or no earth)
Slow average	Seamy (many open seams in horizontal and vertical positions)
Slow	Blocky (wide open seams in all directions and filled with earth or shattered or fissured)

Source: Wilbur, 1982.

To obtain the drillability of a particular rock mass, the points for each characteristic are added together (Table 22.5). In extreme cases of drilling conditions, judgment should be made cautiously. If three characteristics are fast and one (e.g., formation) is slow, the three fast ones would be revised to average, or to a total of 10 (3 + 3 + 3 + 1) points, correcting a fast condition to an average condition. On the other hand, if three characteristics are slow and one (e.g., formation) is fast, the fast one would be revised to an average, or the three slow ones would be revised to a slow average.

Drillability, in other words, may be measured by the drilling speed (centimeter per minute) at which a drill bit penetrates the rock mass. A drillability factor has been determined for all drilling conditions from a performance study of rock drilling jobs both in the field and in the laboratory (Table 22.6). The drillability factor of each condition has subsequently been correlated with drilling speed (Table 22.6); therefore, Table 22.6 can be used to figure out the drilling speed once the drilling condition is known.

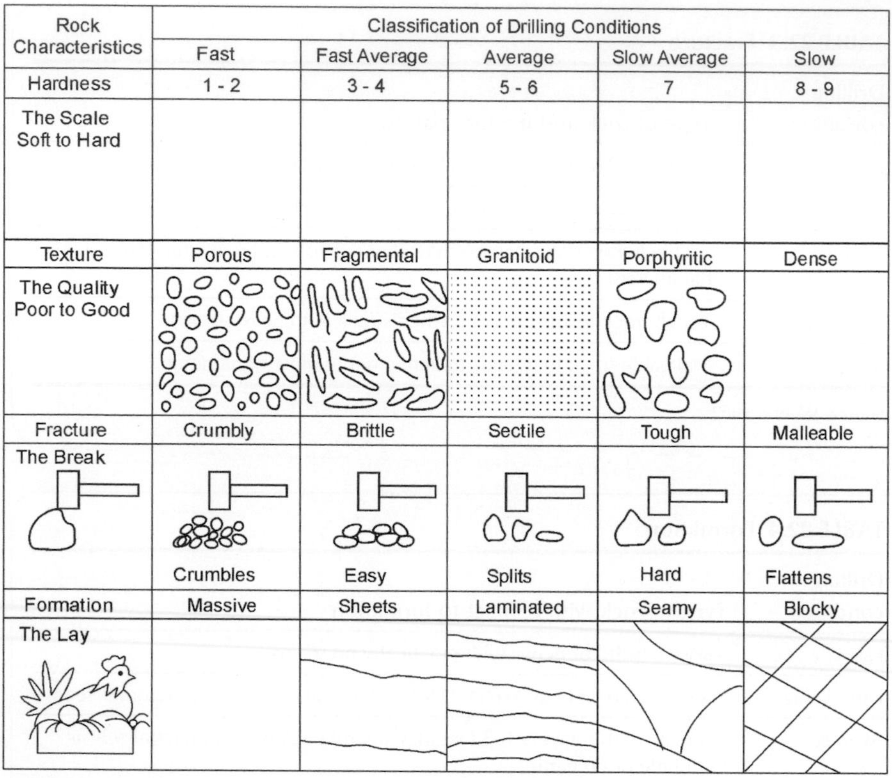

Rock Characteristics	Classification of Drilling Conditions				
	Fast	Fast Average	Average	Slow Average	Slow
Hardness	1 - 2	3 - 4	5 - 6	7	8 - 9
The Scale Soft to Hard					
Texture	Porous	Fragmental	Granitoid	Porphyritic	Dense
The Quality Poor to Good					
Fracture	Crumbly	Brittle	Sectile	Tough	Malleable
The Break	Crumbles	Easy	Splits	Hard	Flattens
Formation	Massive	Sheets	Laminated	Seamy	Blocky
The Lay					

FIGURE 22.1 Rock drilling characteristics. *(From Nast, 1955)*

TABLE 22.5 Drilling Condition Point System Chart

Nature of rock	Fast	Fast average	Average	Slow average	Slow
Hardness	8	4	3	2	1
Texture	8	4	3	2	1
Fracture	8	4	3	2	1
Formation	8	4	3	2	1
Total	32	16	12	8	4

Source: Nast, 1955.

TABLE 22.6 Drillability versus Drilling Speed

Drilling condition	Fast	Fast average	Average	Slow average	Slow
Drillability factor	2.67	1.33	1.0	0.67	0.33
Drilling speed (centimeter/minute)	50	25	18	12	6

Source: Nast, 1955.

OTHER APPROACHES

Scleroscope hardness reading (SHR), as used by the Joy Manufacturing Company in its laboratory, gives more definitive results in determining drillability of rocks (Bateman, 1967). In this method, a small diamond pointer hammer is dropped from a height of 25 cm through a thin glass tube to strike rock samples and the height of rebound is measured. The harder the sample, the higher the rebound of the diamond pointer hammer. The typical observations of rebound height for several rock types are shown in Table 22.7. Soft rocks are crushed to powder by the hammer, while hard rocks are partly shattered, with most of the energy returned in the rebound. This action is analogous to the percussion drill and provides useful information on the drillability of rock masses.

TABLE 22.7 Typical Values of Diamond Pointer Rebound for Several Rock Types

Minerals		Igneous rocks	
Gypsum	12	Basalt	90
Calcite	45	Diorite	90
Feldspar	90	Rhyolite	100
Quartz	115	Granite	100–110
Sedimentary rocks		**Metamorphic rocks**	
Shale	30–50	Marble	40–50
Limestone	40–60	Slate	50–60
Sandstone	50–60	Schist	60–65
Taconite	90–115	Quartzite	100–115

Source: Bateman, 1967.

REFERENCES

Bateman, W. M. (1967). Rock analysis. Joy/Air Power, Joy Manufacturing Company, March–April. In J. O. Bickel & T. R. Kuesel (Eds.), *Tunnel engineering handbook* (1982, Chap. 7).

Bickel, J. O., & Kuesel, T. R. (1982). *Tunnel engineering handbook* (p. 670). New York: Van Nostrand Reinhold.

Nast, P. H. (1955). Drillers handbook on rock. O'Davey Compressor Company, Kent, Ohio. In J. O. Bickel & T. R. Kuesel (Eds.), *Tunnel engineering handbook* (1982, Chap. 7).

Wilbur, L. D. (1982). Rock tunnels. In J. O. Bickel & T. R. Kuesel (Eds.), *Tunnel engineering handbook* (Chap. 7, pp. 123–207).

Chapter 23

Permeability and Groutability

A hazard foreseen is hazard controlled.

Anonymous

PERMEABILITY

Permeability is defined as a property of porous material that permits passage or seepage of fluids, such as water and or gas, through its interconnecting voids.

The resistance to flow depends upon the type of rock, the geometry of the voids in the rock (size and shape of the voids), and the surface tension of water (temperature and viscosity effects). The coefficient of permeability is a function of rock type, pore size, entrapped air in the pores, rock temperature, and viscosity of water.

Because of rock defects, such as irregularity in the amount of fissures and voids and their distribution, permeability of rocks is non-linear and non-uniform. Non-uniform permeability in rocks may also be caused by contraction and expansion of rock fissures; therefore, the concept of a regular groundwater table is not applicable in complex geological conditions.

PERMEABILITY OF VARIOUS ROCK TYPES

Anisotropic conditions in rocks do permit a permeability chart; however, the approximations in Table 23.1 are just for guidance.

Knill (1969) conducted extensive field studies at 89 concrete dam sites in the United Kingdom. Figure 23.1 illustrates his correlation between velocity ratio and permeability measured by conventional packer tests. Velocity ratio is defined as a ratio between field velocity measured from seismic survey and velocity through rock core measured in the laboratory. It is essential that both the measurements are performed on saturated rocks. In situ permeability increases by ten thousand times with a decrease in the velocity ratio from 1.0 to 0.5 due to fractures.

According to Barton (2008), the permeability (k) of the rock mass is roughly given by Eq. (23.1) at 20°C.

$$k \approx \left(\frac{0.002}{Q}\right)\left(\frac{100}{JCS}\right)\left(\frac{1}{H^{5/3}}\right),\ m/sec \qquad (23.1)$$

where Q = in situ rock mass quality (Q = 0.1 to 100) and $(RQD/J_n)(J_r/J_a)(J_w/SRF)$, JCS = joint wall compressive strength in MPa, and H = depth of a point under consideration below ground surface.

TABLE 23.1 Approximate Coefficient of Permeability of Rocks at 15°C and Porosity

In situ rock	Coefficient of permeability k (cm/sec)	Porosity (%)
Igneous rocks		
Basalt	10^{-4} to 10^{-5}	1 to 3
Diabase	10^{-5} to 10^{-7}	0.1 to 0.5
Gabbro	10^{-5} to 10^{-7}	0.1 to 0.5
Granite	10^{-3} to 10^{-5}	1 to 4
Sedimentary rocks		
Dolomite	4.6×10^{-9} to 1.2×10^{-8}	—
Limestone	10^{-2} to 10^{-4}	5 to 15
Sandstone	10^{-2} to 10^{-4}	4 to 2
Slate	10^{-3} to 10^{-4}	5 to 2
Metamorphic rocks		
Gneiss	10^{-3} to 10^{-4}	—
Marble	10^{-4} to 10^{-5}	2 to 4
Quartzite	10^{-5} to 10^{-7}	0.2 to 0.6
Schist	10^{-4} to 3.0×10^{-4}	—
Slate	10^{-4} to 10^{-7}	0.1 to 1

Source: Jumikis, 1983.

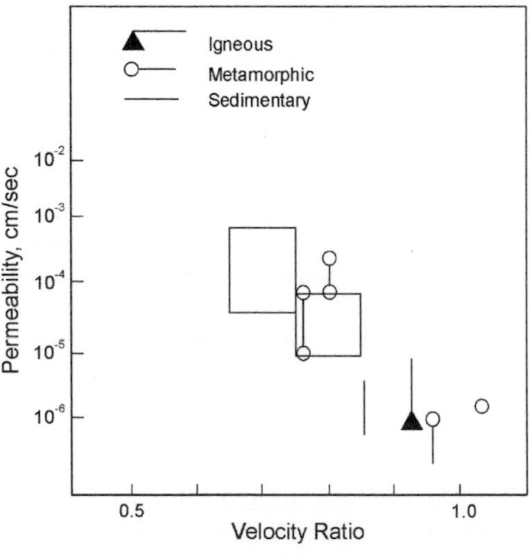

FIGURE 23.1 Correlation between in situ permeability and velocity ratio. *(From Knill, 1969)*

The Dekchu and Bhasmey hydroelectric projects in Sikkim, in the Himalayas, tend to confirm Eq. (23.1).

PERMEABILITY FOR CLASSIFYING ROCK MASSES

Houlsby (1977) suggested a classification of rock masses according to their permeabilities as per Table 23.2.

PERMEABILITY VERSUS GROUTING

Houlsby (1982) presented a very useful keynote paper on cement grouting in dams. When is grouting warranted? This question is fully answered in Figure 23.2. If permeability is less than 1 lugeon, no grouting is required as the rock is likely to be tightly jointed and of good quality. If permeability is more than 10 lugeons, grouting is required for most types of dams. A permeability of 100 lugeons is encountered in a heavily jointed rock mass with relatively open joints (Table 23.2).

DETERMINATION OF PERMEABILITY

The permeability of in situ soils and rocks are usually determined by a pumping test and/or the water pressure test, which is also called a "Lugeon test."

Lugeon Test

The Lugeon method or water pressure test is done in a drillhole. The test does not give the permeability coefficient, k. The test does, however, give a quantitative comparison of the in situ permeabilities. The Lugeon test is generally performed to establish a criterion for grouting rock masses.

TABLE 23.2 Classification of Rock Masses Based on Permeability (Lugeon Values)

Lugeon value	Strong, massive rock with continuous jointing	Weak, heavily jointed rock
0	Completely tight	Completely tight
1	Sometimes open joints up to about 1 mm	Sometimes open to hair crack size of 0.3 mm
3.5	Occasionally open to 2.5 mm	Occasionally open to 1.2 mm
20	Often open to 1.2 mm	Often open to 1.2 mm
50	Often open to 2.5 mm	Often open to 2.5 mm
100	Often open to 6.2 mm	Often open to 6.2 mm

Joint measurements are in millimeters; 1 lugeon $\approx 10^{-7}$ m/sec. Local variation in permeability is probable due to locally open fractures.

Source: Houlsby, 1977.

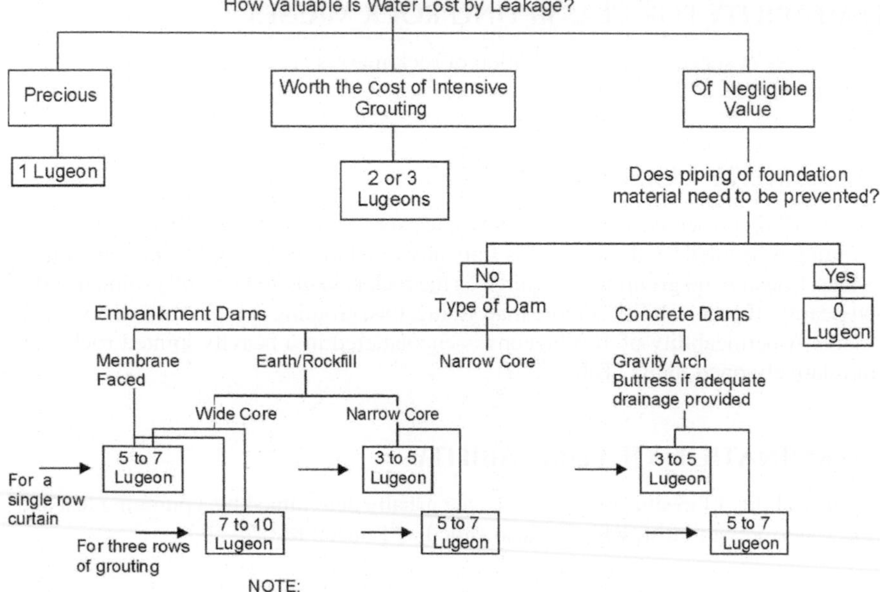

FIGURE 23.2 Guide for deciding when grouting is needed, and if so, to what intensity. *(From Houlsby, 1982)*

The approach, developed by Professor Maurice Lugeon (1933), is based on the lugeon unit, which is obtained from water injection and absorption testing in situ. One lugeon unit corresponds to 1 liter of water absorption at the rate of 1 liter/minute from a 1 meter test length of a borehole when the water in the borehole remains at a pressure of 1 MPa over a period of 10 minutes. Accordingly, a rock mass absorbing less than one lugeon unit of water is considered to be reasonably watertight so no grouting is needed.

GROUTING

If in doubt, do not scream and shout, grout and grout throughout.

Grouting is a process of injecting a slurry of cement or other suitable material under pressure into a rock formation through a borehole to mend fissures and cracks. The purpose of the grouting is to

1. Strengthen the ground or rock mass
2. Make the rock mass watertight
3. Do both at the same time

If the rock mass has poor strength, pre-grouting is aimed at improving its mechanical strength allowing:

- Easier and safer excavation works
- Construction through zones that are difficult to penetrate by traditional methods (e.g., cohesionless or flowing ground, thick shear zones, fault zones, etc.)
- Passage through zones where environmental conditions are difficult

Grouting for waterproofing, on the other hand, is used to form curtains (below dams and around water conductor systems) capable of reducing the underground flow of water. It also provides acceptable tunneling conditions, both for the work and the environment in

- Rocks that are of good structure, however, fissured, fractured, or strongly permeated with water
- Highly permeable grounds that prove unstable

Pre-grouting can be done from ground surface from an adjacent or pre-existing work or directly from a gallery under construction. Consolidation grouting generally has a waterproofing effect. Both types of grouting are often used below groundwater level in underground works.

Grouting increases the modulus of deformation of rock masses. It cuts down the amount of discharge of seepage water, and with a judiciously installed drainage system, grouting may also contribute to reduce uplift pressure on hydraulic structures. All of these improvements in rock properties improve the stability of the rock structure system.

Pre-grouting is grouting done before excavation, and it is done in subway tunnels to reduce the differential settlement of buildings above any overlying clay layer. It is possible to reduce permeability to only 1 lugeon using cement grout with cement particles of the maximum 100–140 μm size. Micro-fine or ultra-fine cements with maximum particle size as small as 30 and 15 μm make it possible to grout fractures with crack openings of about 0.1 and 0.05 cm, respectively. The rule of thumb is that crack openings (aperture) should be about three or four times the maximum particle size (Barton, 2002).

Grouting reduces the degree of anisotropy of jointed rock mass and associated engineering problems (i.e., reduction in the allowable bearing pressure, increase in stress concentration in tunnels, expanding zones of stress relaxations in high walls of caverns, subsidence above mines). Barton (2002) observed that grouting reduced the maximum permeability by 17 times and the minimum permeability by one-tenth. Consequently, there was also rotation of axis of anisotropy. Thus the most permeable and least normally stressed joint set was successfully grouted and, presumably, even the least permeable joint set was well sealed. This may result in significant improvement of the rock mass quality by about one order of magnitude in dry rock masses. The support pressure will be reduced and only a light support system is needed in openings. Pre-grouting is definitely needed in water-charged rock mass or flowing ground conditions. Barton proved why construction engineers often grouted weak zones to improve their quality. The effect of grouting is more in blocky rock mass (Q = 1 to 40) because damage due to blasting is more extensive (Grimstad, 2006). Engineers should adopt smooth blasting for excavation within a pre-grouted rock mass to minimize the damages from blasting.

Goel (2006) showed that the grouting of full-column rock bolts also helps improve rock mass quality in rock mass with Barton's Q between 1 and 40. The improvement is found in rock mass in the vicinity of the rock bolts. This improvement in rock mass quality (Q) is up to three times in poor rock masses, represented by Q = 1, and two times in good rock masses, represented by Q = 20.

Grout Types

There are three main types of grout:

1. Suspension
2. Liquid or solution
3. Special

Suspension Grouts

Suspension grouts are a combination of one or more inert products such as cement, fly-ash, clays, and so on suspended in a liquid (i.e., water). Depending on the dry matter content, suspension grouts are classified as either stable or unstable.

Unstable suspensions are a mixture of pure cement with water. This mixture is homogenized by an agitation process. A sedimentation of suspended particles occurs rapidly when agitation stops.

Stable suspensions are generally obtained by using the following methods:

- Increasing the total dry matter content
- Incorporating a mineral or colloidal component, often from the bentonite family
- Incorporating sodium silicate in cement and clay/cement suspensions

Stability depends on the dosage of various components and on the agitation process. Stability is relative because sedimentation occurs more or less rapidly when agitation ceases.

Liquid Grouts

Liquid grouts consist of chemical products in a solution or emulsion form and their reagents. The most frequently used products are sodium silicate and certain resins. Hydrocarbon emulsions can also be used in specific cases.

Special Grouts

Special grouts have one or more special features. These grouts include quick-setting grouts, cellular type grouts (expanding or swelling grout and expanded or aerated grout), and grouts with improved special properties.

Quick-Setting Grouts

Setting times for these grouts have been modified, and in some cases the setting time may be reduced to a few seconds. The products used for quick-setting grouts include:

- Pure cement-based grout: Among additives, the most common are accelerators such as calcium chloride and sodium silicate. Portland cements and aluminous cement mixes are also used.
- Bentonite/cement grout: The most common accelerator is sodium silicate.

Expanding or Swelling Cellular Type Grout

The volume of this type of grout increases after the grout is placed. Swelling of the grout is obtained through the formation of gas inside the grout. Expansion is generally more than 100%. These grouts are used for filling large solution cavities in soluble rocks such as limestones.

The cells are most often obtained by the formation of hydrogen caused by the action of lime element in cement on aluminum powder incorporated in the grout at mixing time. Immediate stability of the grout can be improved by adding small quantities of sodium silicate. The quantity of aluminum powder in the grout may be up to 2 kg/m^3. At many projects, rock anchors are installed using cement grout without aluminum powder. Consequently, cement grout shrinks after setting and the pull-out capacity of anchors decreases to miserably low values; thus quality control of grout materials used in ground/rock anchors is necessary.

Expanded or Aerated Cellular Type Grouts

The volume of these grouts is increased before use by introducing a certain volume of air. Air is added by introducing a wetting agent when the grout is mixed. This operation can be made easier by blowing air into the grout during preparation. The objective with aerated grout is to increase the grout volume by forming bubbles. The volume generally increases by 30 to 50% before the grout is injected. These types of grouts are used to fill cavities so that a compacting effect occurs in a closed space.

Grouts with Improved Special Properties

Grout with improved penetrability: This type of grout is capable of penetrating voids smaller than those usually filled and also to reach even farther, if necessary. Various methods are used to increase cement grout penetrability:

- Decreasing viscosity and shearing strength using additives with a fluidifying action in the constant presence of dry matter. The additives are used to deflocculate bunches of grains that form in the usual grouts. These products can be derived from natural organic products such as sodium bicarbonate.
- Increasing resistance to filtering effects using activators that reduce grout filtration. This is obtained by dispersion of grout grains (or peptizing agents) or through the action of water retaining polymers on intergranular water.
- Decreasing the dimensions of the grains suspended in grouts. This is a costly alternative that involves regrinding of material.

Grouts with improved mechanical strength: These types of grouts are used to obtain an increased final strength of grouts, either by applying a treatment that does not modify certain other characteristics, such as dry matter content or viscosity, or by using additives that are cheaper than the constructive products of the original grout.

Grouts with improved resistance to washing-out: These types of grouts are used to avoid any washing-out processes when the grouts are applied in largely open spaces filled with water, and particularly when flowing water is present. This is achieved:

- By using hardened grouts that are almost instantaneous and in some cases halt the washing-out process. Controlling the hardening time also permits penetrability to be controlled.
- By improving resistance through the use of flocculating, coagulating, or thickening types of organic additives. These additives improve the resistance to washing-out tendencies and also increase viscosity and cohesion which, in turn, tend to modify grout rheology as well as the behavior at the grout-water separation surface.

Details on grouts can be obtained from the International Tunneling Association (ITA) Special Report on Grouting of Underground Works (1991) and International Society for Rock Mechanics (ISRM) Commission on Rock Grouting (1996).

Grouting Parameters

Three main parameters must be considered when controlling the grout injection process:

1. Grout volume (V) per pass
2. Injection pressure (P)
3. Rate of injection output (Q)

These parameters are determined by a set of injection points and relate to one injection phase. Time of injection (t) for one pass, where t = V/Q average, which must be in accordance with the setting time, is the fourth parameter to be checked.

Volume (V) depends upon the volumetric ratio, defined as grout volume/volume of treated ground, which integrates the porosity of the ground, the filling coefficient of voids for the phase under consideration, and the geometry of treatment given by spacing between holes and length of injection pass.

The speed (Q) must be limited so that the injection pressure (P) remains lower than the ground fracturing pressure, which depends on in situ stresses. An experimental approach with regard to P and Q parameters is recommended to assure that the treatment is accomplished correctly.

Figure 23.3 shows a correlation between grout-take, field velocity, and velocity ratio for grout curtains. This is done by using a pound of cement or cement plus filler per square foot of cut-off. Knill (1969) pointed out that correlations for other countries differ and data may be too scattered. Nevertheless, the advantage of classifying rock masses is clearly seen.

For consolidation grouting, limited available data suggest the following correlation (Figure 23.3):

$$\% \text{ voids infilling} = (0.04) \cdot \text{grout-take} \tag{23.2}$$

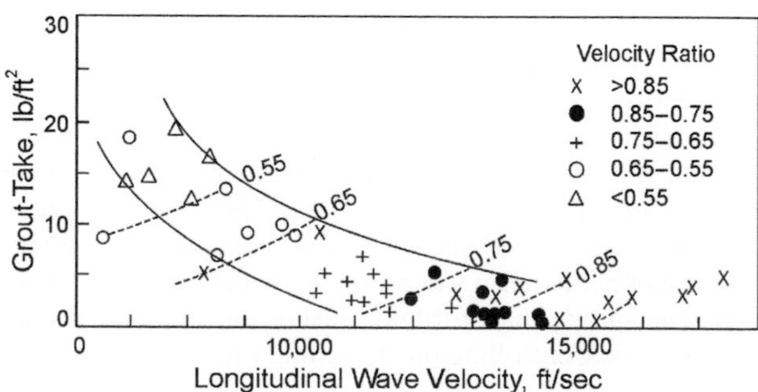

FIGURE 23.3 Correlation between grout-take, longitudinal wave velocity, and velocity ratio. *(From Knill, 1969)*

The grout-take depends upon field wave velocity. If a rock mass is not fully saturated, some allowances must be made for recording a lower velocity. Velocities may be observed to be higher in the area of tectonic stresses. Other factors affecting the velocity are anisotropy, joint system, and the presence of wave guide. There are limitations to a classification system based solely upon velocity ratio. Field studies are needed to update trends observed by Knill (1969).

The effectiveness of consolidation grouting may be checked by observing improvements in rock quality designation (RQD) and field velocity after grouting. For example, if the velocity ratio is raised to a value of more than 0.85 and field velocity becomes more than 13,000 feet/sec (4300 m/sec), the grouting operation is successful.

Effectiveness of Grouting

Effectiveness of grouting may be better checked by measuring the permeability in new drillholes. If the permeability of a rock mass at shallow depths is reduced to the extent shown in Figure 23.2, no further grouting is required.

For grout pressure, the well-known rule of thumb of 1 psi per foot is usually a good compromise for a rock mass of poor quality. Figure 23.4 illustrates the current trend

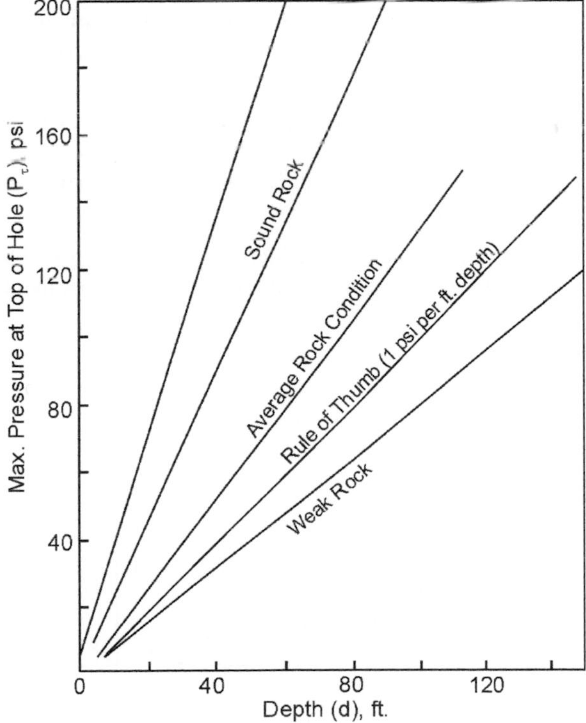

FIGURE 23.4 Recommended maximum grout pressures.

(1 MPa = 150 psi; 1 m = 3.3 feet). One disadvantage of working with grout is "working blind." Since there is little control of where the grout is moving, it is impossible to ensure complete filling of all rock voids. Based on the characteristics of the time-pressure diagrams plotted during grout injection (Figure 23.5a, c), Jahde (1937) suggested a way to identify whether grouting is successful or not.

Figure 23.5a shows that pressure increases slowly and uniformly until the pump capacity or the allowable injection pressure is attained. This may be interpreted as a successful injection.

Figure 23.5b indicates that the pressure drops after an initial increase. This may mean that the grout has "broken out", for example, a clay gauge, filling a crack that might have ended in the free atmosphere, has been expelled out of the crack. This can also be thought of as a successful injection.

Figure 23.5c conveys the idea that after an initial increase in pressure, the pressure drops and again increases slowly. After the occurrence resulting in Figure 23.5b, it can be interpreted that the crack, seam, or joint subsequently closed and the injection is successful. The effectiveness of a grouting operation is usually verified by making check borings in the grouted zone and examining rock cores extracted from these boreholes.

Details on how the grouting will work, assumptions, and discussions on improving various parameters of rock mass quality (Q) because of pre-grouting are described by Barton et al. (2001) as shown in Table 23.3.

The increase or decrease in each parameter of Q seems to be small (Table 23.3), but the combined effect is remarkable. This implies that the following effects can be noted in the Q-system:

- Where there are dry conditions, pre-grouting may improve the rock mass quality one rock quality class.
- Where there are wet conditions, pre-grouting may improve the rock mass quality two or even three rock quality classes.

Barton et al. (2001) observed that the Q-value increased from 0.8 to 16.7 because of pre-grouting, that is, a very poor rock mass was converted to a good rock mass (see Example 23.1 and Table 23.3).

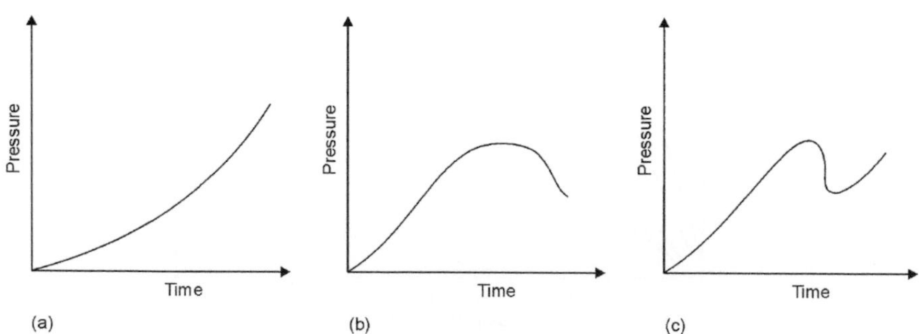

FIGURE 23.5 Plots to check the success of grouting. *(From Jahde, 1937)*

TABLE 23.3 Improvement of Rock Mass Properties with Pre-Grouting

Effective	RQD	Increases	30 to 50%	
Effective	J_n	Reduces	9 to 6	
	J_r	*Increases	1 to 2	(Changed set)
	J_a	*Reduces	2 to 1	(Changed set)
	J_w	Increases	0.5 to 1.0	
	SRF	Reduces	2.5 to 1.0	

**The critical joint set may even change after the grouting.*

Source: Barton et al., 2001.

Example 23.1

The average rock mass quality (Q) of a 10 m diameter tunnel in quartzite under 200 m overburden is 0.8 (very poor). It is proposed to pre-grout the rock mass with a fine cement grout to improve Q up to 16.7. Predict the various engineering parameters and details of support with and without pre-grouting. $q_c = 50$ MPa, $J_r = 1$, $J_a = 4$, $\gamma = 2.5$g/cc, $J_w = 1$, and SRF $= 1.0$.

The approximate estimates of rock parameters are presented in Table 23.4. Since overburden is less than 320 m, so f = 1, and also less than $350Q^{1/3}$ [$=350(0.8/2.5)^{1/3}$], f′ is made equal to 1.0 because no squeezing is likely to take place (see Eqs. 8.6 and 8.7 in Chapter 8).

Heaving of Foundation upon Grouting

Grouting is injurious to a rock mass if it heaves due to an injecting pressure that is more than the overburden pressure. Heaving should be monitored to control the injecting pressure. A practical approach is to undertake grouting in different stages, the first stage at a low pressure and subsequent stages at stepped up pressures, reaching the final pressure at the end. Grouting of dam abutments may destabilize rock slopes and cause a landslide because effective normal pressure across the plane of sliding is reduced. In light of these issues, grouting should be done very carefully and cautiously with adequate supervision. This aspect could be critical when joints are open and dip toward the slope.

TABLE 23.4 Improvement in Rock Parameters after Grouting

S. No.	Engineering parameters	Before grouting	After grouting	Correlations used
1	Rock mass quality Q	0.8 (very poor)	16.7 (good)	As observed
2	Normalized rock mass quality Q_c	0.4	8.3	$Q_c = (Q \cdot q_c)/100$
3	Modulus of deformation E_d in GPa	7	20	$E_d \approx 10\, Q_c^{1/3}$
4	P-wave velocity V_p in km/sec	3.1	4.4	$V_p = \log Q_c + 3.5$
5	Rock mass strength q_{cmass} in MPa	12	33	$q_{cmass} = 7\, \gamma Q^{1/3}$ SRF = 2.5
6	Angle of internal friction ϕ_p in degrees	19	40 (see Table 6.10 for RMR = 70)	$\tan\phi_p = (J_r\, J_w/J_a) + 0$
7	Cohesion of rock mass c_p in MPa	4.3	7.7	$c_p = \frac{q_{cmass}\left(1 - \sin\phi_p\right)}{2\, \cos\phi_p}$
8	Parameter A	1.0	3.6	$A = \frac{2\, \sin\phi_p}{1 - \sin\phi_p}$
9	Suggested strength criterion	$\sigma_1 - \sigma_3 = 12 + 0.5(\sigma_2 + \sigma_3)$	$\sigma_1 - \sigma_3 = 33 + 1.8(\sigma_2 + \sigma_3)$	$\sigma_1 - \sigma_3 = q_{cmass} + \frac{A(\sigma_2 + \sigma_3)}{2}$
10	Tensile strength of rock mass q_{tmass} in MPa	0.07	0.17	$q_{tmass} = 0.029\, \gamma Q^{0.31}$
11	Residual cohesion c_r in MPa	0.1	0.1	Art. 13.10
12	Residual angle of internal friction ϕ_r in degrees	14	30	$\phi_r = \phi_p - 10° \geq 14°$

13	Angle of dilatancy Δ in degrees	2	5	$\Delta = (\phi_p - \phi_r)/2$, beyond failure
14	Permeability k	7×10^{-7} m/sec	3.5×10^{-8} m/sec	Eq. (23.1); $JCS = q_c$
15	Vertical in situ stress P_v in MPa	5	5	$P_v = \gamma H$
16	Major horizontal in situ stress P_H in MPa	7.5	7.5	$P_H = 1.5 + 1.2\ P_v$
17	Convergence of tunnel roof Δ_v in mm	40	2	$\Delta_v = \frac{B}{100\ Q}\sqrt{\frac{P_v}{q_c}}$ B = 10,000 mm
18	Convergence of tunnel roof Δ_h in mm	48	2	$\Delta_h = \frac{H_t}{100\ Q}\sqrt{\frac{P_H}{q_c}}$ H_t = 10,000 mm
19	Critical strain ε_{cr} in %	0.2	0.2	$\varepsilon_{cr} = \frac{q_{cmass}}{E_d \text{ in MPa}}$
20	Construction problems	Anticipated $u_a/a >> \varepsilon_{cr}$	No	u_a = radial displacement, a = radius of tunnel
21	Self-supporting size of tunnel D_e in meters	1.8	6	$D_e = 2\ ESR\ Q^{0.4}$ ESR = 1.0
22	Support pressure p_{roof} in MPa	0.2	0.08	$p_{roof} = (0.2/J_r)Q^{-1/3} \cdot f \cdot f'$
23	Thickness of SFRS in mm	100 mm	None	Figure 8.5
24	Spacing of rock bolts	1.6 m c/c	2.4 m c/c	Figure 8.5
25	Length of rock bolts	3 m	3 m	Figure 8.5

It is recommended that smooth blasting be adopted for excavation within pre-grouted rock mass. Conventional blasting may reduce Q-value significantly due to the damage by blasting.
Further research work is needed to develop correlations for grouted rock masses.

REFERENCES

Barton, N. (2002). Some new Q-Value correlations to assist in site characterisation and tunnel design. *International Journal of Rock Mechanics and Mining Sciences*, *39*, 185–216.

Barton, N. (2008). *Training course on rock engineering* (p. 502). Organized by ISRMTT & CSMRS, Course Coordinator Rajbal Singh, December 10–12, New Delhi, India.

Barton, N., Buen, B., & Roald, S. (2001). Strengthening the case for grouting. *Tunnels and Tunnelling International*, December, 34–39.

Goel, R. K. (2006). Full-column grouted rock bolts and support pressure. In *Proceedings of the ISRM International Symposium and 4th Asian Rock Mechanics Symposium on Rock Mechanics in Underground Construction* (p. 251). Singapore, November.

Grimstad, E. (2006). Personal communication with R. K. Goel.

Houlsby, A. C. (1977). Engineering of grout curtains to standards. *ASCE*, *103*, GT 9, 53–70.

Houlsby, A. C. (1982). Cement grouting for dams, Keynote Paper. In W. H. Baker (Ed.), *ASCE Symposium on Grouting in Geotechnical Engineering* (pp. 1–33), New Orleans.

ISRM. (1996). ISRM Commission on Rock Grouting. *International Journal of Rock Mechanics and Mining Science—Geomechanics Abstracts*, *33*(8), 803–847.

ITA Special Report. (1991). Grouting underground works: Recommendations on grouting for underground works, Association Francaise des Travaux Souterrain, ITA. *Tunnelling and Underground Space Technology*, *6*(4), 383–461.

Jahde, H. (1937). *Die Abdichtung des Untergrundes beim Tals perrenbau, Beton und Eisen*, No. 12, p. 193. In *Rock mechanics* by A. R. Jumikis, 1983.

Jumikis, A. R. (1983). *Rock mechanics* (2nd ed., p. 163). Zurich, Switzerland: Trans Tech Publications.

Knill, J. L. (1969). The application of seismic methods in the prediction of grout take in rock. In *Proceedings of the Conference on In Situ Investigations in Soils and Rocks* (pp. 93–99), London.

Lugeon, M. (1933). Barrages et Geology Methods des Recherches. In *Terrassement et Impermeabilization Lausanne: Librarie de l'universite' F. Rouge et cie* (p. 87).

Chapter 24

Gouge Material

Science may set limits to knowledge, but should not set limits to imagination.

Bertrand Russell

GOUGE

Gouge is a fine graded material occurring between the walls of a fault, a joint, a discontinuity, and so on as a result of the grinding action of rock joint walls. In other words, gouge is a filling material such as silt, clay, rock flour, and other kinds of geological debris in joints, cracks, fissures, faults, and other discontinuities in rocks. The study of gouge material is important from the point of the stability of underground openings, slopes, and foundations.

Brekke and Howard (1972) in Hoek and Brown (1980) presented seven groups of discontinuity infillings or gouges that have significant influence upon the engineering behavior of rock masses.

1. Joints, seams, and sometimes even minor faults may be healed through precipitation from solutions of quartz or calcite. In this instance, the discontinuity may be "welded" together. Such discontinuities might, however, have broken up again, forming new surfaces. It should be emphasized that quartz and calcite may be present in a discontinuity but may not always be healing it.
2. Clean discontinuities include those without fillings or coatings. Many of the rough joints or partings have a favorable character. Close to the surface, however, it is imperative not to confuse clean discontinuities with "empty" discontinuities from where filling material has been leached and washed away from surface weathering.
3. Calcite fillings may dissolve due to seepage during the lifetime of an underground opening, particularly when they are porous or flaky. Their contribution to the strength of the rock mass then disappears. This is a long-term stability (and sometimes fluid flow) problem easily overlooked during design and construction. Gypsum fillings may behave the same way.
4. Coatings or fillings of chlorite, talc, and graphite make very slippery (i.e., low strength) joints, seams, or faults, particularly when wet, due to the loss of cohesion.
5. Inactive clay material in seams and faults naturally represents a very weak material that may squeeze or wash out.

6. Swelling clay gouge may cause serious problems through free swell and consequent loss of strength or through considerable swelling pressure when confined by a tunnel lining.
7. Material that has been altered to a more cohesionless material (sand-like) may run or flow into a tunnel immediately after excavation.

Influence of Gouge Material

Brekke and Howard (1972) summarized the consequences of encountering filled discontinuities during tunnel excavation as shown in Table 24.1.

If the gouge consists of montmorillonite clay mineral, variation in its moisture content may bring about catastrophic instability of the rock slope. Any clay gouge in a sloped discontinuity makes the rock mass slide easily, and when such a gouge becomes wet it promotes sliding of the rock blocks. In either case, the presence of a significant thickness of gouge has a major influence on the stability of a rock mass (Hoek & Bray, 1981). Figure 24.1 shows an idealized picture of rough undulating joints (Barton, 1974), which have these four types of fillings:

TABLE 24.1 Influence of Discontinuity Infilling upon the Behavior of Tunnels

Dominant material in gouge	Potential behavior of gouge material	
	Near face of tunnel	*Later*
Swelling clay	Free swelling, sloughing; swelling pressure and squeezing pressure on shield	Swelling pressure and squeezing pressure against support or lining, free swell with down-fall or wash-in if lining is inadequate
Inactive clay	Slaking and sloughing caused by squeezing pressure; heavy squeezing pressure under extreme conditions	Squeezing pressure on supports of lining where there is unprotected slaking and sloughing due to environmental changes
Chlorite, talc, graphite, or serpentine	Raveling	Heavy loads may develop on tunnel supports due to low strength, particularly when wet
Crushed rock fragments; sand-like	Raveling or running; stand-up time may be extremely short	Loosening loads on lining, running, and raveling, if unconfined
Porous or flaky calcite, gypsum	Favorable conditions	May dissolve, leading to instability of rock mass

Source: Brekke and Howard, 1972.

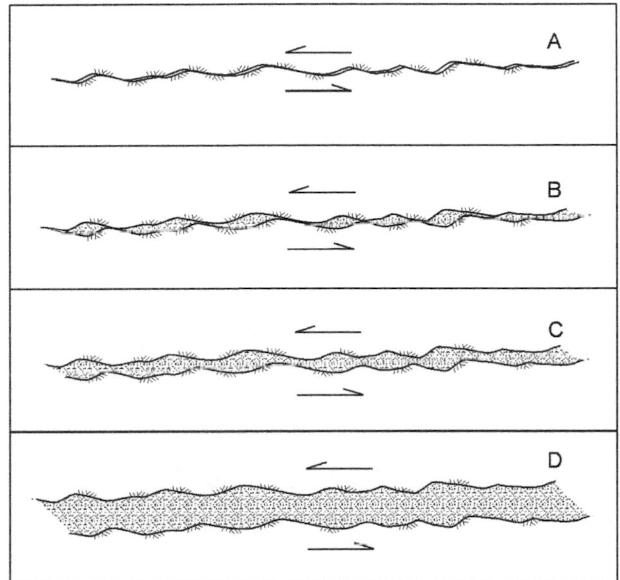

FIGURE 24.1 Categories of discontinuities according to the filling thickness. *(From Barton, 1974)*

1. Category A indicates direct rock/rock asperity contact. The shear strength differs little from the unfilled strength because the rock/rock contact area at peak strength is always small. Dilation due to rock/rock contact causes negative pore pressures to be developed in filling if the shearing rate is fast due to a nearby high-intensity earthquake.
2. Category B may develop the same amount of rock/rock asperity contact as in category A, but the required displacement may be larger. The dilation component of peak shear strength is greatly reduced since the peak strength is similar to the residual strength for unfilled joints. There will be fewer tendencies for negative pore pressures due to reduced dilation.
3. Category C does not show an occurrence of rock/rock contact, but there will be a buildup of stress in the filling where the adjacent rock asperities come close together. If the shearing rate is fast there is an increase in pore pressures in these highly stressed zones and the shear strength is low. If, on the other hand, the shearing rate is low, consolidation and drainage occur. Drainage toward the low stress pockets on either side of the consolidation zones results in a marked increase in shear strength when compared to that under the fast shearing rate.
4. Category D indicates that when the discontinuity filling has a thickness several times that of the asperity amplitude, the influence of the rock walls disappears provided the filling is uniformly graded and predominantly clay or silt. The strength behavior is governed by usual principles of geotechnical engineering.

Goodman (1970) demonstrated the importance of joint infillings in a series of tests in which artificially created saw-tooth joint surfaces were coated with crushed mica. The decrease in shear strength with the increase in filling thickness is shown in Figure 24.2,

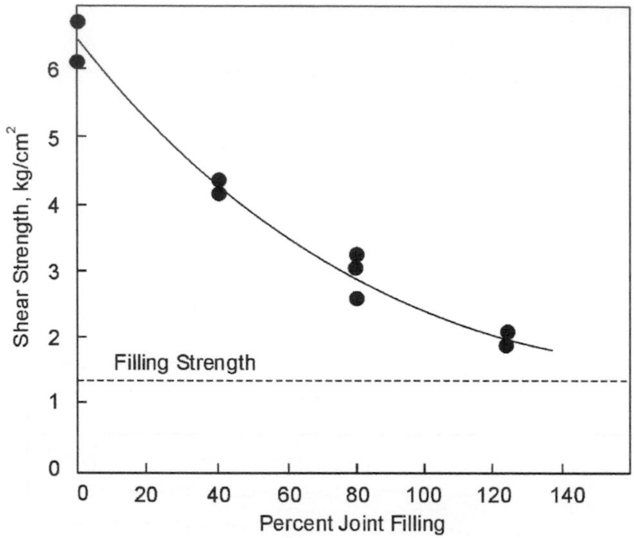

FIGURE 24.2 Effect of joint filling thickness on shear strength. *(From Goodman, 1970)*

which indicates that once the filling thickness (t) exceeds the amplitude (a) of the surface projections, the strength of the joint is controlled by the strength of the filling material.

Goodman, Heuze, and Ohnishi (1972) examined the influence of the thickness (t) of the filling material (kaolinite clay) in granite and sandstone joints. They reported that for a very small thickness of filling material, there is an augmentation of the strength by virtue of the geometry of the rough joint walls. As the thickness increases, the clay filling reduces its strength. At a ratio of thickness and amplitude (t/a) of 3, the shear strength is reduced to that of the filling material.

SHEAR STRENGTH OF FILLED DISCONTINUITIES (SILTY TO CLAYEY GOUGE)

Sinha and Singh (2000) successfully simulated the filled discontinuity in a slope in triaxial tests on two 38 mm ϕ Perspex cylinders with inclined saw-tooth joints that were filled with remolded gouge. The study by Sinha (1993) brought out the following strength criteria for a thick gouge.

1. Deviator stress, which controls the shear failure, is a better criterion for evaluating shear strength of a joint with a thick gouge (t/a >1.25). The following modifications in Eq. (15.6) (Barton, 1974, 1987) were made for the evaluation of the shear strength of a rock joint with a clay gouge and t/a > 1.25.

   (a) For undulating joints

$$\frac{\sigma_1 - \sigma_3}{2} = \sigma_n' \cdot f_t \cdot \tan\left[\text{JRC} \log_{10} \frac{\sigma_1 - \sigma_3}{\sigma_n'} + \phi_b'\right] \tag{24.1}$$

(b) For planar joints

$$\frac{\sigma_1 - \sigma_3}{2} = \sigma_n' \cdot f_t \tan \phi_b' \quad (24.2)$$

where σ_n' = effective normal stress on joint plane; f_t = correction factor due to thickness of gouge (t/a) and $= 0.98 + 0.96\, e^{(-t/a)}$ for undulating joints and $= 0.80 + 0.61\, e^{(-t)}$ for planar joints; t = thickness of gouge in meters; JRC = joint roughness coefficient as shown in Chapter 15 (range 0 to 20); ϕ_b' = basic frictional angle; and $(\sigma_1 - \sigma_3)/2$ = maximum shear stress as obtained after conducting triaxial tests on joints filled with gouge; and β = angle between joint plane and major principal stress plane ($\beta > \phi_b'$ for failure to occur).

Further, it is observed by Sinha (1993) that at a higher thickness of gouge (t > 20 mm), σ_n' becomes less than $\sigma_1 - \sigma_3$, resulting in compaction (negative dilation) of the gouge.

2. On the basis of experimental data, a non-linear relationship for the shear modulus of gouge in joints is found to be

$$\frac{G}{G_o} = 1.46 + 7.13\ c^{-(t/a)\tan\beta} \quad \text{undulating joints} \quad (24.3)$$

$$\frac{G}{G_o} = 1.10 + 3.48\ e^{-(t/4)\tan\beta} \quad \text{planar joints} \quad (24.4)$$

where G/G_o = normalized shear modulus, G = shear modulus, G_o = shear modulus of gouge of very large thickness (t >> a), t/a = thickness-amplitude ratio, β = dip angle (angle between joint plane and major principal plane), and t = thickness of gouge in millimeters.

This testing technique has been appreciated by Norwegian Geotechnical Institute (NGI) scientists. There is a need for further studies on over-consolidated clayey gouge and samples of larger diameter (d/t). The dynamic shear modulus will be much higher than the static modulus because dynamic strain is very small.

DYNAMIC STRENGTH

Shear zones near slopes may have over-consolidated clayey gouge due to erosion of the overburden; thus, there may be some cohesive resistance, particularly in joints having over-consolidated clayey gouge. Under seismic loading the dynamic cohesion may increase enormously because of negative pore water pressure (PI > 5):

$$c_{dyn} = c_{consolidated\ undrained} \quad (24.5)$$

Particles of soil and rock take some time to slip with respect to each other due to inertial forces of particles and lack of time for creep during seismic loading, so a much higher dynamic stress is needed to develop failure strain. Consequently, dynamic strength enhancement in cohesion is likely to be very high along discontinuities filled with over-consolidated clayey gouge (PI > 5) under impulsive seismic loading due to a high intensity earthquake with a nearby epicenter. Further research is needed on dynamic behavior of filled discontinuities.

REFERENCES

Barton, N. (1974). *A review of the shear strength of filled discontinuities in rock*. NGI Publication No. 105 (pp. 1–48), Oslo.

Barton, N. (1987). The shear strength of rock and rock joints. In A. Singh & M. L. Ohri (Eds.), *Current practices in geotechnical engineering* (Vol. 4, pp. 149–202). Associated Publishers IBT & Geo-Environ Academica.

Brekke, T. L., & Howard, T. (1972). Stability problems caused by seams and faults. In *Proceedings of the First North American Rapid Excavation and Tunnelling Conference* (pp. 25–41), New York: AIME.

Goodman, R. E. (1970). Deformability of joints, determination of the in situ modulus of deformation of rock. In *Symposium in Denver, Colorado*, 1969 (pp. 174–196). ASTM, Special Technical Publication 477.

Goodman, R. E., Heuze, F. E., & Ohnishi, Y. (1972). *Research on Strength, Deformability, Water Pressure Relationship for Faults in Direct Shear*. Berkeley: University of California.

Hoek, E., & Bray, J. W. (1981). *Rock slope engineering* (Chap. 5, pp. 83–126). Institute of Mining and Metallurgy. London: Maney Publishing.

Hoek, E., & Brown, E. T. (1980). *Underground excavations in rocks. Institution of Mining and Metallurgy* (Chap. 2, pp. 20–25). London: Maney Publishing.

Sinha, U. N. (1993). *Behaviour of clayey gouge material along discontinuity surfaces in rock mass* (p. 290). Ph.D. Thesis. IIT Roorkee, Uttarakhand, India.

Sinha, U. N., & Singh, B. (2000). Testing of rock joints filled with gouge using a triaxial apparatus. *International Journal of Rock Mechanics and Mining Sciences*, *37*, 963–981.

Chapter 25

Engineering Properties of Hard Rock Masses

Good judgement comes from experience. But where does experience come from? Experience comes from bad judgement.

Mark Twain

HARD ROCK MASSES

Hard rock masses are encountered in a majority of countries and extensive underground excavation work is being carried out through such rocks. The engineering properties of hard rock masses are discussed in this chapter separately for ready reference.

The properties of hard rock masses are required for designing engineering structures. Hard rock is defined as rock material having a uniaxial compressive strength (UCS) of more than 100 MPa. Hard rocks are geologically very old and have well-developed and highly weathered joints; therefore, there may be serious problems with rock falls and seepage in tunnels due to these joints if left unsupported. Hard rock is a misnomer as engineers may believe that it will not pose any instability problems. The deceptively nice appearance of hard rock has created many construction problems in the tunnels of South India, in the upper Himalayas, the Alps, and the United States.

MODULUS OF DEFORMATION

For rock foundations, knowledge of modulus of deformation of rock masses is of prime importance. Geomechanics classification is a useful method for estimating in situ deformability of rock masses (Bieniawski, 1978). As shown in Figure 6.3, the following correlation is obtained:

$$E_d = 2\ \text{RMR} - 100,\ \text{GPa} \tag{25.1}$$

where E_d is in situ modulus of deformation in GPa for RMR $>$ 50 and RMR is as discussed in Chapter 6.

UCS

Barton (2002) proposed the following correlation for mobilized UCS for good and massive rock masses in tunnels based on the correlation of Singh et al. (1998):

$$q_{cmass} = 5\ \gamma\ Q_c^{1/3}\ \text{MPa} \tag{25.2a}$$

where $Q_c = Q \cdot \frac{q_c}{100}$ and γ is unit weight of the rock mass in gm/cc.

Laubscher (1984) found UCS for hard rock masses in mines (Eq. 25.2b), which are nearly the same as Eq. (25.2a).

$$q_{cmass} = q_c \cdot \frac{(\text{RMR} - \text{rating for } q_c)}{106} \tag{25.2b}$$

UNIAXIAL TENSILE STRENGTH

Uniaxial tensile strength (UTS) of a rock mass (q_{tmass}) is obtained by using Eq. (25.3), which is a suggested extension of Eq. (13.21) for hard rocks

$$q_{tmass} = 0.029 \cdot\ \gamma \cdot\ f_c \cdot\ Q^{0.3}\ \text{MPa} \tag{25.3}$$

where $f_c = \frac{q_c}{100}$ and γ = unit weight of the rock mass in g/cc or T/m^3.

STRENGTH CRITERION

The UCS of massive hard rock mass is approximately the same as that of its rock material. However, a small size correction in q_c is needed as shown in Eq. (10.4). The shear strength of hard rock masses proposed by Hoek and Brown (1980) is proportional to the average value of UCS of the rock material q_c (after size correction).

$$\sigma_1 = \sigma_3 + [\text{m} \cdot q_c \cdot \sigma_3 + \text{s} \cdot q_c^2]^{1/2} \tag{25.4}$$

For massive rock masses, s = 1. For tunnels/caverns, $s^{1/2} = \frac{5 \cdot \gamma \cdot Q_c^{1/3}}{q_c}$ = strength reduction factor and $\frac{m}{m_r} = s^{1/3}$.

For slopes, rock parameters (m) and (s) are related to the Geological Strength Index (GSI) in Chapter 26, which may be used for slopes, dam abutments, and foundations. The Hoek and Brown criterion, which assumes isotropic rock and rock mass behavior, should only be applied where there are many sets of closely spaced joints with similar properties. Therefore, rock mass behaves as isotropic mass, if the joint spacing is much less than the size of the structure or opening.

When one joint set is significantly weaker than the others, Eq. (25.4) should not be used, because the rock mass behaves as an anisotropic mass. In these cases stability of the structure should be analyzed by considering a natural wedge failure between two or three intersecting joints (Hoek, 2007). Discontinuous joints should also be considered in stability analysis of wedges with higher cohesion for shorter joints. Singh and Goel (2002) presented the software SASW and WEDGE for the analysis of rock slope wedge and UWEDGE for tunnel and cavern wedge. The factor of safety of a wedge hardly depends upon the strength of rock mass within the weakest and unfavorably oriented discontinuities.

With overstressed dry massive hard rocks, sudden failure by rock bursts may take place such as in Kolar gold mines in India and hard rock mines in South Africa. Chances of rock burst are increased if a hard rock is of Class II type (Chapter 7 and the section Rock Burst in Brittle Rocks in Chapter 13). In weak rock masses, squeezing may take place instead of violent failure ($J_r/J_a < 0.5$).

Reservoir induced seismicity (RIS) is more pronounced due to dam reservoirs in hard rocks, for example, the Koyna hydroelectric project in India. In weak rock masses, RIS is low because of its high damping characteristics.

SUPPORT PRESSURE IN NON-SQUEEZING/NON-ROCK BURST CONDITIONS ($H < 350\ Q^{1/3}$)

The ultimate support pressure in underground caverns with overburden (H) in meters may be found from Eq. (8.9), which is also produced here as Eq. (25.5):

$$p_{ult} = \frac{0.2}{J_r} \cdot f \cdot Q^{-1/3},\ \text{MPa} \tag{25.5}$$

where $f = 1 + (H - 320)/800 \geq 1$.

Table 8.6 gives new values of SRF (50 to 400) for rock bursts in hard rocks.

A tunnel may be self-supporting where its width or diameter (B) is less than the self-supporting span B_s given by

$$\begin{aligned} B_s &= 2 \cdot Q^{0.4},\ \text{meters} \\ \text{if } H &\ll 23.4\ N^{0.88}\ B_s^{-0.1},\ \text{meters (Eq. 7.8)} \end{aligned} \tag{25.6}$$

The general requirement for permanently unsupported openings is

(a) $J_n < 9$, $J_r > 1.0$, $J_a < 1.0$, $J_w = 1.0$, $SRF < 2.5$

Further conditional requirements for permanently unsupported openings are given below.

(b) If $RQD < 40$, need $J_n < 2$
(c) If $J_n = 9$, need $J_r > 1.5$ and $RQD > 90$
(d) If $J_r = 1.0$, need $J_w < 4$
(e) If $SRF > 1$, need $J_r > 1.5$
(f) If span > 10 m, need $J_n < 9$
(g) If span > 20 m, need $J_n < 4$ and $SRF < 1$

In the geologically old and matured hard rock masses, joints may be highly weathered due to very long periods of weathering. Thus, small wedge failures in unsupported tunnels are common. Further, water-charged rock masses may also be encountered, particularly during heavy rainy seasons.

HALF-TUNNELS

Generally, half-tunnels are excavated along hill roads passing through steep hills in hard rocks (Figure 25.1). Such tunnels are most common in Himachal Pradesh, India. The top width, B_{ht} (Figure 25.2) is estimated from 11 cases of half-tunnels as per Eq. (25.7) (Anbalagan, Singh, & Bhargava, 2003).

$$B_{ht} = 1.7\ Q^{0.4}\ \text{meters} \tag{25.7}$$

FIGURE 25.1 Photograph of a half-tunnel along hill roads in hard rocks.

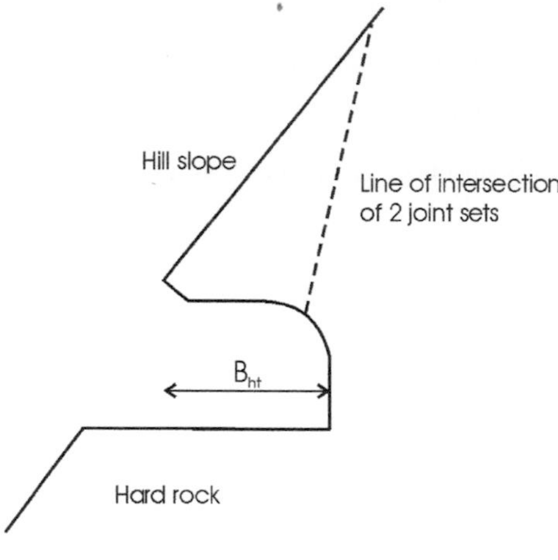

FIGURE 25.2 Line diagram of a half-tunnel along hill roads in hard rocks.

Joints at these sites were discontinuous and the number of joint sets was no more than two with $Q > 18$ ($SRF = 2.5$). These unsupported half-tunnels have been stable for more than 30 years. The factors of safety of wedges, formed by two joint sets and slope, were found to be more than 3 as opposed to sliding along inclined lines of joint intersection planes (Figure 25.2). Anbalagan et al. (2003) presented a detailed study. These half-tunnels saved ecological disturbance because near vertical cut-slopes would be very costly and ecologically unsound. The half-tunnels are also tourist attractions and considered an engineering marvel.

REFERENCES

Anbalagan, R., Singh, B., & Bhargava, P. (2003). Half tunnels along hill roads of Himalaya—An innovative approach. *Tunnelling and Underground Space Technology*, *18*, 411–419.

Barton, N. (2002). Some new Q-Value correlations to assist in site characterization and tunnel design. *International Journal of Rock Mechanics and Mining Sciences*, *39*, 185–216.

Bieniawski, Z. T. (1978). Determining rock mass deformability: experience from case histories. *International Journal of Rock Mechanics and Mining Sciences—Geomechanics Abstracts*, *15*, 237–247.

Hoek, E. (2007). *Practical rock engineering* (Chapter 12). www.rocscience.com/hoek.

Hoek, E. & Brown, E. T. (1980). *Underground excavations in rock* (p. 527). Institute of Mining and Metallurgy. London: Maney Publishing.

Laubscher, D. H. (1984). Design aspects and effectiveness of support system in different mining conditions. *Transactions of the Institute of Mining and Metallurgy*, *93*, A70–A81.

Singh, B., & Goel, R. K. (2002). *Software for engineering control of landslide and tunnelling hazards* (p. 344). Rotterdam: A. A. Balkema (Swets & Zeitlinger).

Singh, B., Viladkar, M. N., Samadhiya, N. K., & Mehrotra, V. K. (1998). Rock mass strength parameters mobilized in tunnels. *Tunnelling and Underground Space Technology*, *12*(1), 47–54.

Chapter 26

Geological Strength Index

The function of Rock Mechanics engineers is not to compute accurately but to judge soundly.
Hoek and Londe

GEOLOGICAL STRENGTH INDEX

Hoek and Brown (1997) introduced the Geological Strength Index (GSI) for both hard and weak rock masses. Experienced field engineers and geologists generally show a liking for a simple, fast, yet reliable classification that is based on visual inspection of geological conditions. A classification system should be non-linear for poor rocks as strength deteriorates rapidly with weathering. Further, increased applications of computer modelling have created an urgent need for a classification system tuned to a computer simulation of rock structures. To meet these needs, Hoek and Brown (1997) devised simple charts for estimating GSI based on the following two correlations:

$$GSI = RMR'_{89} - 5 \text{ for } GSI \geq 18 \text{ or } RMR \geq 23 \quad (26.1)$$

$$= 9 \ln Q' + 44 \text{ for } GSI < 18 \quad (26.2)$$

where Q' = modified rock mass quality,

$$Q' = [RQD/Jn] \cdot [Jr/Ja], \text{ and} \quad (26.3)$$

RMR'_{89} = rock mass rating according to Bieniawski (1989) when the groundwater rating = 15 and joint adjustment rating = 0.

Sometimes, it is difficult to obtain RMR in poor rock masses, and Q' may be used more often because it is relatively more reliable than RMR, especially in openings in weak rocks.

Hoek (Roclab, 2006) and Marinos and Hoek (2000) proposed a chart for GSI (Figure 26.1) so experts can classify a rock mass by visual inspection alone. In this classification, there are six main qualitative rock classes, mainly adopted from Terzaghi's classification (Table 5.2).

1. Intact or massive
2. Blocky
3. Very blocky
4. Blocky/folded
5. Crushed
6. Laminated/sheared

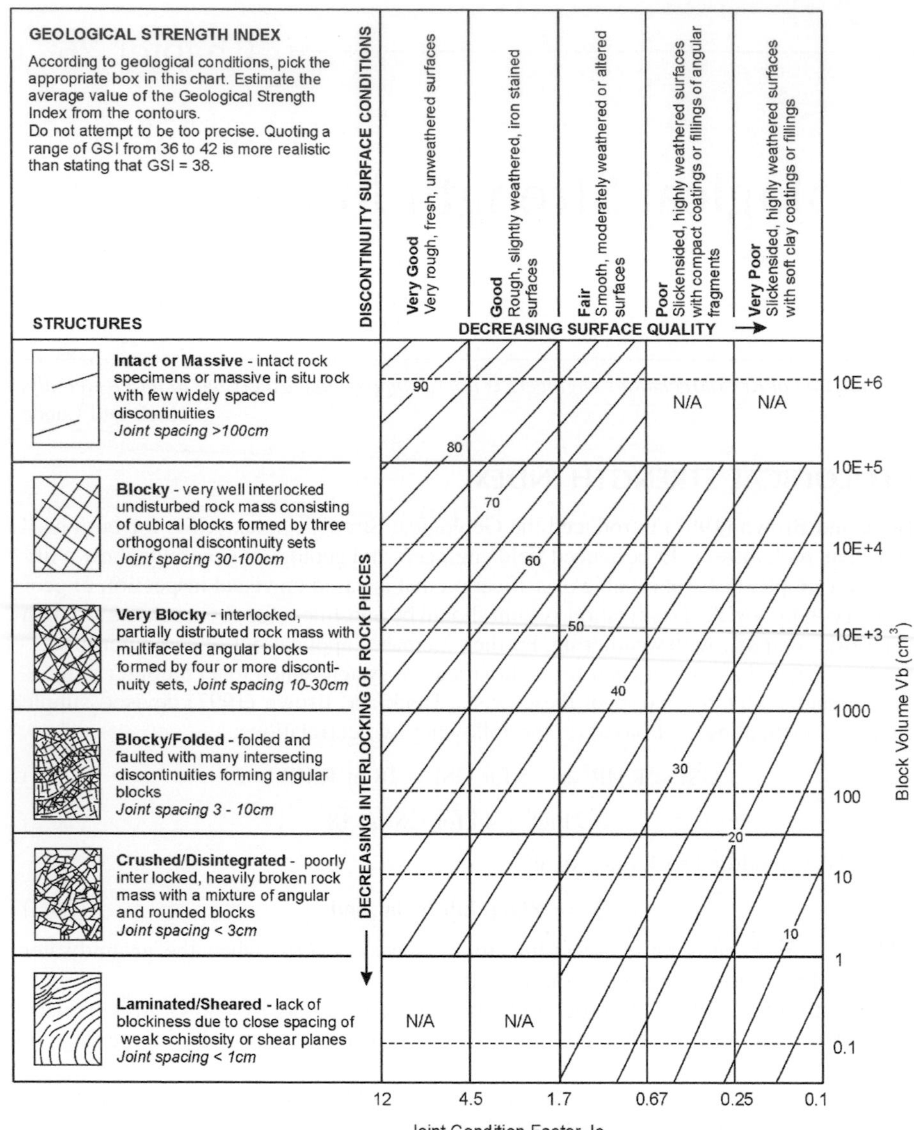

FIGURE 26.1 Estimate of GSI based on visual inspection of geological conditions. *(From Roclab, 2006; Marinos and Hoek, 2000)* Modification by Cai et al. (2004) in terms of its quantification by block volume and joint condition factor is also shown on the right side.

These classifications have been available to engineers and geologists for 60 years. Discontinuities are classified into five surface conditions that are similar to joint conditions in RMR (Chapter 6).

1. Very good
2. Good

3. Fair
4. Poor
5. Very poor

A 6 × 5 block in the matrix of Figure 26.1 is picked up first according to actual and undisturbed rock mass classification and discontinuity surface condition. Then a corresponding GSI is read. According to Hoek (1998) and Marinos and Hoek (2000), a range of values of GSI (or RMR) should be estimated instead of just a single value. This practice has a significant impact on the design of slopes and excavations in rocks. Drastic degradation in GSI, RMR, and Q-values is found to occur in openings after squeezing and rock bursts. This is also seen in openings, hence the need for evaluating the GSI of rock mass in the undisturbed condition (D = 0). Back analysis of both a model (polyaxial strength criterion) and its parameters (from the observed behavior of rock structures) is an ideal method of the rock mass characterization, and GSI is the first step in this direction.

Figure 26.1 is used judiciously for crushed/disintegrated and laminated/sheared rocks. Similarly, hard, thick laminated rocks in the last row of Figure 26.1 may not be applicable, because they may have a higher strength classification (see Table 5.2, Class II).

The GSI chart has been subsequently quantified by Cai et al. (2004) by incorporating the rock block volume (V_b) formed by the joints or discontinuities and the joint condition factor J_C (see Table 4.6). The suggested quantification is also shown in Figure 26.1. The block volume (V_b), affected by the joint set spacing and persistence, can broadly be known by the joint spacing given for six different rock classes in Figure 26.1. The value of joint condition factor, J_C, controlled by joint roughness, weathering, and infilling material, can be obtained by Eq. (26.4) from Cai et al. (2004).

$$J_C = \frac{J_W \, J_S}{J_A} \tag{26.4}$$

where J_W = large-scale joint or discontinuity waviness in meters from 1 to 10 m (Table 26.1), J_S = small-scale smoothness in centimeters from 1 to 20 cm (Table 26.2), and J_A = joint alteration factor (Table 26.3).

Cai and Kaiser (2006), based on the proposed quantitative chart (Figure 26.1), and using surface fitting techniques, suggested the following equation to calculate GSI from J_C and V_b:

$$\mathrm{GSI}(V_b, J_C) = \frac{26.5 + 8.79 \ln J_C + 0.9 \ln V_b}{1 + 0.0151 \ln J_C - 0.0253 \ln V_b} \tag{26.5}$$

where J_C is a dimensionless factor defined by Eq. (26.4) and block volume V_b is in cm^3 (see the section Calibration of RMi from Known Rock Mass Strength Data in Chapter 10).

To avoid double-accounting, groundwater condition and in situ stresses are not considered in GSI because they are accounted for in computer models. GSI assumes that the rock mass is isotropic; therefore, only cores without weak planes should be tested in triaxial cells to determine q_c and m_r as GSI downgrades strength according to schistocity. This classification reduces many uncertainties in rock mass characterization. An undisturbed rock mass should be inspected for classification; however, heavy blasting creates new fractures.

TABLE 26.1 Terms to Describe Large-Scale Waviness (J_W)

Waviness terms	Undulation	Rating for waviness (J_W)
Interlocking (large-scale)		3
Stepped		2.5
Large undulation	>3%	2
Small to moderate undulation	0.3–3%	1.5
Planar	<0.3%	1

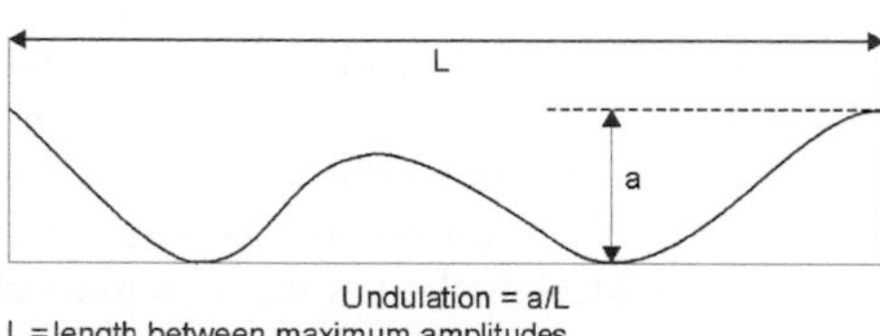

Sources: Palmstrom, 1995; Cai et al., 2004.

TABLE 26.2 Terms to Describe Small-Scale Smoothness (J_S)

Smoothness terms	Description	Rating for smoothness (J_S)
Very rough	Near vertical steps and ridges occur with interlocking effect on the joint surface	3
Rough	Some ridges and side-angles are evident; asperities are clearly visible; discontinuity surface feels very abrasive (rougher than sandpaper grade 30)	2
Slightly rough	Asperities on the discontinuity surfaces are distinguishable and can be felt (like sandpaper grade 30–300)	1.5
Smooth	Surface appears smooth and feels so to the touch (smoother than sandpaper grade 300)	1
Polished	Visual evidence of polishing exists; this is often seen in coating of chlorite and especially talc	0.75
Slickensided	Polished and striated surface that results from sliding along a fault surface or other movement surface	0.6–1.5

Sources: Palmstrom, 1995; Cai et al., 2004.

TABLE 26.3 Rating for Joint Alteration Factor (J_A)

	Term	Description	J_A
Rock wall contact		*Clear joints*	
	Healed or "welded" joints (unweathered)	Softening, impermeable filling (quartz, epidote, etc.)	0.75
	Fresh rock walls (unweathered)	No coating or filling on joint surface, except for staining	1
	Alteration of joint wall: slightly to moderately weathered	The joint surface exhibits one class higher alteration than the rock	2
	Alteration of joint wall: highly weathered	The joint surface exhibits two classes higher alteration than the rock	4
		Coating or thin filling	
	Sand, silt, calcite, talc, etc.	Coating of frictional material without clay	3
	Clay, chlorite, talc, etc.	Coating of softening and cohesive minerals	4
Filled joints with partial or no contact between the rock wall surfaces	Sand, silt, calcite, etc.	Filling of frictional material without clay	4
	Compacted clay materials	"Hard" filling of softening and cohesive materials	6
	Soft clay materials	Medium to low over-consolidation of filling	8
	Swelling clay materials	Filling material exhibits swelling properties	8–12

Sources: Palmstrom, 1995; Cai et al., 2004.

GENERALIZED STRENGTH CRITERION

Hoek, Carranza-Torres, and Corkum (2002) suggested the following generalized Hoek-Brown strength criterion for undisturbed rock masses:

$$\sigma_1 = \sigma_3 + q_c[m_b \frac{\sigma_3}{q_c} + s]^n \tag{26.6}$$

where σ_1 = maximum effective principal stress, σ_3 = minimum effective principal stress, q_c = uniaxial compressive strength (UCS) of rock material (intact) for standard NX size core (see Table 8.13 after Palmstrom, 2000), m_b = reduced value of the material constant m_r, and

$$m_b = m_r \cdot \exp\left[\frac{GSI - 100}{28 - 14D}\right] \tag{26.7}$$

where m_r = Hoek-Brown rock material constant to be found from triaxial tests on rock cores.

In Eqs. (26.8) and (26.9), s and n are Hoek-Brown constants for the rock mass given by the following relationships:

$$s = \exp\left[\frac{GSI - 100}{9 - 3D}\right] \tag{26.8}$$

$$n = \frac{1}{2} + \frac{1}{6}\left(e^{-GSI/15} - e^{-20/3}\right) \tag{26.9}$$

D is a disturbance factor that depends upon the degree of disturbance to which the rock mass has been subjected by blast damage and stress relaxation. It varies from 0 for undisturbed in situ rock masses to 1 for very disturbed rock masses (Table 26.4). Cheng and Liu (1990) found that a zone of blast damage extended for a distance of approximately 2.0 m with D = 0.7 around all large excavations (caverns). While using the disturbance factor D, its values given in Table 26.4 are selected judiciously. The actual value of D is a function of rock mass quality and blasting practices.

Experience in the design of slopes in very large open pit mines has shown that the Hoek-Brown criterion for undisturbed in situ rock masses (D = 0) results in shear strength parameters that are too optimistic. The effects of heavy blast damage as well

TABLE 26.4 Guidelines for Estimating Disturbance Factor D

Appearance of rock mass	Description of rock mass	Suggested value of D
	Excellent quality controlled blasting or excavation by tunnel boring machine results in minimal disturbance to the confined rock mass surrounding a tunnel.	D = 0
	Mechanical or hand excavation in poor quality rock masses (no blasting) results in minimal disturbance to the surrounding rock mass. Where squeezing problems result in significant floor heave, disturbance can be severe unless a temporary invert, as shown in the photograph, is placed.	D = 0 D = 0.5 No invert

TABLE 26.4—Cont'd

Appearance of rock mass	Description of rock mass	Suggested value of D
	Very poor quality blasting in a hard rock tunnel results in severe local damage, extending 2 or 3 m, in the surrounding rock mass.	D = 0.8
	Small-scale blasting in civil engineering slopes results in modest rock mass damage, particularly if controlled blasting is used as shown on the left-hand side of the photograph. However, stress relief results in some disturbance.	D = 0.7 Good blasting D = 1.0 Poor blasting
	Very large open pit mine slopes suffer significant disturbance due to heavy production blasting and also due to stress relief from overburden removal. In some softer rocks, excavation can be carried out by ripping and dozing and the degree of damage to the slopes is less.	D = 1.0 Production blasting D = 0.7 Mechanical excavation

Sources: Hoek et al., 2002; Hoek, 2007.

as stress relief due to removal of the overburden of the rock mass results in disturbance of the rock mass. It is considered that the "disturbed" rock mass parameters with D = 1 in Eqs. (26.7) and (26.8) are more appropriate for these rock masses (Hoek et al., 2002).

Thus, UCS of a rock mass obtained from Eq. (26.6) is

$$q_{cmass} = q_c \cdot s^n \tag{26.10}$$

and uniaxial tensile strength (UTS) of a good rock mass is

$$q_{tmass} = -\frac{s\, q_c}{m_b} \tag{26.11}$$

Equation (26.11) is obtained by setting $\sigma_1 = \sigma_3 = q_{tmass}$ in Eq. (26.6). This represents a condition of biaxial tension. Hoek (1983) showed that the UTS is equal to the biaxial tensile strength for brittle materials.

Hoek (2007) proposed Eq. (13.5) for estimating rock mass strength (q_{cmass}) from laboratory strength of intact rock material (q_c) and GSI for D = 0.

MOHR-COULOMB STRENGTH PARAMETERS

Mohr-Coulomb's strength criterion for a rock mass is expressed as

$$\sigma_1 - \sigma_3 = q_{cmass} + A\sigma_3 \tag{26.12}$$

where q_{cmass} = UCS of the rock mass, which = 2 c cosϕ/(1 − sinϕ); c = cohesion of the rock mass; A = 2 sinϕ/(1 − sinϕ); and ϕ = angle of internal friction of the rock mass.

Hoek and Brown (1997) made extensive calculations on the linear approximation of non-linear strength criterion (Eq. 26.6). They found that strength parameters c and ϕ depend upon σ_3; thus, they plotted charts for average values of c (Figure 26.2) and ϕ (Figure 26.3) with D = 0 for a quick assessment. It may be noted that c and ϕ decrease non-linearly with GSI unlike RMR (Table 6.10). The rock parameter m_r may be guessed from ϕ_p of a rock material at GSI of 90, if adequate triaxial tests are not done. Table 26.5 lists typical values of m_r for various types of rock materials.

The angle of dilatancy of a rock mass after failure is recommended approximately as

$$\begin{aligned} \Delta &= (\phi/4) \quad \text{for GSI} = 75 \\ &= (\phi/8) \quad \text{for GSI} = 50 \\ &= 0 \qquad\quad \text{for GSI} \leq 30 \end{aligned} \tag{26.13}$$

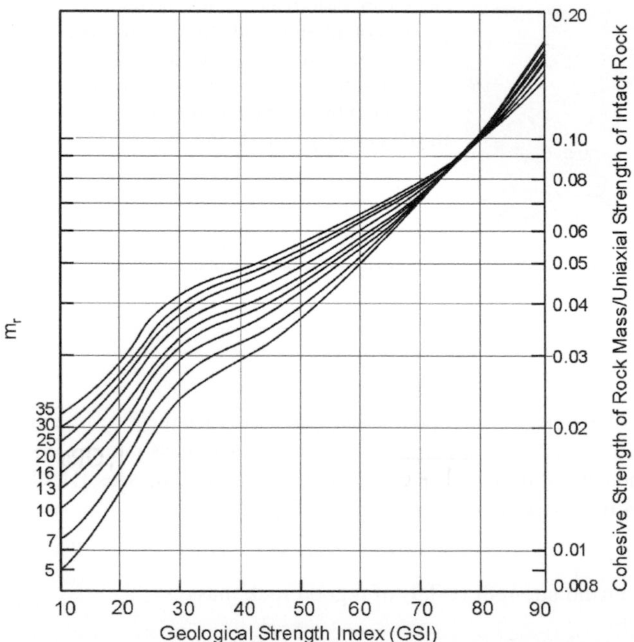

FIGURE 26.2 Relationship between ratio of cohesive strength of rock mass to UCS of intact rock (c/q_c) and GSI for different m_r values for D = 0. *(From Hoek and Brown, 1997)*

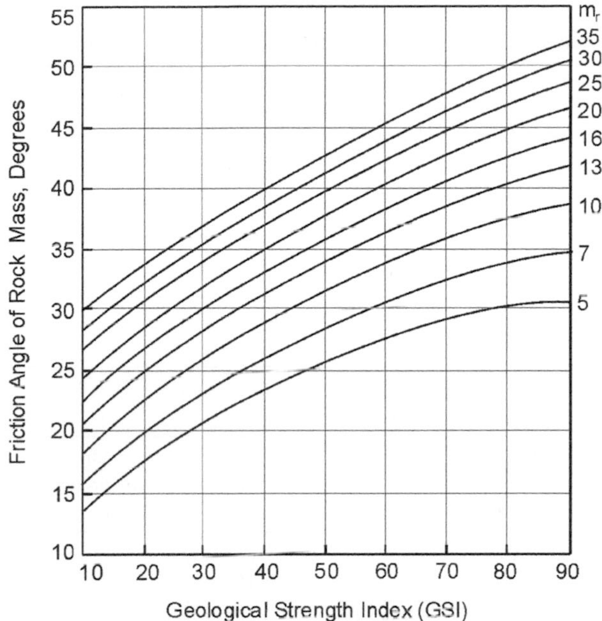

FIGURE 26.3 Friction angle (ϕ) of rock mass for D = 0 for different GSI and m_r values. *(From Hoek and Brown, 1997)*

The Hoek et al. (2002) correlations for s are valid for rock slopes and open pit mines, but not for structurally controlled rock slopes and transported rockfill slopes. For tunnels and caverns, there is an enormous strength enhancement (Chapter 13).

MODULUS OF DEFORMATION

Hoek and Diederichs (2006) found a useful correlation for modulus of deformation (E_d) of rock mass based on approximately 496 in situ tests.

$$E_d = E_r\left[0.02 + \frac{1 - D/2}{1 + \exp((60 + 15D - GSI)/11)}\right], \text{GPa} \tag{26.14}$$

where E_r = modulus of elasticity of intact rock in GPa.

The elastic modulus (E_e) is obtained from the unloading cycles of the uniaxial jacking tests. It is correlated for both dry and saturated rock mass as follows (Chapter 8 and Eq. 8.19):

$$E_e = 1.5Q^{0.6}\, E_r^{0.14}, \text{GPa} \tag{26.15}$$

where Q = rock mass quality.

Equation (26.15) is suggested for the dynamic analyses of concrete dams during a major earthquake and machine (generator) foundations on the rock masses.

The original equation proposed by Hoek and Brown (1997) has been modified by the inclusion of factor D to allow for the effects of blast damage and stress relaxation. The strength and deformation parameters estimated from the GSI system are very close to those obtained from in situ tests (Cai et al., 2004). Back analysis of observed displacements in openings may give more realistic values of the design parameters including the disturbance factor by trial and error.

TABLE 26.5 Values of the Constant m_r for Intact Rock Material by Rock Group

Rock type	Class	Group	Texture: *Coarse*	*Medium*	*Fine*	*Very fine*
Sedimentary	Clastic		Conglomerate (22)	Sandstone 19	Siltstone 9	Claystone 4
			Greywacke (18)			
	Non-clastic	Organic	---------- Chalk ---------- 7			
			---------- Coal ---------- (8–21)			
		Carbonate	Breccia (20)	Sparitic limestone (10)	Micritic limestone 8	—
		Chemical	—	Gypstone 16	Anhydrite 13	—
Metamorphic	Non-foliated		Marble 9	Hornfels (19)	Quartzite 24	—
	Slightly foliated		Migmatite (30)	Amphibolite 25–31	Mylonites (6)	—
	Foliated*		Gneiss 33	Schists 4–8	Phyllites (10)	Slates 9
Igneous	Light		Granite 33	—	Rhyolite (16)	Obsidian (19)
			Granodiorite (30)	—	Dacite (17)	—
			Diorite (28)	—	Andesite 19	—
	Dark		Gabbro 27	Dolerite (19)	Basalt (17)	—
			Norite 22	—	—	—
	Extrusive pyroclastic type		Agglomerate (20)	Breccia (18)	Tuff (15)	—

The values given are estimates. It is suggested to get the m_r values from triaxial test data.

**These values are for intact rock specimens tested normal to bedding or foliation. The value of m_r will be significantly different if failure occurs along a weakness plane.*

Source: Hoek, Marinos, and Benissi, 1998.

ROCK PARAMETERS FOR INTACT SCHISTOSE

In argillaceous or anisotropic rocks (shales, phyllites, schists, gneisses, etc.), the UCS of rock material q_c depends upon the orientation of the plane of weakness. Both GSI and RMR take into account the orientation of joints. To avoid double-accounting for joint orientation in both UCS and GSI, it is a common engineering practice to use the upper bound value of q_c and corresponding m_r for rock cores with nearly horizontal planes of weakness for estimating m_b, s, and E_d for jointed rock masses.

Cohesion along joints is needed for wedge analysis or computer modeling. Cohesion along bedding planes or planar continuous joints (longer than 10 m) may be negligible. However, cohesion along discontinuous joints (assumed continuous in the wedge analysis) may be the same as cohesion (c) of the rock mass. The cohesion of the rock mass is due to the cohesion of the discontinuous joints. The ratio of c and cohesion of rock material (Figure 26.2) may be of the same order as the area of intact rock bridges per unit area of discontinuous joints.

ESTIMATION OF RESIDUAL STRENGTH OF ROCK MASSES

To extend the GSI system for estimation of rock mass residual strength, Cai et al. (2007) proposed an adjustment of the original GSI value based on the two major controlling factors in the GSI system, block volume (V_b) and joint condition factor (J_C), to reach the residual values.

The difference between the peak and residual strength of a rock mass with non-persistent joints is larger than that of a rock mass with persistent joints. The implication is that a drop of GSI from peak to residual values is larger for rock masses with non-persistent joints. Besides rock bridges, rock asperity interlocking also contributes to the difference between peak and residual strengths.

Residual Block Volume

If a rock experiences post-peak deformation, the rock in the broken zone is fractured and consequently turned into a poor and eventually "very poor" rock (Figure 7.2). The properties of a rock mass after extensive straining should be derived from the rock class of "very poor rock mass" in the RMR system (Chapter 6) or "disintegrated" in the GSI system.

For the residual block volume, it is observed that the post-peak block volume is small because the rock mass has experienced tensile and shear fracturing. After the peak load, the rock mass becomes less interlocked and is heavily broken with a mixture of angular and partly rounded rock pieces.

Detailed examination on the rock mass damage state (before and after the in situ block shear tests at some underground cavern sites in Japan) revealed that in areas not covered by concrete, the failed rock mass blocks are 1–5 cm in size. The rock mass is disintegrated along a shear zone in these tests. As such, Cai et al. (2007) suggested the following residual block volume V_b^r:

- If $V_b > 10\ cm^3$, V_b^r (in disintegrated category) $= 10\ cm^3$
- If $V_b < 10\ cm^3$, $V_b^r = V_b$

Residual Joint Condition Factor

The residual joint surface condition factor J_C^r is calculated now from Eq. (26.16).

$$J_C^r = \frac{J_W^r J_S^r}{J_A^r} \tag{26.16}$$

where J_W^r, J_S^r, and J_A^r are residual values of large-scale waviness, small-scale smoothness, and joint alteration factor, respectively. The reduction of J_W^r and J_S^r is based on the concept of mobilized joint roughness and the equations are given as

$$\text{If } \frac{J_W}{2} < 1,\ J_W^r = 1;\ \text{Else } J_W^r = \frac{J_W}{2} \tag{26.17}$$

$$\text{If } \frac{J_S}{2} < 0.75,\ J_S^r = 0.75;\ \text{Else } J_S^r = \frac{JS}{2} \tag{26.18}$$

There is no reduction in J_A.

Residual GSI Value and Strength Parameters

The residual GSI_r is a function of V_b^r and J_C^r, which can be estimated using Eq. (26.5).

Fracturing and shearing do not weaken the intact rocks (even if they are broken into smaller pieces) so the mechanical parameters (q_c and m_r) should be unchanged. Therefore the generalized non-linear criterion for the residual strength of jointed rock masses can be written as

$$\sigma_1 = \sigma_3 + q_c\left[m_{br}\frac{\sigma_3}{q_c} + s_r\right]^{n_r} \tag{26.19}$$

where m_{br}, s_r, and n_r are the residual constants for the rock mass. These constants can be determined from a residual GSI_r (Cai et al., 2007).

$$m_{br} = m_r \cdot \exp\left[\frac{GSI_r - 100}{28}\right] \tag{26.20}$$

$$s_r = \exp\left[\frac{GSI_r - 100}{9}\right] \tag{26.21}$$

$$n_r = \frac{1}{2} + \frac{1}{6}\left(e^{-GSI_r/15} - e^{-20/3}\right) \tag{26.22}$$

Because the rock masses are in a damaged, residual state, $D = 0$ is used for the residual strength parameter calculation.

CLASSIFICATION OF SQUEEZING GROUND CONDITION

Hoek (2001) classified squeezing ground conditions based on tunnel strain (u_a/a) or the ratio between rock mass strength and in situ stress (γH), as shown in Figure 26.4. In very severe squeezing ground ($u_a/a > 5\%$), the tunnel face may exhibit plastic extrusion due to the failure of rock mass all around the tunnel and the face has to be stabilized. For a rock mass strength (q_{cmass}) of 1.5 MPa and in situ stress of 13.5 MPa (γH), the ratio ($q_{cmass}/\gamma H$) = 0.11. Figure 26.4 shows that this corresponds to a tunnel strain of approximately 10% and very severe squeezing ground condition should be anticipated.

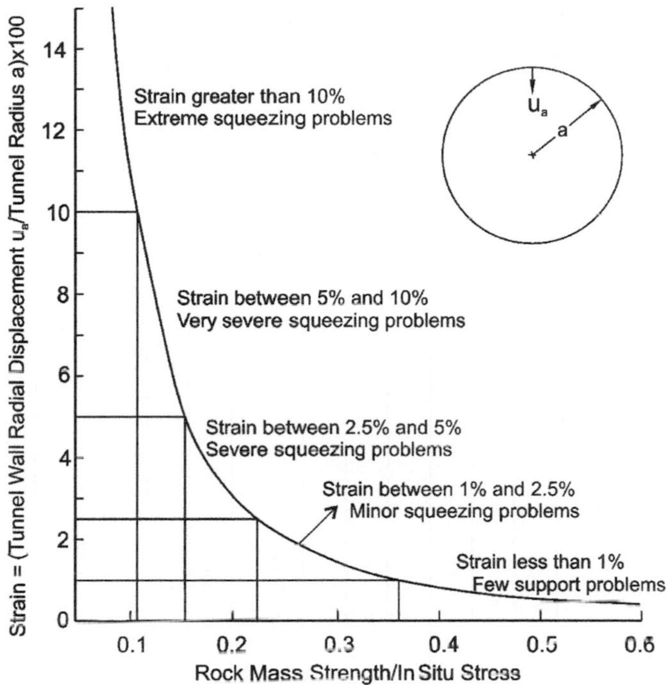

FIGURE 26.4 Tunneling problems associated with different levels of strain. *(From Hoek, 2001; Singh and Goel, 2006)*

Example 26.1

In a major hydroelectric project in dry quartzitic phyllite, the rock mass quality (Q) is in the range of 6 to 10. The joint roughness number J_r is 1.5 and joint alteration number J_a is 1.0 for critically oriented joints in the underground machine hall. The unit weight of phyllite rock is 2.78 gm/cc. The upper bound strength envelope between σ_1 and σ_3 from triaxial tests gave UCS (q_c) = 80 MPa, $\phi_p = 32°$, $m_r = 5.3$, and $E_r = 11.6$ GPa when the plane of schistocity is horizontal. The average UCS for various angles of schistocity is 40 MPa. The GSI is estimated to be about 55 as rock mass is micro-folded and joints are very rough and unweathered. With these values, it is required to consider the engineering parameters of the undisturbed (D = 0) rock mass for the machine hall cavity (width 24 m and height 47 m).

The average rock mass quality is $\sqrt{(6 \times 10)} = 8$ (approximately). Other calculations are presented in Table 26.6 for the undisturbed rock mass. The peak angle of internal friction works out to be 27° from Figure 26.3 and 32° from triaxial tests and 56° from the J_r/J_a value. Thus, a value of $\phi_p = 32°$ appears to be realistic. A blast damaged zone of about 2 m depth may be assumed in the computer modeling all around the cavity with half the values of c_p, q_{cmass}, E_d, and G.

It may be emphasized that Table 26.6 suggests parameters for the first iteration only in the computer modeling. The more realistic model and parameters may be back calculated from the observed displacements of the cavity during upper half excavation.

TABLE 26.6 Recommended Engineering Parameters of Undisturbed Rock Mass

S. No.	Rock mass parameter	Reference	Recommended value	Remarks
1	n	Eq. (26.9)	0.5	—
	m_b	Eq. (26.7)	1.1	D = 0
	s	Eq. (26.8)	6.7×10^{-3}	D = 0
2	c_p	Figure 26.2	3.6 MPa	$q_c = 80$ MPa
3	ϕ_p	—	32°	Same as that of rock material
4	UCS q_{cmass}	$2\ c_p \cos\phi_p/(1 - \sin\phi_p)$	13 MPa	Intercept on σ_1 and σ_3 envelope
5	UTS q_{tmass}	$0.029\ \gamma\ Q^{0.31}$	0.15 MPa	Q = 8
6	Angle of dilatancy Δ	$(\phi_p - \phi_r)/2$	5°	—
7	ϕ_r	$\phi_p - 10 \geq 14°$	22°	—
8	Residual cohesion c_r	Chapter 13; see the section Tensile Strength Across Discontinuous Joints	0.1 MPa	—
9	Residual UCS	$2\ c_r \cos\phi_r/(1 - \sin\phi_r)$	0.3 MPa	—
10	Modulus of deformation E_d	Uniaxial jacking test	7.5 MPa	Pressure dependency not observed
11	Poisson's ratio	—	0.20	
12	Shear modulus	$E_d/10$	0.75 MPa	Axis of anisotropy along bedding plane
13	Suggested model for peak strength	Eq. (13.14)	$13 + 2.2(\sigma_2 + \sigma_3)/2$ MPa	—
14	Model for residual strength	Mohr-Coulomb's theory	$0.3 + 1.2\ \sigma_3$ MPa	

Example 26.2

Given the strength of rock material (q_c) = 50 MPa, Hoek-Brown parameters for rock material (m_r) = 10, GSI = 45, and overburden above tunnel (H) = 100 m. Estimate the shear strength parameters of both undisturbed and heavily blasted rock mass (D = 1.0) using Hoek's computer program Roclab (2006).

For an undisturbed in situ rock mass surrounding a tunnel at a depth of 100 m, with a disturbance factor D = 0, the equivalent friction angle ϕ = 47° while the cohesive strength is c = 0.58 MPa. A rock mass with the same basic parameters but in a highly disturbed slope of 100 m height, with a disturbance factor of D = 1, has an equivalent friction angle of ϕ = 28° and a cohesive strength of c = 0.35 MPa.

REFERENCES

Bieniawski, Z. T. (1989). *Engineering rock mass classifications* (p. 251). New York: John Wiley.

Cai, M., & Kaiser, P. K. (2006). Visualization of rock mass classification systems. *Geotechnical and Geological Engineering*, *24*(4), 1089–1102.

Cai, M., Kaiser, P. K., Tasaka, Y., & Minami, M. (2007). Determination of residual strength parameters of jointed rock masses using GSI system. *International Journal of Rock Mechanics and Mining Sciences*, *44*, 247–265.

Cai, M., Kaiser, P. K., Uno, H., Tasaka, Y., & Minami, M. (2004). Estimation of rock mass deformation modulus and strength of jointed hard rock masses using the GSI system. *International Journal of Rock Mechanics and Mining Sciences*, *41*, 3–19.

Cheng, Y., & Liu, S. (1990). Power caverns of the Mingtan pumped storage project, Taiwan. In J. A. Hudson (Ed.), *Comprehensive Rock Engineering* (Vol. 5, pp. 111–132). Oxford, UK: Pergamon.

Hoek, E. (1983). Strength of jointed rock masses. 23rd Rankine Lecture. Institution of Civil Engineers. *Geotechnique*, *33*(3), 187–223.

Hoek, E. (1998). Reliability of Hoek-Brown estimates of rock mass properties and their impact on design. Technical Note. *International Journal of Rock Mechanics and Mining Sciences*, *35*(1), 63–68.

Hoek, E. (2001). Big tunnels in bad rock, 36th Terzaghi Lecture. *Journal of Geotechnical and Geoenvironmental Engineering, ASCE*, *127*(9), 725–740. *http://150.217.9.3/geotecnica/hoek_badrock.pdf.*

Hoek, E. (2007). *Practical rock engineering*. Chap. 12. www.rocscience.com.

Hoek, E., & Brown, E. T. (1997). Practical estimates of rock mass strength. *International Journal of Rock Mechanics and Mining Sciences*, *34*(8), 1165–1186.

Hoek, E., Carranza-Torres, C., & Corkum, B. (2002). Hoek-Brown Failure Criterion — 2002 edition. In *5th North American rock mechanics Symposium* (Vol. 1, pp. 267–273). 17th Tunnel Association of Canada, NARMS-TAC Conference, Toronto.

Hoek, E., & Diederichs, M. S. (2006). Empirical estimation of rock mass modulus. *International Journal of Rock Mechanics and Mining Sciences*, *43*, 203–215.

Hoek, E., Marinos, P., & Benissi, M. (1998). Applicability of the Geological Strength Index (GSI) classification for very weak and sheared rock masses—The case of Athens schist formation. *Bulletin of Engineering Geology and Environment*, *57*, 151–160.

Marinos, P., & Hoek, E. (2000). GSI — A geologically friendly tool for rock mass strength estimation. In *Proceedings of the GeoEngineering 2000 Conference*. Melbourne, Australia.

Palmstrom, A. (1995). RMi — A system for characterising rock mass strength for use in rock engineering. *Journal of Rock Mechanics and Tunnelling Technology*, *1*(2), 69–108.

Palmstrom, A. (2000). Recent developments in rock support estimates by the RMi. *Journal of Rock Mechanics and Tunnelling Technology*, *2*(1), 1–24.

Roclab, A. (2006). *Computer program 'Roclab' downloaded from Rocscience web site. www.rocscience.com.*

Singh, B., & Goel, R. K. (2006). *Tunnelling in weak rocks* (p. 488). Amsterdam: Elsevier.

Chapter 27

Evaluation of Critical Rock Parameters

The foundation of all concepts is simple unsophisticated experience. The personal experience is everything, and logical consistency is not final.

D. T. Suzuki, Professor of Philosophy, Otani University, Japan

INTRODUCTION

A list of all rock parameters and an understanding of all rock properties and rock mechanics are necessary before the start of any rock engineering project. Then an objective-based method of planning should be undertaken. A procedure for identifying the mechanics and rock properties most relevant to the project within the scope of the objective is next, and finally the ability to select relevant engineering techniques rounds out the process. Taking these steps, we utilize existing knowledge in an optimal way to develop site investigation, design, construction, and monitoring procedures for any project. The Rock Engineering System (RES) for selecting site-specific critical rock parameters (Hudson, 1992) is presented in this chapter. The sequence of critical rock parameters should be determined and then checked and confirmed by ratings of various classification systems. This process should minimize judgment errors.

CRITICAL PARAMETERS

There is some degree of coupling between joints, stress, flow, and construction, which is why this concept of interaction matrix was developed by Hudson (1992). The parameters in question are placed along the leading diagonal. The twelve leading diagonal terms for slopes and underground excavations considered by Hudson (1992) are given in the tables in the next two sections.

Slopes

	Parameters (P_i)	Representing
1.	Overall environment	Geology, climate, seismic risk, etc.
2.	Intact rock quality	Strong, weak, weathering susceptibility
3.	Discontinuity geometry	Sets, orientations, apertures, roughness

4.	Discontinuity properties	Stiffness, cohesion, friction
5.	Rock mass properties	Deformability, strength, failure
6.	In situ rock stress	Principal stress magnitudes/ directions
7.	Hydraulic conditions	Permeability, etc.
8.	Slope orientations, etc.	Dip, dip direction, location
9.	Slope dimensions	Bench height/width and overall slope
10.	Proximate engineering	Adjacent blasting, etc.
11.	Support/maintenance	Bolts, cables, grouting, etc.
12.	Construction	Excavation method, sequencing, etc.

Underground Excavations

1.	Excavation dimensions	Excavation size and geometry
2.	Rock support	Rock bolts, concrete liner, etc.
3.	Depth of excavations	Deep or shallow
4.	Excavation methods	Tunnel boring machines, blasting
5.	Rock mass quality	Poor, fair, good
6.	Discontinuity geometry	Roughness, sets, orientations, distributions, etc.
7.	Rock mass structure	Intact rock and discontinuities
8.	In situ rock stress	Principal stress magnitude and direction
9.	Intact rock quality	Hard rocks or soft rocks
10.	Rock behavior	Responses of rocks to engineering activities
11.	Discontinuity aperture	Wide or narrow
12.	Hydraulic conditions	Permeabilities, water tables, etc. (after commissioning of hydro projects)

PARAMETER INTENSITY AND DOMINANCE

We know that some parameters have a greater effect on a rock structure system than others and that the system has a greater effect on some parameters than others. The approach for quantifying the intensity and dominance of parameters is presented in this section. This is achieved by Hudson (1992) by coding the interaction matrices and studying the interaction intensity and dominance of each parameter.

Generic Matrix Coding

There are five categories into which the mechanism can be classified: no, 0; weak, 2; medium, 3; strong, 4; and critical, 5. This coding method is viable for any matrix and serves to demonstrate how the simple systems approach is developed.

The Cause-Effect Plot

The cause refers to the influence of a parameter on the system and the effect refers to the influence of the system on the parameter. Consider Figure 27.1, which shows the generation of the cause and effect coordinates. The main parameters (P_i) are listed along the leading diagonal with parameter construction as the last box. We intercept the meaning of the rows and the columns of the matrix, as highlighted in Figure 27.1 by the row and the column through P_i. From the construction of the matrix, it is clear that the row passing through P_i represents the influence of P_i on all the other parameters in the system.

Conversely, the column through P_i represents the influence of the other parameters, that is, the rest of the system on P_i. Once the matrix has been coded approximately, the sum of each row and each column can be found. Now, think of the influence of P_i on the system; the sum of the row values is called the *cause* and the sum of the column is called the *effect*, designated as coordinates (C, E). Thus, C represents the way in which P affects the system and E represents the effect that the system has on P. Note that construction has (C, E) coordinates that represent the post- and pre-construction mechanisms, respectively.

It is important to note that the dual nature of rock parameters is accounted for in this approach. Strength and weakness go together. Poor rock masses are likely to be less brittle, impervious in some cases, and have high damping characteristics — unlike hard rocks. The long life of a support system and drainage system is essential in civil engineering projects unlike in mining projects where the support system is temporary and associated with very large deformation rates.

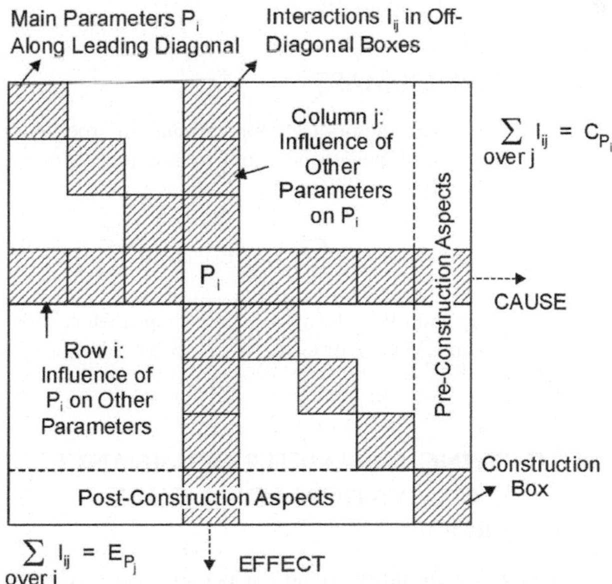

FIGURE 27.1 Summation of coding values in the row and column through each parameter to establish the cause and effect coordinates. *(From Hudson, 1992)*

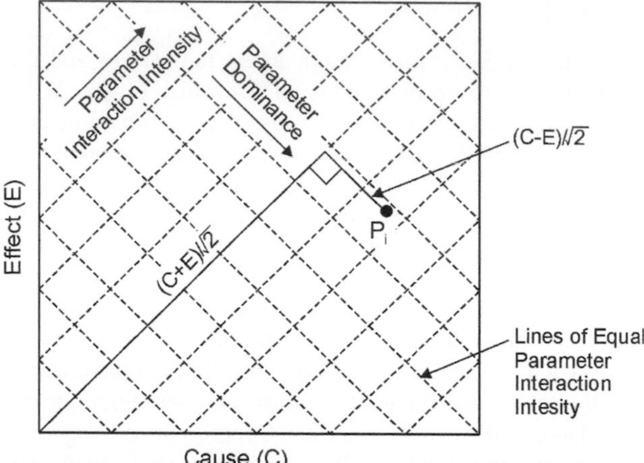

FIGURE 27.2 Lines of equal parameter interaction intensity and dominance. *(From Hudson, 1992)*

Interpretation of Cause-Effect Plot

The parameter *interaction intensity* and the parameter *dominance characteristics* are shown in Figure 27.2. The two sets of 45 degree lines in the plot indicate contours of equal value for each of the two characteristics. It is particularly important to note that, while the parameter interaction intensity increases from zero to the maximum parameter interaction, the associated maximum possible parameter dominance value rises from zero to a maximum of 50% of the parameter interaction intensity and then reduces back to zero at a maximum parameter intensity value. The specific numerical values of the two characteristics are $(C + E)/\sqrt{2}$ and $(C - E)/\sqrt{2}$, as indicated in Figure 27.2.

CLASSIFICATION OF ROCK MASS

It is necessary to evolve weightage factors (w_i) for various "m" rock parameters separately for underground openings, slopes, mines, and foundations. Hudson (1992) suggested the following rock classification index:

$$\text{Rock Classification Index} = \sum_{i=1}^{m} (C_i+E_i) \cdot w_i \Big/ \sum_{i=1}^{m} (C_i+E_i) \tag{27.1}$$

where C_i and E_i are the cause and effect rating of the *ith* parameter. This rock classification index may be better than RMR, Q, or GSI, which do not account for the important site-specific parameters.

EXAMPLE FOR STUDYING PARAMETER DOMINANCE IN UNDERGROUND EXCAVATION FOR A COAL MINE WITH A FLAT ROOF

The 12 leading parameters for an underground excavation matrix were listed earlier in this chapter. A 12 × 12 matrix keeping these 12 parameters in the leading diagonal has been prepared with numerical coding from 0 to 4 for parameter interaction as shown in

Figure 27.3. To explain the coding method, we can highlight some of the extreme values. For example, Box 1, 9 (first row and ninth column of the matrix in Figure 27.3) is coded as 0. This is the influence of cavern dimensions on intact rock quality. There could be some minor effect such as larger caverns might cause a greater degradation of the intact rock quality but, within the resolution of the coding, we would assign this box a value of 0. On the other hand, Box 2, 10 has been assigned a maximum value of 4; this is a critical interaction because it influences the rock support on rock behavior. The whole purpose of rock support is to control the rock behavior as illustrated in Box 2, 10, so the coding must be 4.

The associated cause-effect plot in the lower part of Figure 27.3 shows that the mean interaction intensity is higher and the parameter dominance and subordinancy is stronger. The cause-effect plot for underground excavations is clarified in Figure 27.4 with the individual parameter identifiable. In this plot, the most interactive parameter is number 3, the depth of excavation. The least interactive parameter is number 6, the discontinuity geometry. The most dominant parameter is number 7, the rock mass structure, and the most subordinate (least dominant) parameter is number 10, rock material behavior, which we would expect because this is conditioned by all the other parameters.

It is emphasized that these are general conclusions about the nature of underground excavations as determined from the generic matrix. If faced with a specific rock type, a specific site, and a specific project objective, the generic matrix could be coded accordingly. Naturally this would change the critical parameters.

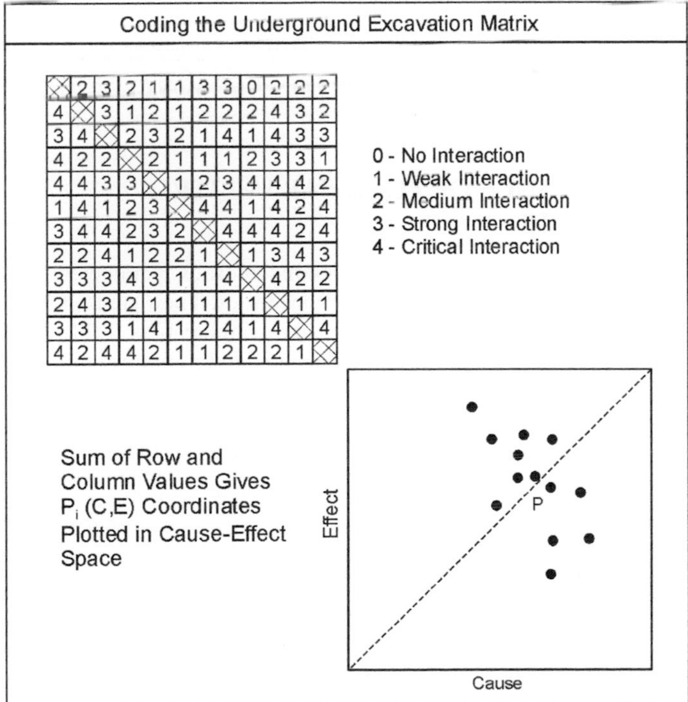

FIGURE 27.3 Coding values for the generic underground excavations interaction matrix and the associated cause versus effect plot. *(From Hudson, 1992)*

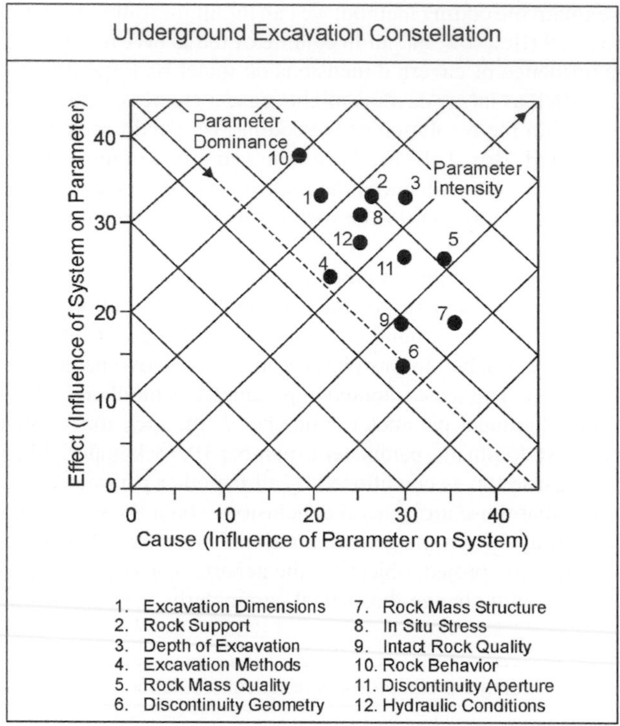

FIGURE 27.4 Cause versus effect plot for the generic 12 × 12 underground excavations for the coding values given in Figure 27.3. *(From Hudson, 1992)*

RELATIVE IMPORTANCE OF ROCK PARAMETERS IN MAJOR PROJECTS

Hudson and Harrison (1997) reported histograms of rock parameters for pressure tunnels, large caverns, and radioactive waste repositories. Their study is based on current practice, recommended practice, and over 320 research papers. Table 27.1 lists their relative importance for site-specific planning, testing, and monitoring of projects. Further, there is no need for hoop reinforcement in the concrete lining of water pressure tunnels as plain cement concrete (PCC) may be allowed to crack. The PCC lining has been working satisfactorily since 1980 (Singh et al., 1988) in hydroelectric projects in India.

INTERACTION BETWEEN ROCK PARAMETERS

The real-world response of rock masses is often highly coupled or interacting. There is a non-linear complex relationship between mechanical properties and rock parameters, especially in weak argillaceous rock masses. Hudson (1992) schematically showed this

TABLE 27.1 Relative Importance of Rock Engineering Parameters in Rock Structures

Water pressure tunnels in hydroelectric projects	Large underground caverns	Radioactive waste repositories
In situ stress	Depth of cavern	In situ stress
Discontinuity persistence	Discontinuity orientation	Induced displacement
Topographic factors	In situ stress	Thermal aspects
Presence of faults/folds	Presence of faults	Discontinuity geometry
Location of tunnel	Rock type	Permeability
Discontinuity aperture	Discontinuity frequency	Time-dependent properties
Rock mass geometry	Discontinuity aperture	Elastic modulus
Discontinuity fill	Preexisting water conditions	Compressive strength
Tunnel water pressure	Intact rock elastic modulus	Porosity
Pre-existing water conditions	Rock mass elastic modulus	Density

Source: Hudson and Harrison, 1997.

complex interaction for tunneling (Figure. 27.5). Hudson identified the following 12 rock parameters affecting the tunneling conditions.

1.	Excavation dimensions	Excavation size and geometry
2.	Rock support	Rock bolts, concrete liner, etc.
3.	Depth of excavations	Deep or shallow
4.	Excavation methods	Tunnel boring machines, blasting
5.	Rock mass quality	Poor, fair, good
6.	Discontinuity geometry	Sets, orientations, distributions, etc.
7.	Rock mass structure	Intact rock and discontinuities
8.	In situ rock stress	Principal stress magnitudes and directions
9.	Intact rock quality	Hard rocks or soft rocks
10.	Rock behavior	Responses of rocks to engineering activities
11.	Discontinuity aperture	Wide or narrow
12.	Hydraulic conditions	Permeabilities, water tables, etc.

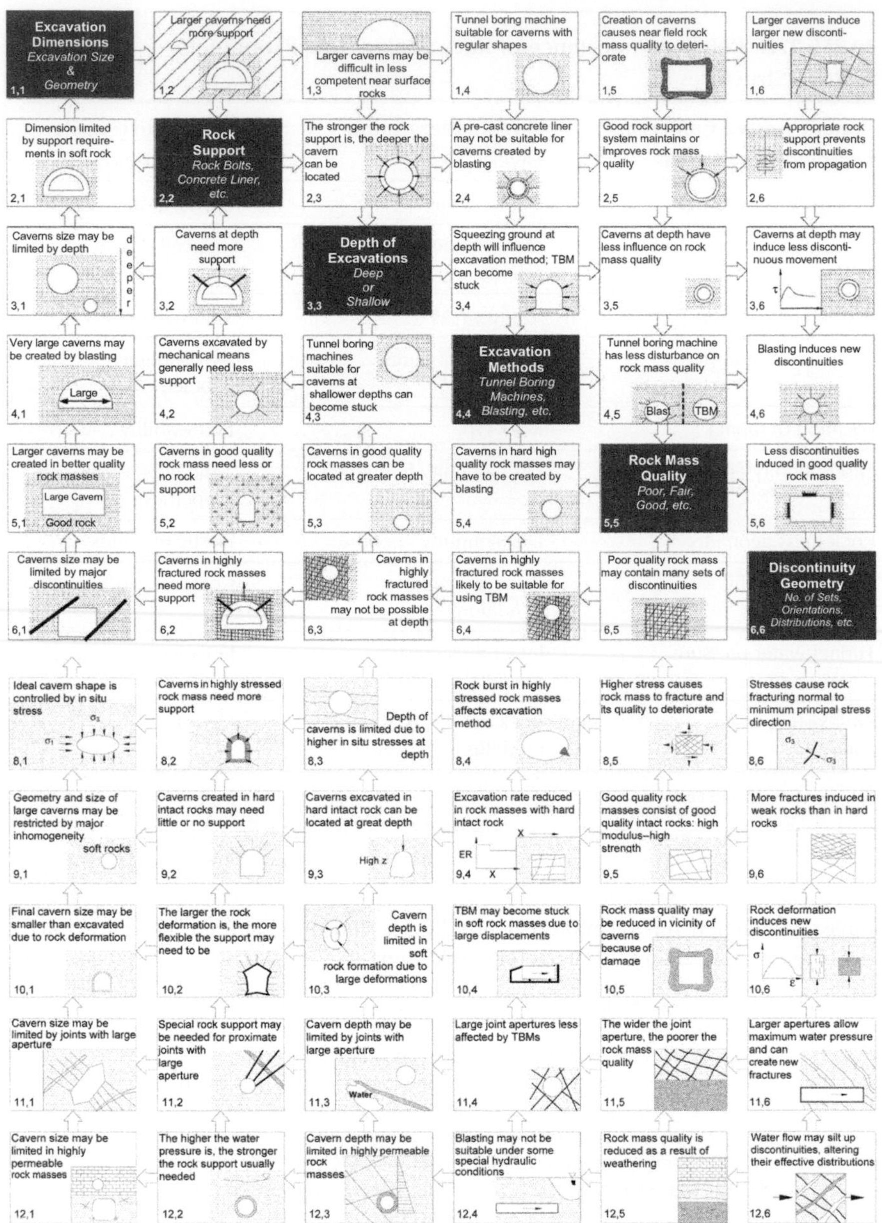

FIGURE 27.5 Interaction of rock parameters in underground excavations. *(From Hudson, 1992)*

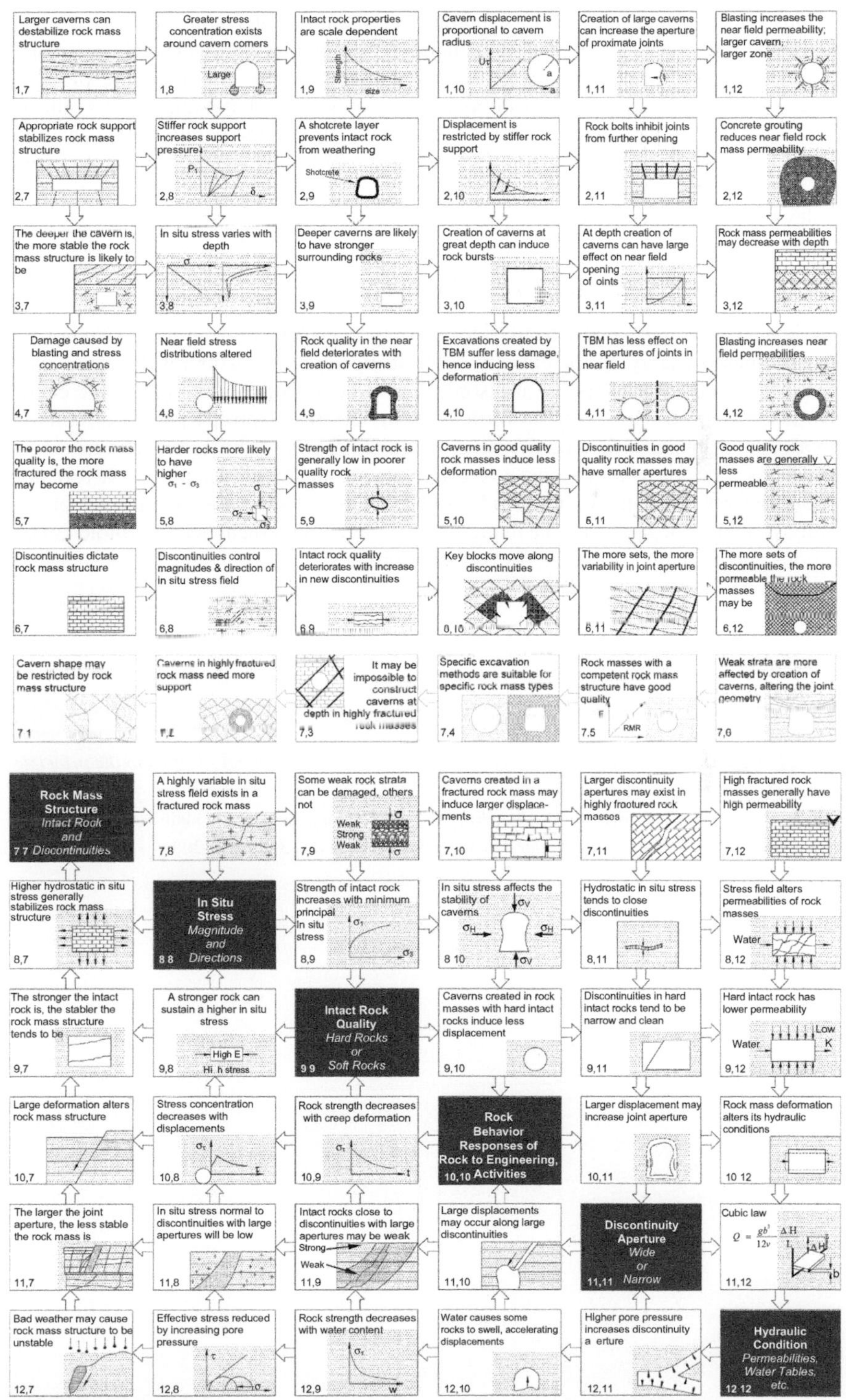

FIGURE 27.5—Cont'd

Hudson (1992) made the system's approach very simple, interesting, and based on the actual experiences and judgments of tunneling experts. His approach makes decision making very easy when planning geotechnical investigations for tunneling projects. Figure 27.5 for underground excavations is self-explanatory. For example (7,1) means the effect of the 7th parameter (rock mass structure) on the first parameter (excavation dimensions). The problem is a coupled coordinate (1,7), which means they are the effect or excavation dimensions on the rock mass structure such as opening or discontinuities and development of new fractures.

APPLICATION IN ENTROPY MANAGEMENT

Generic matrix coding can also be used for entropy management of a project. Today, the effect of unused energy on the entropy is blissfully forgotten. This results in an ever-increasing entropy or side effects or disorderliness, confusion, noise, unhygienic conditions, toxic gases, diseases, and so forth. The anxiety from entropy can be effectively decreased by planting a micro-ecosystem around the project, road network, and landslide-prone areas. Entropy within a house or office can be decreased by placing a few pots of indoor plants inside the rooms. Hudson (1992) noted that change in entropy of live healthy systems is negative, unlike matter; hence there is an urgent need for biotechnical solutions and reduction of the inefficient technologies.

REFERENCES

Hudson, J. A. (1992). *Rock engineering systems—Theory and practice* (p. 185). London: Ellis Horwood Limited.

Hudson, J. A., & Harrison, J. P. (1997). *Engineering rock mechanics—An introduction to the principles* (p. 144). Amsterdam: Elsevier Science.

Singh, B., Nayak, G. C., Kumar, R., & Chandra, G. (1988). Design criteria for plain concrete lining in power tunnels. *Tunnelling and Underground Space Technology*, *3*(2), 201–208.

Chapter 28

In Situ Stresses

Everything should be made as simple as possible, but not simpler.

Albert Einstein

THE NEED FOR IN SITU STRESS MEASUREMENT

In situ stresses are generally measured by the hydro-fracturing method, which is economical, and faster and simpler than other methods. The magnitude and orientation of in situ stresses could be a major influence on planning and design of underground openings in hydroelectric projects, mining, and underground space technology. The orientation of in situ stresses is controlled by geological structures like folds, faults, and intrusions.

CLASSIFICATION OF GEOLOGICAL CONDITIONS AND STRESS REGIMES

Ramsay and Hubber (1988) showed how types of faults rotate principal in situ stresses (Figure 28.1).

Normal Fault Area (Figure 28.1a)

These are steeply dipping faults where slip occurs more often along the dip direction than along its strike, and the hanging wall is moved downward. Normal faults are formed due to tensional forces. The mechanics of failure suggest that the vertical stress (σ_v) is the major principal stress and the minimum horizontal stress (σ_h) acts along the dip direction. As such, the order of in situ stresses is given here:

$$\sigma_v > \sigma_H > \sigma_h$$

In a subducting boundary plate, normal faults are more common as the downward bending of this plate reduces horizontal stresses along the dip direction. However, in the upper boundary plate, thrust faults are generally seen because of the tectonic thrust, so there is an urgent need for stress analysis of the interaction of plate boundaries (Nedoma, 1997).

Thrust Fault Area (Figure 28.1b)

Thrusts have mild dip with major slip along the dip direction compared to along the strike, and the hanging wall is moved upward. Normal faults are formed due to compressional forces. The mechanics of brittle failure indicate that the vertical stress in this case

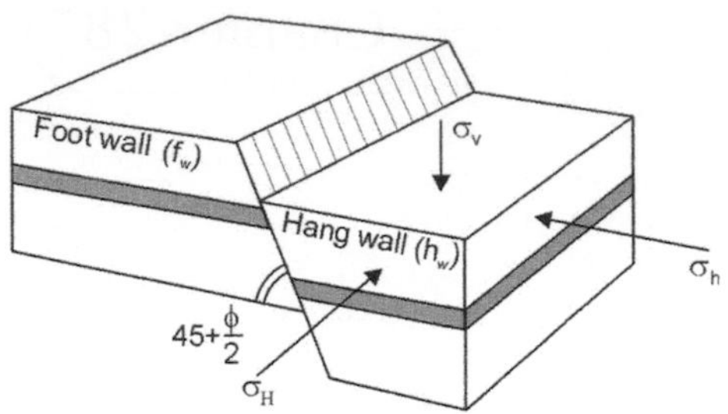

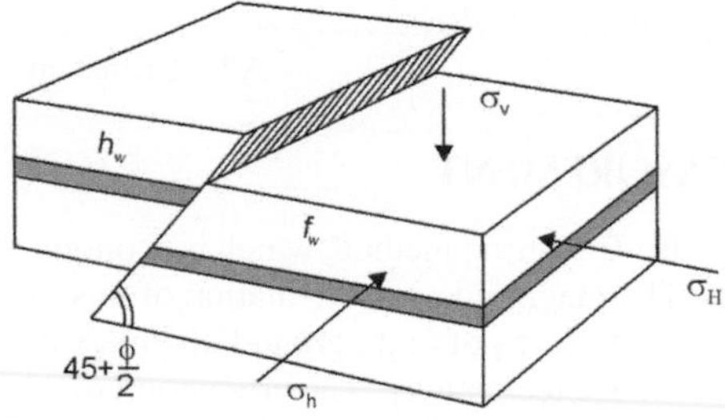

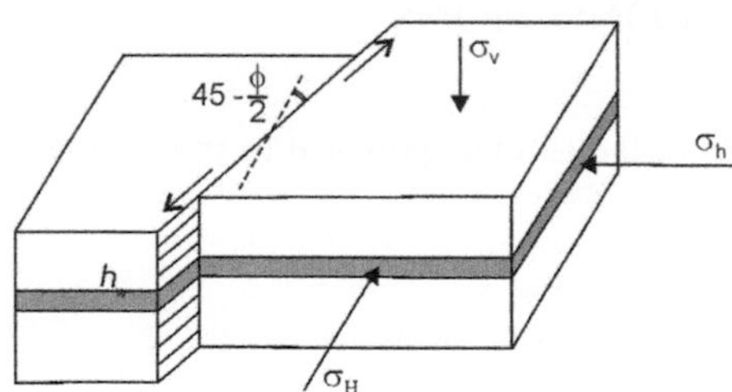

FIGURE 28.1 Orientation of in situ stresses in various geological conditions. *(From Ramsay and Hubber, 1988)*

should be the minimum principal in situ stress and the horizontal stress along the dip direction is the maximum principal. The order of the in situ stresses in the thrust fault area is as follows:

$$\sigma_H > \sigma_h > \sigma_v$$

The correlations developed in India refer to the geological region of the upper boundary plate with frequent thrust and strike-slip faults in the Himalayas.

Strike-Slip Fault Area (Figure 28.1c)

Such faults are steeply oriented and usually vertical. The slip occurs more often along the strike than along the dip direction. In a strike-slip fault, the major principal stress and minor principal stress are oriented as shown in Figure 28.1c. The order of the in situ stresses is given in the following equation:

$$\sigma_H > \sigma_v > \sigma_h$$

Both magnitude and orientation of horizontal in situ stresses change with erosion and tectonic movements, especially in hilly regions. The regional horizontal in situ stresses are relaxed in steep mountainous regions. These stresses are relaxed closer to the slope face. The gradient of the horizontal stress with depth (or vertical stress) may be more in steeply inclined mountainous terrain compared to plane terrain. Vertical stress just below the valley may be much higher than the overburden pressure due to the stress concentration at the bottom of the valley; thus the in situ stresses are different at a given depth in three fault areas and vary locally near faults, folds, and thermic regions.

VARIATION OF IN SITU STRESSES WITH DEPTH

In soils, the in situ horizontal stress is given by the condition of zero lateral strain, thus, we get

$$\sigma_H = \sigma_h = v \cdot \sigma_v/(1 - v) \tag{28.1}$$

where v is Poisson's ratio of soil mass.

Rock masses have significant horizontal stresses even near ground surface due to the non-uniform cooling of Earth's crust. Tectonic stresses also significantly affect the in situ stresses. Hoek and Brown (1980) analyzed worldwide data measured from in situ stresses. They found that the vertical stress is approximately equal to the overburden stress.

The regional stresses vary in a wide range as follows (depth $z < 2000$ m):

$$\sigma_H < 40 + 0.5\,\sigma_v \text{ MPa} \tag{28.2}$$

$$\sigma_h > 2.7 + 0.3\,\sigma_v \text{ MPa} \tag{28.3}$$

$$\sigma_v \cong \gamma Z \tag{28.4}$$

where γ is the unit weight of the rock mass ($\gamma = 2.7$ gm/cc or T/m^3) and z is the depth of the point of reference in meters below the ground surface.

According to McCutchin (1982), the tectonic stress component (at ground level) depends upon the modulus of deformation of the rock mass as given in Eq. (28.5):

$$\sigma_{av} = 7\gamma E_d + \sigma_v(0.25 + 0.007\,E_d), \text{T/m}^2 \tag{28.5}$$

where E_d is the modulus of deformation in GPa.

Stephansson (1993) reported the following trend for in situ horizontal stresses at shallow depth ($z < 1000$ m) from hydro-fracturing tests:

$$\sigma_H = 2.8 + 1.48\,\sigma_v \text{ MPa} \tag{28.6}$$

$$\begin{aligned} \sigma_h &= 2.2 + 0.89\,\sigma_v \text{ Pa} \\ \sigma_v &= \gamma Z \end{aligned} \tag{28.7}$$

Hydro-fracturing tests done by Sharma (1999) showed that previously mentioned trends apply to the thrust area regime (Figure 28.1b). It is also noted that $\sigma_H > \sigma_h > \sigma_v$ (Eqs. 28.6 and 28.7). Sharma also showed that the measured in situ stresses depended significantly on the method of testing.

Sengupta (1998) performed a large number of hydro-fracturing tests within weak rocks in the Himalayan region. The in situ stress data of Sengupta (1998) and Sheorey et al. (2001) are plotted in Figure 28.2. It is heartening to see a good correlation

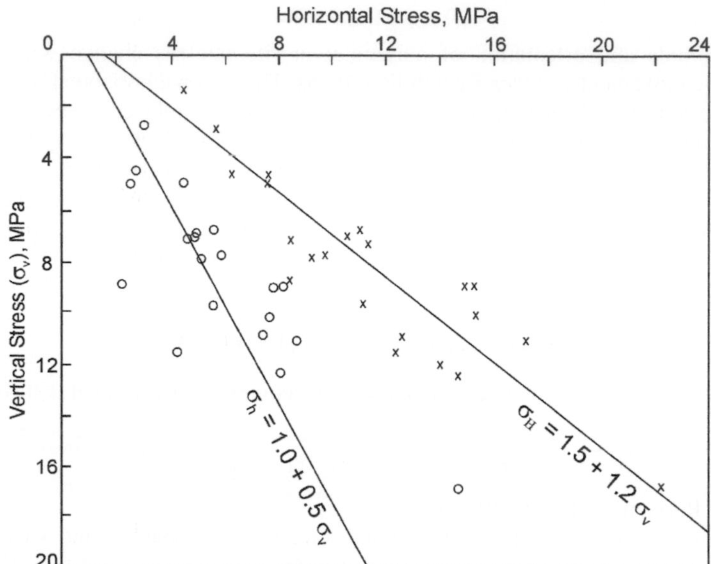

FIGURE 28.2 Variation of in situ stresses near the Himalayan region.

between σ_H and σ_v. The correlation between σ_h and σ_v is not good, perhaps due to mountainous terrain; however, it is inferred that for z < 400 m

$$\sigma_H = 1.5 + 1.2\ \sigma_v \text{ MPa} \tag{28.8}$$

$$\sigma_h = 1.0 + 0.5\ \sigma_v \text{ MPa} \tag{28.9}$$

It appears that Stephansson's correlations (Eqs. 28.6 and 28.7) predict higher values, whereas Sengupta's correlations predict lower values of the actual in situ stresses. In steeply inclined mountainous terrain, Sengupta's correlations (Eqs. 28.8 and 28.9) may be applicable in the stress region ($\sigma_H > \sigma_v > \sigma_h$) as the in situ horizontal stresses are likely to be significantly relaxed. In the upper Himalayas, vertical stress is the intermediate principal stress, and horizontal stresses are major and minor principal stresses.

In other stress regimes, separate correlations need to be developed. In major projects, a statistically significant number of hydro-fracturing tests should be conducted to determine how rotation of in situ stresses takes place along folds and across faults at a site. This may help in mine planning locally as well as in the design of a support system or selection of support strategy in major underground projects.

Sheorey et al. (2001) proposed an equation for estimating the mean horizontal in situ stress, which depends on the elastic constants of rock mass, the geothermal gradient due to cooling of crust, and the coefficient of thermal expansion. According to the theory, the equation for the mean horizontal stress (σ_h) is as follows for isotropic rock mass:

$$\sigma_h = \frac{\nu}{1-\nu}\gamma H + \frac{\beta E G}{1-\nu}(H + 1000) \tag{28.10}$$

where, H = depth of cover; E = elastic modulus; ν = Poisson's ratio; γ = unit weight of the rock mass; β = coefficient of linear thermal expansion of rock (8×10^{-6}/°C appears to be a reasonable representative value for different rock types but not for coal); and G = geothermal gradient (for crustal rocks = 0.024°C/m, for coal measure rocks = 0.03°C/m).

It is interesting to note that the higher the geothermal gradient, the higher the tectonic stress according to Eq. (28.10).

The results obtained from Eq. (28.10) closely matched the observed values of in situ stresses in India in geologically undisturbed areas such as coal measure rock formations, deccan trap mountains caused by lava flow, and the Aravali range system of Precambrian age with folds and thrusts. The estimations are also comparable to North American data, UK coal measures rock data, and the data from Japan. The stress measurements from Italy and Austria in the Alps mountainous region show a scatter that may occur because of the influence of topography, major geological features, and tectonics. Equation (28.10) should be used cautiously in areas of tectonic activities, especially in the Himalayan region.

EFFECTS OF IN SITU STRESS ON ROCK MASS PROPERTIES

In situ stresses affect rock mass properties in a number of ways. Some of these effects, shown earlier in Figure 27.5, are stated in the following list (Hudson, 1992):

- Stronger rock can sustain higher in situ stress.
- Stress concentration decreases with displacements.
- In situ stresses normal to discontinuities with large aperture will be low.
- Effective stress is reduced by increasing pore water pressure.
- Hydrostatic in situ stresses tend to close discontinuities.
- Stress field alters permeability of rock mass.
- Stresses cause normal to minimum rock fracturing at principal stress directions.
- High stress causes rock mass to fracture and its quality to deteriorate.
- Rock bursts in highly stressed rock masses affect excavation methods.
- Ideal cavern shape is controlled by an in situ stress field.
- Tectonic stresses, erosion, topography, and other factors also affect the stress field.
- In situ stress varies with depth.
- Discontinuities control magnitudes and directions of in situ stress fields.
- A highly variable in situ stress field exists in fractured rock mass.

CORE DISCING

Extraction of diamond-drilled core from high stress environments can result in the core breaking into discs. Rock breakage during the coring process is a well-known phenomenon. Often this fragmentation is called "core discing" when it involves formation of fractures that crosscut the core axis to form relatively thin discs. Sometimes core discing is seen as an impediment to sampling or to over-coring measurements, but it has also been repeatedly suggested as a potential method for in situ stress estimation.

The analysis of core discing for in situ stress estimation relies on the stress concentration near the base of the advancing coring bit and depends on both the in situ stress state and the length of the unbroken core stub, which has already advanced into the core barrel. Given a typical disc length (i.e., thickness), constraints can, in principle, be placed on the stress field.

The occurrence of discing has been investigated by many researchers. They suggested that a projecting core stub broke off over a curved surface when a lateral stress was applied to the rock. They also found from experimental drilling into stressed rock that, the higher the applied lateral stress, the thinner the resulting discs. They also observed that the fracture surfaces appeared clean and unsheared, suggesting tension failure, and that failure may start near the center of the core. Readers interested in knowing more about core discing are referred to the literature on the subject.

"A scientist should also be a good businessman in the future."

REFERENCES

Hoek, E. & Brown, E. T. (1980). *Underground excavations in rock* (p. 527). Institution of Mining and Metallurgy. London: Maney Publishing.

Hudson, J. A. (1992). *Rock engineering systems—Theory and practice* (p. 185). London: Ellis Horwood Ltd.

McCutchin, W. R. (1982). Some elements of a theory of in situ stresses. *International Journal of Rock Mechanics and Mining Sciences—Geomechanics Abstracts, 19*(4), 201–203.

Nedoma, J. (1997). *Part I — Geodynamic analysis of the Himalayas and Part II — Geodynamic analysis.* Technical Report, No. 721, September, p. 44. Institute of Computer Science. Prague: Academy of Sciences of the Czech Republic.

Ramsay, G., & Hubber, M. I. (1988). The techniques of modern structural geology. In *Folds and Fractures* (Vol. 2, pp. 564–566). San Diego: Academic Press.

Sengupta, S. (1998). *Influence of geological structures on in situ stresses.* Ph.D. Thesis, p. 275. Department of Civil Engineering, IIT, Uttarakhand, India.

Sharma, S. K. (1999). *In situ stress measurements by hydro-fracturing—Some case studies.* M.E. Thesis, p. 104. Roorkee: WRDTC, IIT, Uttarakhand, India.

Sheorey, P. R., Mohan, G. M., & Sinha, A. (2001). Influence of elastic constants on the horizontal in situ stress. Technical Note. *International Journal of Rock Mechanics and Mining Sciences, 38*, 1211–1216.

Stephansson, O. (1993). Rock stress in the Fennoscandian Shield. In *Comprehensive rock engineering* (Vol. 3, Chap. 17, pp. 445–459). New York: Pergamon.

Appendix I

Shear and Normal Stiffness of Rock Joints

Humanity is acquiring all the right technology for all the wrong reasons.

R. Buckminster Fuller

The normal stiffness k_n of an unweathered rock joint is estimated as follows:

$$k_n = \frac{\text{normal stress } (\sigma)}{\text{joint closure } (\delta)}$$

$$\delta \propto \frac{\sigma}{E_r}$$

or $k_N \propto E_r$

$$\therefore \quad \frac{E_r}{k_n} = \text{constant for a given joint profile } (A_j) \qquad (I.1)$$

where E_r = modulus of elasticity of asperities/rock material. The physical significance of parameter A_j is that k_n is equal to the stiffness of intact rock layer of the thickness A_j.

The manual of the U.S. Corps of Engineers (1997) and Singh and Goel (1982) summarized the typical values of parameter A_j based on results of uniaxial jacking tests in the United States and India, which are marked by an asterisk in Table I.1.

Based on the back analysis of underground powerhouses at the Sardar Sarovar and Tehri Dam projects, Samadhiya (1998) suggested values for the normal and shear stiffness of joints, which are summarized in Table I.1 for various kinds of joints. Normal stiffness during unloading (relaxation of normal stresses) is much higher than during loading, as expected.

PRESSURE DEPENDENT MODULUS

In highly jointed rock masses, the modulus of deformation is significantly dependent upon the confining pressure. The effect of confining pressure on modulus of deformation is very significant for soft rock materials like shales, slates, claystones, and so forth (Janbu, 1963). The effect of pressure dependency is accounted for by Eq. (I.2) in which the effects of intermediate principal stresses σ_2 and σ_3 have been included.

$$E_r = E_o \left[\frac{\sigma_2 + \sigma_3}{2p_a}\right]^{\alpha} > E_o \qquad (I.2)$$

TABLE I.1 Back Analysis Values of Normal and Shear Stiffness of Rock Joints

S. No.	Joint type	$A_j = E_r/k_n$ Loading (cm)	$A_j = E_r/k_n$ Unloading (cm)	k_s/k_n Loading
1	Continuous joint or loose bedding plane in weathered rock mass	115–125	16–18	1/10
2	Continuous joint or loose bedding plane in unweathered rock mass	60*	12	1/10
3	Discontinuous joints in unweathered rock mass	15–25*	5–7	1/10
4	Unweathered cleavage planes but separated	5*	2	1/10
5**	Joint with gouge	t_g	—	1/10

**Adopted from Singh and Goel (1982) and U.S. Corps of Engineers Manual (1997).*
***Stiffness of gouge is of the order of E_g/t_g where E_g and t_g are average modulus of deformation and thickness of the gouge, respectively.*

where E_r = pressure dependent modulus of elasticity of a rock material in triaxial condition; E_o = modulus of deformation corresponding to atmospheric confining pressure (which may be taken equal to the modulus of deformation from uniaxial compressive strength tests); σ_2, σ_3 = effective intermediate and minor principal stresses; p_a = atmospheric pressure; and α = the modulus exponent obtained from triaxial tests conducted at different confining pressures, that is, 0.15 for hard rocks, 0.30 for medium rocks, and 0.50 for very soft rocks.

It may be noted that the increase in modulus of elasticity due to confining pressure also results in a corresponding increase in the stiffness of the joints. Stiffness k_n and k_s may also be increased in the same proportion as the modulus of elasticity of a rock material.

With the distinct element model, from Table I.1 we get

$$k_{n\ model} = \frac{E_r}{A_j \cdot x} \tag{I.3}$$

$$k_{s\ model} = \frac{k_{n\ model}}{10} \tag{I.4}$$

$$E_{r\ model} = E_r \tag{I.5}$$

where x = ratio of "spacing of rock joints in the model" to "actual spacing of joints."

Back analysis is more powerful in guessing probable rock parameters from monitored displacements. 3DEC software seems to be ideal for dynamic analysis of rock structures. This software realistically simulates the pre-stressing effect of intermediate principal stress along the axis of opening on rock wedges (according to Eq. 13.14). This software provides insight into the mechanics of interaction between openings with the rock slope. According to Samadhiya (1998), most of the displacements in rock mass take place because of the displacement of rock blocks along critical joints, not because of the displacement within the rock blocks ($k_s \approx k_n/10$).

REFERENCES

Janbu, N. (1963). Soil compressibility as determined by odometer and triaxial tests. In *European Conference on Soil Mechanics and Foundation Engineering* (Vol. 1, pp. 19–25). Wiesbaden, Germany.

Samadhiya, N. K. (1998). *Influence of anisotropy and shear zones on stability of caverns* (p. 334). Ph.D. Thesis. Uttarakhand, India: Department of Civil Engineering, IIT Roorkee.

Singh, B., & Goel, P. K. (1982). Estimation of elastic modulus of jointed rock masses from field wave velocity. In B. B. S. Singhal & S. Prakashan (Eds.), *R.S. Mithal commemorative volume on engineering geosciences* (pp. 156–172). Sarita Prakashan, Meerut, India.

U.S. Army Corps of Engineers. (1997). *Tunnels and shafts in rock, engineering and design manual.* (No. EM 1110-2-2901). Dept. of Army, May (available on Internet).

Appendix II

Bond Shear Strength of Grouted Bolts

Kindly dig happiness in your self. There is a lot of happiness within all of us.

Unknown Saint

Software used to design support systems requires knowledge of allowable bond shear strength of grouted rock bolts (q_g). The rock bolts are often pulled out and q_g is found (see Chapter 12) as:

$$q_g = \frac{P_{bolt}}{\pi d_g l_g F_g} \tag{II.1}$$

TABLE II.1 Allowable Rock–Grout Bond Shear Strength in Cement Grouted Rock Bolts

Rock description	Compressive strength (q_c) range (MPa)	Allowable bond stress (q_g) (MPa)
Strong rock	>100	1.05–1.40
Medium rock	50–100	0.7–1.05
Weak rock	20–50	0.35–0.7
	Rock type	
Granite, basalt		0.55–1.0
Dolomitic limestone		0.45–0.70
Soft limestone		0.35–0.50
Slates, strong shales		0.30–0.45
Weak shales		0.05–0.30
Sandstone		0.30–0.60
Concrete		0.45–0.90

Source: Wyllie, 1999.

where q_g = allowable bond shear strength of grout–rock mass interface, d_g = diameter of grouted bolt hole, l_g = grouted length of rock bolt, F_g = factor of safety of rock bolt (e.g., 3), and P_{bolt} = pull-out capacity of rock bolt.

A suitable expansion agent (e.g., aluminum powder, etc.) must be added to the cement grout, otherwise pull-out capacity is found to be very low. According to Littlejohn and Bruce (1977), the allowable bond shear strength for cement grouted bolt is

$$q_g = \frac{q_c}{30} \tag{II.2}$$

where q_c is the UCS of the rock material adjacent to the bolt. Table II.1 provides a list of values of q_g for different rocks (Wyllie, 1999; Wyllie & Mah, 2004).

REFERENCES

Littlejohn, G. S., & Bruce, D. A. (1977). *Rock anchors—State of the art*. Essex, UK: Foundations Publications Ltd.

Wyllie, D. C. (1999). *Foundations on rock* (2nd ed., p. 401). London: Taylor & Francis.

Wyllie, D. C., & Mah, C. W. (2004). *Rock slope engineering—Civil and mining* (4th ed., p. 431), based on the 3rd ed. by E. Hoek & J. Bray. London and New York: Spon Press and Taylor & Francis Group.

Index

Note: Page numbers followed by *b* indicate boxes, *f* indicate figures, and *t* indicate tables.

H

I

S

T

U

V

W

Z